T0340703

FINANCIAL MANAGEMENT AND ACCOUNTING FUNDAMENTALS FOR CONSTRUCTION

DANIEL W. HALPIN, ***PURDUE UNIVERSITY***
BOLIVAR A. SENIOR, ***COLORADO STATE UNIVERSITY***

JOHN WILEY & SONS, INC.

This book is printed on acid-free paper. ♾

Published by John Wiley & Sons, Inc., Hoboken, New Jersey.
Published simultaneously in Canada.

For general information about our other products and services, please contact our Customer Care Department within the United States at (800) 762-2974, outside the United States at (317) 572-3993 or fax (317) 572-4002.

Wiley also publishes its books in a variety of electronic formats. Some content that appears in print may not be available in electronic books. For more information about Wiley products, visit our web site at www.wiley.com.

Library of Congress Cataloging-in-Publication Data:

Halpin, Daniel W.
Financial management and accounting fundamentals for construction / Daniel W. Halpin and Bolivar A. Senior.
p. cm.
Includes index.
ISBN 978-0-470-18271-0 (cloth)
1. Building–Estimates. 2. Construction industry–Accounting. 3. Building–Cost control. 4. Construction industry–Finance. I. Senior, Bolivar A. II. Title.
TH435.H323 2009
690.068′1–dc22 2009009715

Printed in the United States of America

SKY10059698_111023

CONTENTS

PREFACE

It has been noted that the construction industry is the "engine" that drives the national economy. Estimates of the of the total annual construction volume in the United States range as high as a trillion (1,000,000,000,000) dollars annually. Recent problems generated by a downturn in the real estate and construction markets have triggered a period of financial turmoil and government intervention to shore up economies worldwide. Financial management of construction revenues and cash flow in this sector obviously has a major role to play in world markets and economic stability in general.

Financial management and the measurement of financial activity in the construction industry is unique, since most revenues are generated in the context of projects that are designed and constructed. That is, the basic production unit of this gigantic industry is the *project*. This is in contrast to other industries, which produce units such as automobiles and electrical appliances or services to individuals such as medical care or foodservice. The number of projects that even a large contractor has at any given point is substantially smaller than the number of cars, refrigerators, or patients that businesses in the other industries produce or service. Moreover, the time frame involved in realizing a construction project ranges from a few weeks up to several years, compared to the few days, hours, or minutes typical of production or service activity in other industries. As will be discussed in this text, the small number of unique projects and the extended period of time required for the completion of each one place special requirements on the accounting and financial management systems typical of the construction industry.

This book takes advantage of experience gained from using a previous text by one of the coauthors, both in the classroom and as a professional reference. This original book by Halpin, entitled *Financial and Cost Concepts for*

Construction Management, was also published by John Wiley and Sons, Inc. (Wiley, 1985). The stated objective of that text was "to present both company and project levels of revenue and expense management in an introductory but integrated format. . . ." The Halpin text emphasized how financial activity at the project site is collected and reported to the company level to provide data for financial reporting, control, and management purposes. The present book draws upon the strengths of this original text by reflecting present-day practice and adding information regarding business taxation, project control, engineering economy, and financial forecasting.

The audience for the present book is primarily practitioners and students who may have a strong technology and engineering background, but relatively limited training in the areas of financial management and accounting. It is hoped that this book will help these individuals to become more aware of the way in which fiscal topics and the ebb and flow of revenues and expenses impact the generation of income and profit. In addition, for those individuals who are familiar with conventional accounting methods in manufacturing and service industries, this book will, hopefully, provide insights into how the project format of construction changes the way in which financial management is exercised in this large and specialized industry.

Chapters 1 to 4 provide an introduction to company-level financial management and accounting topics such as financial reporting, analysis of financial data, and the rudiments of the accounting procedures required to generate company Balance Sheets, Income Statements, and summary documents.

Chapter 5 addresses the importance of cost control systems and the establishment of cost accounts at the project level. Chapter 6 looks at the issues involved in forecasting cash flow and controlling overhead as well as concepts such as breakeven analysis.

Chapters 7 and 8 deal with the way in which borrowing, interest, and the time value of money influence financial decision making. Chapter 9 presents an introductory treatment of taxation and its impact on company operations and fiscal management.

The appendices provide support material covering the structure of a typical Chart of Accounts, the flow of various types of transactions through a construction accounting system, and tables required for mathematical analysis of transactions in which interest and time value of money must be considered.

Each chapter has been designed to be as self-contained as possible. Many college construction programs resort to the use of multiple generic courses in accounting, engineering economy, and other similar topics to address the subject matter covered here. It is hoped that the relative autonomy of each chapter provides teaching flexibility and adequate subject coverage within the context of a single-course format.

Review questions and exercises are included at the end of each chapter. They emphasize open ended thinking and reasoning rather than mechanical responses. The most common keywords found, even in numerical questions, are *why*, *explain*, and *discuss*. It is hoped this approach will promote freeform responses and a better understanding of the chapter's content. The construction-specific nature of the book is designed to provide a practice-oriented context to support better understanding of the generic (e.g., accounting, etc.) subject matter.

ACKNOWLEDGMENTS

Many people have made important contributions to this book. Material covered in this text has been taught by the authors to classes at Colorado State University, Georgia Institute of Technology, Purdue University, and the University of Maryland, College Park, over the past 20 years. Feedback from students has been incorporated into this to reworking of the classroom material. The authors would also like to thank colleagues who have used this text and provided feedback on improvements and modifications. In particular, the authors would like to recognize E. Paul Hitter, Jr. of Messer Construction Co. in Cincinnati, Ohio, for reviewing some of the chapters and providing comments from a practice-oriented viewpoint.

Finally, we would like to recognize the support and patience of our wives, Maria and Ana, during the preparation this text.

Daniel W. Halpin, Crestview Hills, Kentucky
Bolivar A. Senior, Ft. Collins, Colorado

CHAPTER 1

INTRODUCTION

THE BIG PARADOX

> "A construction manager is like an Olympic decathlon athlete who must show great competence in a multitude of areas ranging from design of construction operations to labor relations."*

Notwithstanding the multi-faceted nature of construction management, construction professionals are forced to focus heavily on the technical side of their work. Each project is a unique technological and organizational puzzle. A construction manager is in a race against time and money to reach targets relating to cost and required completion deadlines. Surprisingly, business objectives such as making a profit often take a back seat to the complex interplay of technology and organization. Bringing a project in on time and at bid price is like landing a jet fighter on an aircraft carrier in heavy seas.

Financial and business issues are often foreign to the interests of the field personnel who are locked in combat, on a day by day basis, with the solution of practice oriented problems in the field. It is almost as if making a profit is a secondary issue—a necessary evil. And yet, without profit, businesses fail. Small mistakes in judging the financial landscape often lead to big losses.

---

*Halpin, Daniel W., (2006), *Construction Management, 3rd Edition,* John Wiley and Sons, Inc. New York.

WHAT IS FINANCIAL MANAGEMENT?

Financial management concerns all the decisions involving money that a company must take every day. Some financial choices, such as deciding to stop building condominiums in order to free up resources, can have a substantial impact on a company. Others may be of much smaller scope, such as deciding to take advantage of vendor discounts available by paying invoices in a timely fashion. Regardless of size or impact, financial decisions can be made using a rational analysis of relevant factors just on the basis of intuition. A main proposition of this book is that rational and informed decisions will prevail in the long run over intuitive but uninformed choices.

Financial management finds its way into almost every corner of human activity (think of how many things in life involve money). It would be nearly impossible to address all the issues within its scope. Taxes, for example, are of relevance for almost everyone. Computing a project's profit to date, however, is much more relevant for a construction professional than to a stock trader. Optimizing a stock portfolio, on the other hand, is of little direct significance in construction, but it is of utmost importance for a stock trader. Consequently, this book—like any other specialty-focused book—is a subset of all the topics that we could address in financial management. Its topics are not only a collection of standard areas found in most construction oriented financial textbooks but are also a selection of what, in the judgment of the authors, will be useful to you throughout your career.

As a construction professional, you need to know accounting fundamentals, project-related financial matters, and company-level financial issues. Each one of these three areas has a substantial impact on your ability to succeed in your career. Let us take a bird's-eye view of these topics with some attention to the issues that they comprise.

First Stop: Financial Accounting

Financial accounting involves the capture of information regarding the purchase and sale of effort and products (e.g., TV sets, bicycles, real estate, construction of concrete footers, etc.). The information of interest is the revenue derived from sale and the expense involved in producing work and products for sale. The history of accounting is as old as commerce in society. It led to early forms of mathematics so that a system of measures could be used to keep track of value and the transfer of value between individuals. Businesses exist to produce a profit, and accounting allows for the determination of whether a profit or loss is occurring because of the activities of a given business activity.

Records of purchase and sale offer interesting insights into the operations of society from the time of ancient civilizations up to the present day. We encounter references to bookkeeping or accounting in classical stories such as Charles Dickens famous *A Christmas Carol*. Bob Cratchit, one of the main characters, is the bookkeeper for the firm of Marley and Scrooge. We see him sitting at a high desk writing figures into a ledger book using a quill pen.

Financial records maintained by historical figures tell us a great deal about their life and times. By studying accounting records from the eighteenth century, we can determine how founding fathers such as Washington, Jefferson, and John Adams faired financially throughout their brilliant and hectic careers. We can determine whether Mozart was really as poor as he is often portrayed (Actually, he had an annual income most of his adult life on the order of $250,000.)

We, as individuals, become involved in accounting at an early age as we receive and spend money from parents, aunts, and uncles. At some point, we open a bank account and must deal with a checkbook. We learn to study and reconcile bank statements, comparing how much we have deposited to how much we have spent.

Accounting is founded upon the acquiring, storing, and analyzing of financial information. This implies extensive record keeping and data management. The data captured by accounting systems, when properly displayed and analyzed, tell us something about the financial position or health of a business entity (e.g., Blockbuster Construction Co.) or an individual (e.g., Sarah Smith). Let us take a first look at the main components of financial accounting, which will be addressed in detail in Chapters 2, 3, and 4.

In order to summarize financial activities at a point in time (i.e., December 31, 2010), one report has become the cornerstone document used worldwide to provide a picture of the financial position of a person or a business activity. This report will be described and discussed in great detail in Chapter 2. Suffice it to say, —the *balance sheet*—attempts to capture a snapshot of financial position *at a point in time*. This snapshot is expressed in terms of *assets* and *liabilities*.

Assets are financial entities that have value and are controlled or owned by a firm or individual. Assets are what you have or own. Your bank account, car, and CD player are assets. Even if you owe money on your car or furniture, they are still considered your assets as long your ownership can be established.

Liabilities are what you owe or are committed to pay based on agreements and commitments with other parties. If you borrow money to buy your car and the loan is still not paid off, the amount you owe is a liability. All of this derives broadly from the idea of property, ownership, and legally binding commitments (sometimes formalized as written contracts). Commitments are also referred to as obligations.

The document that attempts to capture and reflect assets held and the obligations of a company or person is called a balance sheet. The balance sheet structure is a reflection of the basic equation of financial accounting. Simply stated, it indicates what a person or company has or owns and what debts or obligations are pending against what is owned. What is owned is referred to as assets. When one subtracts the obligations pending from what one owns, we have calculated the net value or (in financial terms) net worth of the person or company. This can be calculated at any point in time. The balance sheet is a detailed report of what one owns and what one owes at any given point in time. A detailed discussion of the balance sheet and its structure will be presented in the Chapter 2. Chapter 3 centers on the interpretation of financial information, and Chapter 4 covers the mechanics of creating financial reports.

Why Construction Accounting Is Different from Accounting in Other Business Sectors*

Worldwide construction is the largest economic sector of the global economy. Construction ranks number two in the amount of economic activity contributed to the gross national product (GNP) of the United States. It is the largest U.S. industry that focuses on the production of a physical product as opposed to provision of a service (e.g., the health care industry.) The dollar volume of the industry is on the order of one trillion (1,000 billion) dollars annually. The process of realizing a constructed facility such as a road, bridge, or building, however, is quite different from what is involved in manufacturing an automobile, a computer, or a cell phone.

Manufactured products are typically designed and produced without a designated purchaser. In other words, products (e.g., automobiles or TV sets) are produced and then presented for sale to any potential purchaser. The product is produced on the speculation that a purchaser will be found for the item produced. A manufacturer of bicycles, for instance, must determine the size of the market, design a bicycle that appeals to the potential purchaser, and then manufacture the number of units that market studies indicate can be sold. Design and production are done prior to sale. In order to attract possible buyers, marketing and advertising are required and are an important cost center.

Many variables exist in this undertaking, and the manufacturer is "at risk" of failing to recover the money invested once a decision is made to proceed with design and production of the end item. The market may not respond to the product at the price offered. Units may remain unsold or sell at or below the cost of production (i.e., yielding no profit). If the product cannot be sold so as to recover the cost of manufacture, a loss is incurred and the enterprise is unprofitable. When pricing a given product, the manufacturer must not only recover the direct (labor, materials, etc.) cost of manufacturing but also the so-called indirect and general and administrative (G&A) costs such as the cost of management and the implementation of the production process (e.g., legal costs, marketing costs, supervisory costs, etc.) Finally, unless the enterprise is a "nonprofit," the desire of the manufacturer is to increase the

---

*This and the following section is taken with permission from *Construction Management* by Daniel W. Halpin, published by John Wiley and Sons, 2006.

value of the firm. Therefore, profit must be added to the direct, indirect, and G&A costs of manufacturing.

Manufacturers offer their products for sale either directly to individuals (e.g., by mail order or directly over the Web), to wholesalers who purchase in quantity and provide units to specific sales outlets or to retailers who sell directly to the public. This sales network approach has developed as the traditional framework for moving products to the eventual purchaser. (See if you can think of some manufacturers who sell products directly to the end user, sell to wholesalers, and/or sell to retail stores.)

In construction, projects are sold to the client in a different way. The process of purchase begins with a client who has need for a facility. The purchaser typically approaches a design professional to more specifically define the nature of the project. This leads to a conceptual definition of the scope of work required to build the desired facility. Prior to the age of mass production, purchasers presented plans of the end object (e.g., a piece of furniture) to a craftsman for manufacture. The craftsman then proceeded to produce the desired object. For example, if King Louis XIV desired a desk at which he could work, an artisan would design the object, and a craftsman would be selected to complete the construction of the desk. In this situation, the purchaser (King Louis XIV) contracts with a specialist to construct a unique object. The end item is not available for inspection until it is fabricated. That is, since the object is unique, it is not sitting on the showroom floor and must be specially fabricated.

Because of the "one of a kind" unique nature of constructed facilities, this is still the method used for building construction projects. The purchaser approaches a set of potential contractors. Once an agreement is reached among the parties (e.g., clients, the designer, etc.) as to the scope of work to be performed, the details of the project or end item are designed and constructed. The purchase is made based on a graphical and verbal description of the end item, rather than the completed item itself. This is the opposite of the speculative process, where the design and manufacture of the product are done prior to identifying specific purchasers. For instance, it would be hard to imagine building a bridge without having identified the potential buyer. (Can you think of a construction situation where the construction is completed prior to identifying a buyer?)

Who Is at Risk?

The nature of risk is influenced by this process of purchasing construction. For the manufacturer of a refrigerator, risk is related primarily to being able to produce units at a competitive price. For the purchaser of the refrigerator, the risk involves mainly whether the appliance operates as advertised.

In construction, since the item purchased is to be produced (rather than being in a finished state), there are many complex issues that can lead to failure to complete the project in a functional and/or timely manner. The

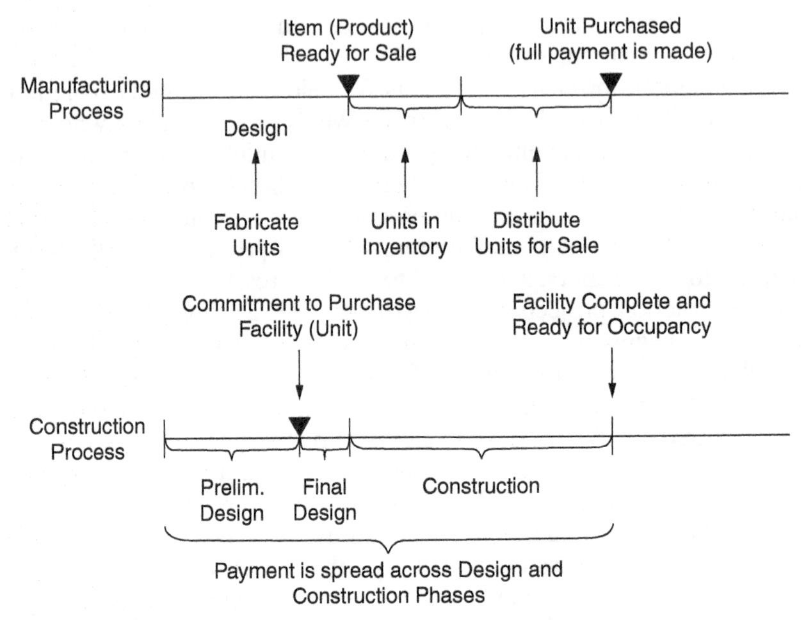

Figure 1.1 Manufacturing vs. construction timeline.

number of stakeholders and issues that must be dealt with prior to project completion lead to a complex level of risk for all parties involved (e.g., the designer, constructors, government authorities, real estate brokers, etc.). A manufactured product is, so to speak, "a bird in the hand." A construction project is a "bird in the bush."

The risks of the manufacturing process to the consumer are somewhat like those incurred when a person goes to the store and buys a music CD. If the recording is good and the disk is serviceable, the risk is reduced to whether the customer is satisfied with the musical group's performance. The client in a construction project is more like a musical director, who must assemble an orchestra and do a live performance, hoping that the performance and the final effect will be pleasing. The risks of a failure in this case are infinitely greater. A chronological diagram of the events involved in the manufacturing process versus those in the construction process are shown schematically in Figure 1.1.

Projects: The Output of the Construction Process

Another aspect that greatly influences the way in which construction is accounted for relates to the project format used for delivering the completed product. As noted previously, the construction industry is generally focused on the production of a single unique end product. That is, the product of the construction industry is a facility that is usually unique in design and method of fabrication. It is a single "one-off" item that is stylized in terms of its function, appearance, and location. In some cases, basically similar units are constructed, as in the case of town houses or fast-food restaurants. But even in this case, the units must be site adapted and stylized to some degree.

Mass production is typical of most manufacturing activities. Some manufacturing sectors make large numbers of similar units or batches of units that are exactly the same. A single item is designed to be fabricated many times. Firms manufacture many repetitions of the same item (e.g., telephone instruments, thermos bottles, etc.) and sell large numbers to achieve a profit. In certain cases, a limited number or batch of units of a product is required. For instance, a specially designed transformer or hydropower turbine may be fabricated in limited numbers to meet the special requirements of a specific client. This production of a limited number of similar units is referred to as batch production.

Mass production and batch production are not typical of the construction industry. Since the industry is oriented toward the production of single unique units, the format in which these one-off units is achieved is called project format. Both the design and production of constructed facilities are realized in the framework of a project.

Construction projects are completed over extended time periods. Even simple construction projects require many weeks or months to complete and can often extend over more than one year. This means that the client typically makes partial or progress payments to the constructor over the life of the project. Therefore, construction is paid for in a "pay as you go" format as opposed to the payment of a single amount at the time ownership is transferred. This requires a totally different method for recognizing the value transferred payment by payment. Methods used to account for the sale of manufactured products (e.g., refrigerators) are not applicable when dealing with projects delivered to the client over an extended period of time. A different form of accounting is needed, and that project form of accounting is a major focus of this text.

PROJECT-LEVEL CONTROLS

Since projects are the main business units for any construction company, a fundamental raison d'être for financial management, as applied in the construction industry, is the development and use of appropriate controls at the project level. We will address the financial planning and control at the project level in Chapters 5 and 6. Chapter 5 emphasizes the planning and control of operations. How can we know whether a project is ahead of or behind schedule, and over or under its budget? We will use the concepts of earned value, scheduled value, and actual value to anchor the simple and ingenious principles applied in modern cost control and analysis. Chapter 6 centers on the estimating of the cash requirements for a project. A contractor can execute a million dollar project with much less than one million dollars invested in the project at any given time. The progress payments paid by the project owner to the contractor, as well as the trade credit that a contractor can procure, serve to reduce the cash requirements needed to build the project. As noted above, the project is sold month by month to its owner as the construction proceeds.

TIME VALUE OF MONEY

No enterprise can survive the modern business environment without a very good grasp of the concepts and techniques related to the time value of money. At the most obvious level, the company must be able to estimate the payments that it will make to repay borrowed money. But, many other, more subtle issues can be equally important. Determining the attractiveness of a business scheme, comparing several alternatives, and finding the true cost of a business proposition when interest is considered are examples of the immediate and critical usefulness of these techniques. Chapter 7 addresses the time value of money, using the techniques of *engineering economy*.

ENTREPRENEURIAL ISSUES

No text on construction financial management would be complete without including information about two crucial aspects of the construction business—the financing process and tax issues. Chapter 8 is about construction loans and credit. How does a company get the financial resources to execute a contract? The role and cost of lines of credit and term loans are critical to the success of a project. How an entrepreneur procures the money to build the project will be examined. This aspect is sometimes underestimated by contractors. The financial merits and attractiveness of a project are of great importance to the contractor. Money is, after all, a cascading resource. If the source of money runs dry at the entrepreneur's level, the contractor and everyone working under him will suffer the consequences. The policy of paying suppliers in a timely manner to receive discounts on the invoiced amounts will also be discussed.

Chapter 9 offers an introduction to tax issues affecting the typical contractor. On one hand, there are opportunities to save money paid in taxes when there is an understanding of the rationale and implementation of the current tax system. On the other hand, the lack of such knowledge can result in missed opportunities at best and imprudent decisions at worst.

REVIEW QUESTIONS AND EXERCISES

1. What attracted you to the construction industry? Discuss the advantages and disadvantages of a career in construction. List in descending order what you like the most about this industry. Develop a similar list with the factors that you dislike.
2. Write a one-page review of the financial challenges for a major ongoing construction project or for a current issue in the construction industry involving the management of finances. You can use sources such as the *Engineering News Record* (ENR), a magazine available in most college and public libraries. A great deal of information is also available on the Internet, including a free, limited version of ENR.
3. Contrast the effort of building a car and building a residential unit. Which one requires more initial capital? What kinds of resources are involved? What are the main differences in their management needs?
4. Would the cost of residential construction benefit from a greater use of prefabricated building components? Why is it that prefabrication is not more widely used in this industry?
5. Interview a construction professional, and report the financial controls that he or she currently uses. In that person's opinion, how could these controls be used more effectively?
6. This chapter discusses some important differences between the manufacturing and construction industries. Can you think of issues that are common to both industries?

CHAPTER 2

UNDERSTANDING FINANCIAL STATEMENTS

INTRODUCTION

It is common to talk of "the bottom line" of an action as its ultimate outcome. This everyday expression comes straight from the accounting world. It refers to the last line of a company's Income Statement, which usually contains the amount of its profit. The bottom line sums up the performance of a company's management in the same way that a competition score reflects an athlete's or team's performance. No athlete can be successful without a good understanding of their game's scoring system. For the same reason, a construction manager needs to know how his or her performance—the bottom line—is being tallied up.

Accounting is about tallying up a company's performance as reflected by its bottom line. We will discuss here the information required to keep score in the construction game, how the scoreboard looks, and what tricks of the trade are useful to make sense of the score. As in sports, the world of business needs the feedback provided by tracking scores to decide whether current performance is effective and to change strategies if necessary.

More formally, accounting can be defined as the process of recording, summarizing, and communicating financial data. Each one of these three primary functions has enormous importance for anyone concerned with a company's performance. Lenders want to know whether the company can be reasonably expected to pay back all loans; investors gauge the company's potential to generate profit, especially compared to other possible investment possibilities; the company owners need to keep track of the actions taken on their behalf by its managers; and the government wants to know the company's bottom line to tax it accordingly.

WHY SHOULD YOU CARE ABOUT ACCOUNTING?

Let's use flying an airplane as an analogy to managing a company. As you fly though a cloud bank, you may be flying upside down without realizing it. The plane's instruments let you know where you are, how fast you are flying, and your altitude and attitude (i.e., whether you are level, turning, banking, or upside down). It is very important to know where you are so that you can plan for where you are going. Without an accounting system, from a financial point of view, you have no "instruments" and you are flying blind.

So that the instruments are reliable, consistency is crucial. That is, the definition of what is up, down, sideways, the basis for measuring air speed, the definition of altitude (i.e., above the terrain, above sea level, etc.) must be consistent and reliable. Therefore, definitions play a key role in keeping score financially. If a debit means one thing in Akron, Ohio, and a different thing in Denver, the definitions are not consistent and we have chaos.

With millions of companies involved in a wide variety of businesses, it is remarkable that conceptually a single system can keep track and report the finances of them all. It is amazing that, in concept, the accounting used in the United States is applicable, with minimal variations, throughout the world. A construction company uses essentially the same accounting principles used by a car manufacturer, the corner grocery store, or a school PTA. The major differences relate to the type of product or service and the framework (i.e., in terms of time, etc.) within which the product or service is delivered.

GENERALLY ACCEPTED ACCOUNTING PRINCIPLES

Modern accounting is the result of a long evolution and trial-and-error process that began centuries ago. The set of principles known as Generally Accepted Accounting Principles (GAAP) synthesize the assumptions underlying all modern accounting. In the United States, the Financial Accounting Standards Board (FASB) translates these principles into actionable rules. FASB is a private institution, but its statements, bulletins, and other pronouncements carry such weight that any formal financial document is virtually guaranteed to follow its rules. The Internal Revenue Service (IRS) and a few other government agencies also provide accounting rules. The IRS and FASB resolve any potential contradiction in their rules before they need to be followed.

There are many Generally Accepted Accounting Principles. The following informal summary includes the most important for a beginner accountant, with a very brief explanation of their meaning.

Conservatism. An accounting system should recognize losses as soon as they are foreseeable but recognize gains only when they are certain.

Consistency. Accounting methods and reports should not change from period to period. If a change is implemented, its effect on the interpretation of the financial status of the company must be disclosed.

Acquisition cost principle. Assets should be valued in the accounting system at their actual, initial cost.

Revenue realization principle. Revenue can be recorded when the sale takes place and there is a reasonable expectation of collecting the money involved in the sale.

Matching principle. The revenues and expenses of any transaction must be reported in the same Income Statement. This is the very basis of double-entry accounting.

Full disclosure principle. Financial reports must be complete and explicit enough as to avoid confusion or mislead its users.

Materiality principle. Any transaction significant enough to influence the judgment of a reasonable person making a financial decision must be included in an appropriate report.

The unified environment offered by GAAP as translated into rules by FASB provides an unambiguous system for analyzing a company's finances. A clear-cut set of rules is extremely important because crafty mangers can be tempted to tweak the accounting system to paint their company in an undeserved favorable light. A notable case was that of Enron Corporation in the late 1990s and early 2000s. To convince investors and shareholders that the company was in good financial health, some of its managers created a number of illegal schemes such as "selling" their losing ventures to phantom companies that would not appear in the company financial statements. This scheme worked largely because the company was able to make its external auditors—essentially, their financial police—look the other way. As soon as an accounting review was performed by concerned investors, the ongoing misdeeds were detected. The system, after all, did work even in this particularly appalling case.

CASH AND ACCRUAL BASES: TWO WAYS TO LOOK AT ACCOUNTING

Let's consider further the metaphor of accounting as a scoring system. Every sport—particularly if it is widely practiced—will have some latitude in its rules, depending on who is playing. Rules for Little League baseball would not make sense for professional adult leagues. Similarly, accounting has rules for small companies that would be as inappropriate for a large corporation as for a big-leaguer to appear in the roster of a T-ball team. We will see these size accommodations in income recognition, taxation, and other practical aspects of accounting. The first one is considered here: When should we account for money? The answer is "When money or value is transferred from one person, group, or financial entity to another." The question now becomes "When should we recognize the transfer of money?"

CASH BASIS OF ACCOUNTING

Let's discuss when and how money or value (measured in terms of money) is transferred from point to point. Consider a contract situation where money is received from a client as a progress payment. The client makes payments over the life of the project. When the payment is received, the contractor can pay expenses that have occurred in construction of the project. A small contractor, perhaps with no permanent employees, may feel that there's no reason to complicate money matters. The contractor would base his accounting on the simple relationship:

$$\text{Profit to date} = \text{Money received to date} - \text{Money paid out to date}$$

When we compute the profit to date with this formula, we are using the Cash basis of accounting. The Cash basis considers only money actually received or paid. If we will receive an amount of money tomorrow, then we will wait until tomorrow before factoring it into our accounting. Similarly, we will account for any payment made only when we physically write the check paying it. Someone looking at our financial statements right now would not be able to know how much we owe or are owed.

The Cash basis is relatively common in the construction industry, since many companies have a small volume of work and lack the resources to implement an accounting system on a more sophisticated basis. Of 802,349 construction-related companies operating in 2004, the U. S. Census Bureau (2004 Economic Census, NAICS 23) reports that 645,669 (80.5%) had 9 or less employees. For these small companies, the simplicity of checkbook-balancing accounting—which is, in essence, the Cash basis of accounting—can be attractive.

A substantial problem with using the Cash basis in the context of the construction industry is due to the slow and uneven rate at which clients pay contractor's progress payment requests. Profit to date will appear smaller if the money received to date doesn't account for uncollected bills, and will appear disproportionately large in the period when delayed payments are received. Moreover, even small contractors usually make purchases or receive services on credit. Profit to date would be inflated by delaying payments to suppliers or other creditors. These biases make the profit computed using the formula above nearly useless. A contractor can be needlessly panicked or dangerously overconfident because of these unbalanced results.

There are other problems with the Cash basis. For example, a company may purchase materials and place them in inventory for weeks or months. Using the Cash basis, this inventory could reflect negatively on the company's profit for months. There could be a long time between the moment the company pays for these materials and the moment it receives a progress payment from the project's owner for their installation. The payment occurs only after these materials have been satisfactorily incorporated into the project.

ACCRUAL BASIS OF ACCOUNTING

The alternative formula most widely used looks similar to the Cash formula. However, it is fundamentally different:

$$\text{Profit to date} = \text{Money earned to date} - \text{Cost to date}$$

An accounting system based on the preceding formula is using the Accrual basis of accounting. Money or value is recognized at the time at which the responsibility for payment is incurred. In this basis, money earned is the money that we have either received or can expect to receive. Cost to date is all the money that we have either paid or can expect to pay. Profit to date is much more accurately reflected in an Accrual basis, because its computation does not depend on the fortuitous date when an invoice is paid or a progress payment is received. The point in time of interest is the point at which responsibility is incurred. If you receive a valid invoice, you have the responsibility to pay it at the time of receipt. A check may be sent later, but responsibility to pay is incurred immediately.

The Internal Revenue Service (IRS) requires U.S. companies with average annual revenues (invoices to clients) of more than one million dollars to use, with some exceptions, the Accrual basis for tax purposes. Some small companies do choose to use the Cash basis to save on taxes, as opposed to just for its simplicity. That would be the case, for example, of a construction company with a large amount of uncollected invoices or billings. Since the uncollected billings would not count for profit purposes, the company would pay less tax than if its accounting used the Accrual basis.

Financial statements following the Cash basis do not have Accounts Receivable or Accounts Payable, since expectations to receive money, in the case of Accounts Receivable, or promises to pay debts, in the case of Accounts Payable, are not considered in the Cash approach. Only actual transfer of cash is of interest. As a result, the Cash basis would make job costing extremely difficult.

Imagine how unreliable the as-built unit cost of, say, a cubic yard of reinforced concrete would be using the Cash basis. Invoices for some components (such as formwork and reinforcement) might have been paid but others (for example, concrete) might still be unpaid. The Cash basis would indicate that all unpaid components cost nothing, and the resulting unit cost would be totally inaccurate. Any practical job cost system assumes that funds will eventually be disbursed for their payment, and therefore, their cost is considered as soon as they are used, not when cash payment is received.

Because of its limited ability to reflect real-world financial transactions, this book will only discuss the Cash basis in the broadest terms. Emphasis will be placed instead on the Accrual basis of accounting.

ACCOUNTS

The building blocks of any accounting system are its accounts. We can imagine an account as a labeled sheet of paper where we record and keep track of how much the company owns, owes, receives and spends related to a particular topic. An account is a tally of how much we have spent or taken in for an activity center (e.g., paving, slab on grade, etc.) Typical account titles are:

Cash
Accounts Receivable – John Q. Dalton
Accounts Receivable – Jane Q. White
Fixed Assets – Equipment – CAT D8 ID 4356
Accounts Payable – My Supplier, Inc.
Owners' Equity – Common Stock
Cost of Materials - Concrete

We need to keep track of how much we owe to My Supplier, Inc., and therefore we keep track of the cost of our purchases and our payments to this supplier. The amount that we currently owe to My Supplier, Inc. is the balance of the account. To be useful, an account must focus on a single financial issue. It would not make sense to have an account called Inventory and Accounts Receivable. Although it is important to know the value of materials in stock and how much the company has in outstanding billings, the sum of these two items (that is, the balance of this strange account) would be of no use. It would be a veritable case of apples and oranges. Deciding how each account is defined and how many accounts to maintain for good financial control are key issues in establishing a functional accounting system. Again using the airplane analogy, depending on the size and complexity of the aircraft, only a few (a small one-engine plane) or many instruments (a Boeing 767) are required.

ACCOUNT HIERARCHY

An accounting system monitors all account balances and summarizes this financial information at several levels. The balances of the individual accounts for Mr. Dalton and Ms. White, mentioned in the previous section, are consolidated in an account called "Accounts Receivable" (This account is frequently referred to by its acronym, A/R). An accounting system always has broad-scope[1] accounts whose balances consist of the sum of the balance of similar, more detailed accounts. In this case, the account "Accounts Receivable" has two subaccounts, A/R John Q. Dalton, and A/R Jane Q. White. In turn, each of these two subaccounts could have its own subaccounts, for example "A/R John Q. Dalton – Project A" and "A/R John Q. Dalton, Project B."

Summarization is crucial for making sense of our finances. We need to know how much we owe to all suppliers as much as we need to know how much we owe to each individual one. Each company has a unique detailed list of accounts (and subaccounts), collectively called its Chart of Accounts. A typical Chart of Accounts for a construction company is given in Appendix A. Various trade and professional organizations publish standard Charts of Accounts that may be adopted by a particular construction company. One characteristic of construction accounting is that much of the revenue and expense activity is generated by projects and, therefore, most expense and revenue accounts are maintained by project. A typical set of project-related expense accounts is shown in Figure 2.1.

It can be observed that the numbers used to designate the accounts have been divided into major groups. For example, all accounts between 100 and 699 refer to project work. The level of detail depends on the specific needs and preference of each company. The makeup of the accounts in Figure 2.1 suggests that this company is interested in tracking its concrete costs in detail. Account 240, Poured concrete, includes 14 subaccounts, from .01 to .90 (or, more properly, 240.01 to 240.90. The initial 240 is omitted to avoid visually cluttering the chart). This allows keeping an individual record of the costs incurred for the concrete in footings, grade beams, and the other subaccounts shown.

FINANCIAL REPORTS

As instruments in an aircraft reflect the speed, altitude, and attitude of the plane, financial reports reflect the financial position of a person or a financial entity (e.g., company, organization, project, etc.). The two main financial reports that give us the financial picture are the Balance Sheet and the Income Statement. Both reports show specific broad-scope accounts and their balances (subaccounts are usually omitted), arranged into groups of similar accounts, such as all accounts that list money that the company owes. There are six account groups, or categories. Three categories are shown in the Balance Sheet and three in the Income Statement.

[1] A broad scope account can be viewed as an account that summarizes or consolidates the balances or values of a number of subaccounts.

MASTER LIST OF PROJECT COST ACCOUNTS Subaccounts of General Ledger Account 80.000 PROJECT EXPENSE					
		Project Work Accounts 100–699			Project Overhead Accounts 700–999
100		Clearing and grubbing	700		Project administration
101		Demolition		.01	Project manager
102		Underpinning		.02	Office engineer
103		Earth excavation	701		Construction supervision
104		Rock excavation		.01	Superintendent
105		Backfill		.02	Carpenter foreman
115		Wood structural piles		.03	Concrete foreman
116		Steel structural piles	702		Project office
117		Concrete structural piles		.01	Move in and move out
121		Steel sheet piling		.02	Furniture
240		Concrete, poured		.03	Supplies
	.01	Footings	703		Timekeeping and
	.05	Grade beams		.01	security
	.07	Slab on grade		.02	Timekeeper
	.08	Beams		.03	Watchmen
	.10	Slab on forms	705		Guards
	.11	Columns		.01	Utilities and services
	.12	Walls		.02	Water
	.16	Stairs		.03	Gas
	.20	Expansion joint		.04	Electricity
	.40	Screeds	710		Telephone
	.50	Float finish	711		Storage facilities
	.51	Trowel finish	712		Temporary fences
	.60	Rubbing	715		Temporary bulkheads
	.90	Curing	717		Storage area renta!
245		Precast concrete	720		Job sign
260		Concrete forms	721		Drinking water
	.01	Footings	722		Sanitary facilities
	.05	Grade beams	725		First-aid facilities
	.07	Slab on grade	726		Temporary lighting
	.08	Beams	730		Temporary stairs
	.10	Slab	740		Load tests
	.11	Columns	750		Small tools
	.12	Walls	755		Permits and fees
270		Reinforcing steel	756		Concrete tests
	.01	Footings	760		Compaction tests
	.12	Walls	761		Photographs
280		Structural steel	765		Surveys
350		Masonry	770		Cutting and patching
	.01	8-in. block	780		Winter operation
	.02	12-in. block	785		Drayage
	.06	Common brick	790		Parking
	.20	Face brick			Protection of adjoining property
	.60	Glazed tile	795		Drawings
400		Carpentry	796		Engineering
440		Millwork	800		Worker transportation
500		Miscellaneous metals	805		Worker housing
	,01	Metal door frames	810		Worker feeding
	.20	Window sash	880		Genera! clean-up
	.50	Toilet partitions	950		Equipment
560		Finish hardware		.01	Move in
620		Paving		.02	Set up
680		Allowances		.03	Dismantling
685		Fencing		.04	Move out

Figure 2.1 List of typical project expense (cost) accounts.

The Balance Sheet shows the value of all the items that a company owns and the money that the company owes, to either lenders or the company owners, at a given point in time. It is composed of these three account categories:

1. ***Assets***. Accounts in this category keep track of everything that the company owns. This can also be referred to as "holdings" or "what you have."
2. ***Liabilities***. These accounts list all the company's debts, except those related to company ownership. This category covers what you owe or are committed to pay or perform.
3. ***Owners' Equity***. This account category records all the money that the company owes to its owners.

The Income Statement shows how much money the company earned and spent over a period of time. The difference between the amounts earned and spent is the profit made by the company during the reported period. If the company spent more than what it earned, then the difference is called a loss for the period. The Income Statement encompasses the other three main account categories:

1. ***Revenues***. Accounts here keep track of the money that the company makes over a period of time.
2. ***Expenses***. This category lists all the money that the company spends over a period of time.
3. ***Profit (or income, earnings)***. This is a short but all-important category, consisting of the difference between revenues and expenses for the period covered by the Income Statement.

Understanding the details and nuances of these two reports, their categories and subtotals, and the accounts composing each category will be the object of our attention throughout much of this book. There are other financial reports that support these two cornerstone reports, which we will introduce later in this chapter.

BOOKKEEPING

We have defined the Balance Sheet and Income Statement as the end products of a summarization process. But, how is this summarization process performed? The process of entering ("posting") the value of financial transactions into a company's various accounts is a specialty within accounting, called bookkeeping. Furthermore, bookkeeping keeps track of all changes in a company's financial composition, and checks the integrity of a financial system. We will discuss bookkeeping fundamentals in Chapter 4.

THE BALANCE SHEET

As previously discussed, a Balance Sheet shows everything that a company owns or owes at a point in time. The last part of this definition is very important. A company's Balance Sheet is dynamic and constantly varies as it acquires new items, sells or depletes them, increases the money it owes (for example, by purchasing an item on credit), pays off some of its debts, or disburses part or all of its profit to its owners.

Since the Balance Sheet shows a company's financial structure at an instant in time, it is frequently said that it is like taking a photograph of the company's financial status. To illustrate the idea, let's consider a TV monitor in an airport terminal showing airplane arrivals and departures. This screen will change frequently, as airplanes arrive or leave the terminal. Imagine that we take a picture of this screen at a point in time of our choosing, say midnight. This picture would be quite useful: It would reveal which airlines were operating in this airport, how many gates were in use, how many airplanes were expected to arrive soon, and the times of departures. However, looking at this picture we would not know if an airplane had just arrived or if a flight was delayed an hour ago. All the information would be true at exactly the instant that we took the picture, but possibly five arrivals and two departures were added in the minute after we took our picture. There would be no way to tell.

Imagine now that our screen shows the accounts and balances of a given company. Similarly, we would watch the screen changing as the company operates. If it pays a bill, the balance of its Accounts Payable decreases[2] and its Cash also decreases by the same amount; if it receives a progress payment, its Accounts Receivable decreases[3] and its cash increases. If we take a picture of this screen at a time of our choosing, it would show the balance of each account. As with the airport screen, much useful information would appear in the picture; however, we would not know anything about events happening after the picture is taken, or even whether an account balance has changed in the previous minute (as we would not know whether a flight was rescheduled a minute or a day before our picture).

While the picture of the airport screen has no special name, we do give a very specific title to the picture showing our account balances. It is called the company Balance Sheet at time *X*, *X* being the moment at which we took the picture. The purpose of this mental exercise is to help you clearly understand the gist of a Balance Sheet and its limitations. A most common confusion is to assume that the balance of a particular account shown on a Balance Sheet gives a reliable indication of its average balance in the past. It does not.

---

[2] Since an account payable indicates we owe a certain amount, if we make a payment the balance of the A/P will decrease.

[3] If we receive as payment, the outstanding balance of the Accounts Receivable will get smaller since someone paid us what we are expecting to receive.

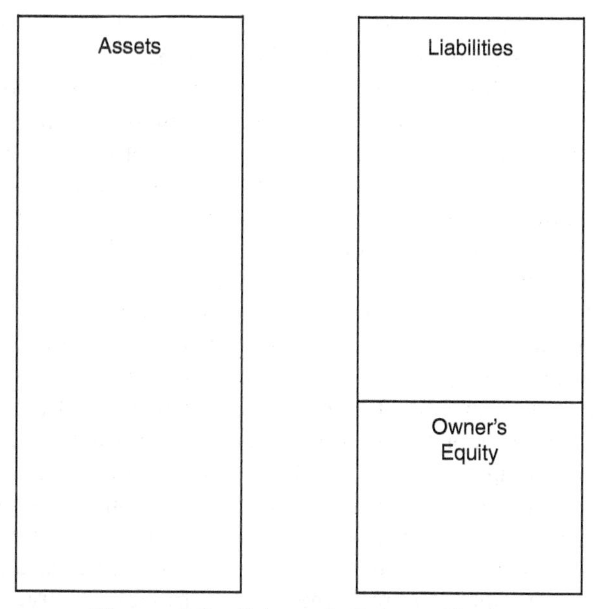

Figure 2.2 Schematic Balance Sheet.

BALANCE SHEET LAYOUT

As mentioned, a Balance Sheet consists of three major account categories, namely Assets, Liabilities, and Owners' Equity. These categories (and their balances) are the broadest level of detail shown in any Balance Sheet, as shown in Figure 2.2.

The Balance Sheet in Figure 2.2 shows Assets on the left and Liabilities and Owners' Equity on the right. Alternatively, they could be shown in one column, beginning with Assets, then Liabilities and finally Owners' Equity. As is true of almost all financial reports, any Balance Sheet (in the United States) must follow one of these two arrangements. Accounting is based upon standardization, and for good reasons: it makes it easier to compare information across companies, and makes it harder to hide information in a maze of accounts.

BALANCE SHEET ACCOUNT CATEGORIES IN DETAIL

Let us revisit, in more detail, the definitions of the three main categories constituting a Balance Sheet. This more detailed view will lead us to the so-called Fundamental Accounting Equation. This equation shows that all accounts are mathematically linked and balanced by a fundamental relationship. Our tour of the Balance Sheet ends by describing the sub-categories within the three main account categories discussed in this section.

Assets are everything of value that the company owns. This definition makes clear that a company's physical possessions such as furniture, equipment, or inventory are assets. Some assets are, however, not physical in the sense of a computer or a bulldozer. These assets pertain to value in terms of cash (e.g., a checking account) or legal obligations on the part of a client of the company (e.g., a bank note documenting a loan). Even more abstract are items representing value such as patents or copyrights. These items are called *intellectual property*, and are assets because they do have value. The company can collect royalties from selling the use of patents and copyright.

Some things may appear to be assets. It is common to hear a company owner state that "Our valued employees are our greatest assets." Although this may be true from one perspective, financially a company does not own its employees. Owning employees went out of fashion following the Civil War. Therefore, employees cannot be viewed as financial assets on the Balance Sheet. In general, if a company does not control an entity for purchase or sale or have authority for its use, that entity cannot be viewed as an asset.

Liabilities encompass commitments (e.g. money, contractually required work performance, etc.) that the company has made to others. This definition extends to all the money that the company must expect to pay. Normally, the receipt of a request for payment or similar document generates recognition of need to pay or perform. In some cases, the obligation is inherent in a contract and is recognized within the liability section of the Balance Sheet even though an invoice for payment has not been received. Businesses to which the company owes money are called *creditors*.

Owners' Equity is the last main category listed in the Balance Sheet. It comprises a special type of liability, namely the money that the company owes its owners. More accurately, Owners' Equity is the residual money amount after deducting the value of all company liabilities from the value of its assets. This definition recognizes that a company must give priority to paying its creditors over any competing request of an owner to remove his or her investment in the company. This perspective of how company owners relate to their investment has resulted in another common name for a company's Owners' Equity, namely *Net Worth.*

Owner's Equity includes two main accounts: the *Initial Capital* that owners provided to start up the company and all the profit that they have chosen to retain and reinvest in it. This reinvested money is registered in an account named *Retained Earnings.*

THE FUNDAMENTAL ACCOUNTING EQUATION

Returning to Figure 2.2, one can see that the area shown for Assets is the same as the areas of the Liabilities plus Owners' Equity. In other words,

$$\text{Assets} = \text{Liabilities} + \text{Owners' Equity}$$

This equality is called the *Fundamental Accounting Equation*.[4]

Why is this equation true? Assets consist of the value of *everything that a company owns*, and Liabilities plus Owner's Equity are *everything a company owes* (owner's equity is a special type of liability). Replacing the two sides of the equation by the words in italics above, the equation means that:

$$\text{Everything a company owns} = \text{Everything a company owes}$$

Could there be anything that a company owns but does not owe to either its creditors or its owners? Think about what happens when a company goes out of business, that is, it ceases to exist either because its owner decides to break it up or it is terminated under bankruptcy. Every single truck, pencil, and the cash in the bank will be claimed by someone. First, all liabilities must be paid off with the money resulting from sale of the company's assets. Then, owners will distribute among themselves every remaining penny.

ASSET VALUES

Assets appearing in a Balance Sheet are valued at their purchase price or a reasonable estimate of their initial value. The purchase price or other initial value (historical cost) of an asset is called its book value. Changing an asset's book value is unusual and complicated (the asset's value must be *impaired*), and for practical purposes, it can be said that assets keep their book value regardless of factors such as inflation and current market price. The exception to this historical cost rule applies to construction equipment and similar assets, whose economic life is intrinsically short. The value of these assets appearing in a Balance Sheet is reduced over time in a process called depreciation. For example, a truck may have a purchase price of $40,000. After one year, it is normally allowable to reduce its book value by 20%. The truck depreciation is then reduced by 20% of $40,000 = $8,000, and its new book value becomes $40,000 – $8,000 = $32,000 (the allowable depreciation percentage is regulated and usually varies for each year of the truck's life).

An asset's book value is usually different from its market value. An asset's market value is determined by how much money the asset would make if it were put up for sale at a given point in time. The market value of an asset can vary from day to day and may have little relation to its book value. The subjectivity of assigning a market value to any asset can easily result in a misleading assessment of the value of a company's assets. This is a main reason for listing assets at their net book value and not their market value.

---

[4] Since owners' equity is also called net worth, the Fundamental Equation is also written as Assets = Liabilities + Net Worth.

THE FUNDAMENTAL EQUATION AND OWNERS' RISK

The Fundamental Equation holds true for any Balance Sheet. It does not matter if the company is large or small, profitable or near bankruptcy, has just received a large amount of money or paid a large sum of money. The accounting balance must be maintained so that Assets (listed at their book value) always equal Liabilities plus Owner's Equity.

The Fundamental Equation ceases to be applicable when a company is closed and its assets are liquidated at their market value. Creditors expect to be paid in full what is owed to them. Owners hope to receive the net worth or residual value of the company based on the equation:

$$\text{Owner's Equity (Net Worth)} = \text{Assets} - \text{Liabilities}$$

What happens if the sale of assets is considerably less that the value shown in the Balance Sheet? In other words, what happens if the market value of the assets is much lower than the book value shown in the Balance Sheet? In such a case, the following equation would hold true:

$$\text{Assets (Market Value)} < \text{Liabilities} + \text{Owner's Equity}$$

Creditors must be paid first, and the remaining value in the company (if any) is distributed to the owners or stockholders (if the company is a corporation). In a worst case scenario, the market value of the assets may not even cover the amount of the liabilities. In this case, the owners receive nothing (e.g., the Enron collapse), and the creditors receive only a percentage of the outstanding liabilities. The amounts paid out have to be determined by a bankruptcy court. Therefore, the creditors and owners of a company are always at risk of not collecting the money owed by the company or invested in the company, respectively.

The risk of not only failing to realize a profit but of losing part or all of their investment, justifies the special role of company owners in a capitalist enterprise. Owners take the glory and financial gains when their company succeeds, but as this section shows, they receive not only the infamy but also the financial consequences of a failure.

BALANCE SHEET FOR FUDD ASSOCIATES, INC.

Fudd Associates, Inc. is a privately held corporation that your favorite uncle, Amadeus T. Fudd, has developed over the past 20 years. The firm specializes in light commercial construction and has been reasonably successful in its market niche. Like most companies, however, it has had its "ups and downs."

Table 2.1 shows the Balance Sheet for Fudd Associates, Inc. It shows accounts typical of most construction companies together with representative (example) balances. In this expanded Balance Sheet, you can see that the

TABLE 2.1 Balance Sheet as of December 31, 20XX for Fudd Associates, Inc.

Fudd Associates, Inc BALANCE SHEET At December 31, 20XX		
Current assets		
Cash and cash equivalents	2,494,298	
Account receivables	8,535,901	
Costs & recog'd earnings in excess of billings	933,165	
Inventories	222,106	
Other current assets	695,080	
Total current assets		12,880,550
Noncurrent assets		
Property, plant, and equipment	4,261,559	
Less accumulated depreciation	(2,344,097)	
Property, plant, and equipment, net	1,917,462	
Other noncurrent assets	1,180,837	
Total noncurrent assets		3,098,299
Total assets		15,978,849
Current liabilities:		
Current maturity on long-term debt	215,714	
Notes payable and notes of credit	356,328	
Accounts payable	6,078,354	
Accrued expenses	913,990	
Billings in excess of costs & recog'd earnings	1,738,499	
Other current liabilities	471,376	
Total current liabilities		9,774,261
Long-term liabilities		
Long-term debt, excluding current maturities	1,017,853	
Other long-term liabilities	453,799	
Total long-term liabilities		1,471,652
Total liabilities		11,245,913
Net Worth:		
Common stock	1,073,779	
Retained earnings	3,659,156	
Total net worth		4,732,935
Total liabilities and net worth		15,978,849

Assets category has two sections. Current Assets and Non-Current Assets. Liabilities are similarly divided into Current Liabilities and Non-Current Liabilities. Current, as used in accounting, usually means near in time or short term. A *current asset* is one that can be generally converted to cash within one year. Accounts Receivable, for example, are considered to be current assets because, in general, companies collect their bills within one year of requesting payment.[5]

[5] The key issue identifying a current asset account is that its underlying assets are intended to be liquidated within a year.

Land and real property, on the other hand, are non-current assets (also called fixed assets) because in many cases companies cannot count on selling real property quickly. Similarly, current liabilities are understood to be due within one year from receipt of request for payment (i.e., an invoice or billing).

The accounts with each asset category are in their approximate order of liquidity. That is, they are listed by how easy it is to convert their value to cash. On the liability side, accounts are listed in their approximate order of payment immediacy, meaning how quickly their balance is due for payment. The Owner's Equity section is subdivided as discussed below.

Although accounts within each section generally follow their order of liquidity or payment immediacy, this sorting is ultimately a matter of tradition. Current Assets is a case in point. Accounts Receivable, for example, is always listed before inventories and construction equipment. This is the case even though certain contractors may contend that the effort to sell a materials inventory or a piece of equipment is less than that required to collect outstanding accounts.

KEY ACCOUNTS

Several accounts are unique to the construction industry and are shown differently in the Balance Sheets of other industries. Moreover, some accounts tend to have more significance, or less significance, than they are afforded in other industries. We will provide a detailed explanation for a few cases of interest.

The following are significant accounts in the Assets side.

Cash and *Accounts Receivable* are the largest asset accounts for most contractors. Construction companies need a lot of cash to operate. Contractors may buy materials and services from as many as 80 to 100 vendors and subcontractors in addition to managing multiple payrolls for in-house personnel. This requires a high degree of skill in cash flow management. Failing to budget cash requirements is one of the most common reasons for the high incidence of business failures in construction.

Inventory, as shown on a construction Balance Sheet, is typically a small account referring to materials inventories.[6] Materials are incorporated into the project immediately and contractors typically maintain very small materials inventories. Inventory in the sense used in manufacturing firms (i.e. finished units ready for shipment and sale) does not exist in construction since the finished unit is the entire project. With the exception of residential home

[6] In manufacturing companies, inventory refers to units of production finished and on hand to be sold. For instance, an electronics manufacturer will have a number of finished TV sets in factory ready to be shipped and considered to be inventory.

builders, building, industrial, and engineering contractors do not build speculative projects and then try to sell once they are completed (see Chapter 1). These contractors do not maintain an inventory of finished projects which are carried as assets on the Balance Sheet.

Cost and Estimated Earnings in Excess of Billings is an account unique to construction. It recognizes work performed or materials installed that have not yet been billed to the client. In manufacturing firms, this account would be called "Work in Progress – Expense" or a similar name. In construction, it is common to have materials on-site which have not been billed. However, these materials are recognized as assets held by the contractor prior to installation. The value of such materials would be carried in this account. Similarly, work performed for rough grading a parking lot on a project were payment is based upon square meter of finished asphalt parking surface would be reflected in this account. Because this account is used to show work performed or materials and services procured but not yet billed, it reflects an "*underbill*" and can be viewed as an "*underbilling*" account.[7]

This account is also utilized to show work performed by the contractor that occurred following the submission of a progress payment request (i.e., billing) at the end of one month in a Balance Sheet prepared at a date prior to the next billing. In other words, work has been performed but not yet billed. Say that a Balance Sheet for the firm is prepared on the 20th of July and the last billing was submitted on June 30th. The value of work performed after June 30th and before July 20th) is posted in this account. The following payment request will include the value of this unbilled work. When this next payment request is submitted, the balance of Cost and Estimated Earnings in Excess of Billings will be reduced by the previously unbilled value of this work, which now will be added to the balance of Accounts Receivable (in turn, the balance of Accounts Receivable will be reduced and Cash will increase when the owner pays this progress payment request).

Fixed Assets are normally smaller in construction than in other industries, since construction firms typically don't maintain a large physical plant to manufacture something or provide services to customers (e.g., a car-manufacturing plant or a chain of fast-food restaurants). Heavy engineering and highway contractors tend to hold large fleets of trucks, cranes, and earth movers. This can lead to a larger amount of value held in the Fixed Asset accounts. Building and industrial contractors tend to rent most of the on-site equipment needed for the construction.

Heavy construction companies, dependent upon fleets of large earth-moving machines and haulers, often establish companies from which they "rent" equipment. Certain advantages (e.g. consolidation of depreciation, etc.) accrue by establishing a subsidiary company from which equipment is rented.

[7] Some contracts, as an exception, allow the contractor to bill for materials purchased for installation stored on-site but not yet installed.

This is particularly profitable for a firm that works mainly on cost reimbursement projects.

Most fixed assets are depreciable and most classes of depreciable assets are shown on the Balance Sheet. The book value, the accumulated depreciation, and the net value of each fixed asset class (e.g., equipment, real estate, etc.) are shown in separate lines. The concepts of depreciation and amortization will be discussed later in this text.

On the Liabilities side, the following accounts are especially important for construction contractors:

Accounts Payable refer to obligations incurred in payroll, subcontractor, and vendor accounts that are due for payment. Since contractors must control project expenditures for labor, equipment, subcontracts and materials these are very large and active accounts moving funds from cash to project expense accounts (this will be described in detail when discussing bookkeeping procedures). A considerable amount of financial management and oversight is required to ensure that these accounts are paid in a timely manner. Cash flow management is a key issue when dealing with Accounts Payable.

Billings in Excess of Costs and Estimated Earnings is the mirror opposite of the asset account entitled *Costs and Estimated Earnings in Excess of Billings*, which we have already addressed in this section. In this case, the company has billed the owner for more than was actually earned. The progress payment request exceeded the earned value of the project. In almost all cases, this overbilling situation results from differences in the estimates by owner and contractor as to the amount of work performed. The overbilled amount usually disappears in the next progress payment request. An overbilling is almost always a matter of estimating and accounting conventions and not of bad faith by the contractor.

Both *underbillings/overbillings* accounts (*Billings in Excess of Costs and Estimated Earnings* and *Costs and Estimated Earnings in Excess of Billings*) may have balances in a contractor's Balance Sheet. This may appear puzzling, since a project cannot have both situations simultaneously. The explanation is that a contractor normally has more than one ongoing project, and some projects may be underbilled and vice versa.

Long-Term Liabilities is an account with a relatively small balance in a construction contractors' Balance Sheet compared to many other industries. Long-Term Liabilities are usually loans used for purchasing capital assets, that is, plant and equipment, which in many industries constitute a large proportion of a company's total assets. As we discussed before, construction contractors tend to have a small amount of fixed assets. The colossal machines and large spaces required for making an automobile or a bulldozer do not have an equivalent in the construction world.

The amount of the long-term debt due within a year of a Balance Sheet's cutoff date is deducted from the balance of Long-Term Liabilities and added to a short-term liability account entitled *Current Portion of Long-term Debt*.

Finally, let's look at some issues concerning the Owner's Equity section of a Balance Sheet. We already discussed some important issues about the special position of owners in claiming their investment in case of bankruptcy.

The *Initial Capital* of a company (also called its *paid-in capital*) has many more nuances than the intended level of coverage of this book. The degree of ownership of each company investor largely depends on how this section is structured. Consequently, this section also determines the percentage of the company's net profit received by each investor, and their authority for decision making.

Corporations are owned by their shareholders. A shareholder's (or stockholder's) level of ownership depends on the type and amount of their shares in the company's stock. *Common shares* usually make up the majority of a corporation's total stock, and provide proportional, full stakes to the company's rights and risks, including their fraction of dividends (money from the company's profit distributed to owners). *Preferred shares*,[8] as the name suggests, give priority to their owners for recovery of their investment in case of bankruptcy. Creditors, however, still have priority in payment even over these types of shares.

Smaller companies are usually proprietorships, which are a one-owner company, and *partnerships*, essentially a proprietorship shared by several partners (and subject to explicit statutory rules). In both cases, the value of any owner(s) contribution is directly incorporated in the Balance Sheet, and the accounts under Initial Capital are much simpler, such as "*Elmer Fudd Initial Contribution.*" In the case of a partnership, the contribution of each partner is registered under a separate account. Chapter 9 addresses in more detail these legal forms of company organization.

Retained Earnings is an account that keeps track of all the money remaining after owners are paid part (or all) of the profit that the company realized during the year. It is, from another perspective, the profit that owners reinvest in the company.

THE INCOME STATEMENT

The second basic report in any financial system is the Income Statement, also known as the *Profit and Loss Statement.* The Income Statement keeps track of the financial activity of a company over a period of time, typically one year or a quarter (i.e., three months). For example, a company's Income Statement shows how much a company spent on subcontracts last year, or how much it received from performing contracts in the same period.

---

[8] Preferred shares normally do not provide voting rights to their owners. They do, however, guarantee that dividends will be paid in years that the company is profitable.

An Income Statement has three main components:

1. *Revenue* is the value earned by the company through the sales of its goods and services during the period being analyzed. Clients accept obligations to pay a company because they expect to get goods or services in exchange.
2. *Expenses* are the cost (to the company) of goods and services that are used up by the company to generate revenue. All expenses have an associated cost, but not all costs are expenses. There are four important types of expenses shown in an Income Statement:
   1. Cost of Goods Sold, frequently addressed in the construction industry as *direct costs*.
   2. *General and Administrative Expenses*, usually referred to in the construction industry as *General Overhead*.
   3. *Interest expenses*, conceptually the "rent" for the money borrowed by the company.
   4. *Tax expenses*, in short, the share of a company's profit that must be paid to the government.
3. *Net Income, Profit or Earnings* (all three are synonymous) is the money left after all expenses have been subtracted from revenue. This leftover belongs to the company owners and is their reward for tying up their money in the company and taking the risk of a loss.

These three components are related by the very definition of net income:

$$\text{Net Income} = \text{Revenue} - \text{Expenses}$$

Figure 2.3 shows these components in their broadest form. In the next section, you will see how this basic layout is expanded in any real-life Income Statement.

What is the difference between a cost and an expense? Many people use the words cost and expense as if they were synonymous. In accounting, however, they have fundamental differences. All expenses have an associated cost, but not all costs are expenses. Cost is the value of the money paid, or promised to be paid, to acquire a good or service. If you pay $10 for a ream of paper, its cost to you is $10. It doesn't matter if you buy the paper on credit: the premise is that you will pay the $10 at some point. Expense is the cost of a good or service that is consumed by the company to generate revenue.

If you store the ream of paper, then it is not considered an expense, because it has not been used to generate revenue. You have exchanged one asset, that is, cash, for another asset, in this case paper. If you use right away 40% of the ream and stock the rest, the expense is $4, and the remaining $6 increases the company's inventory. As the stored paper is taken out from inventory and

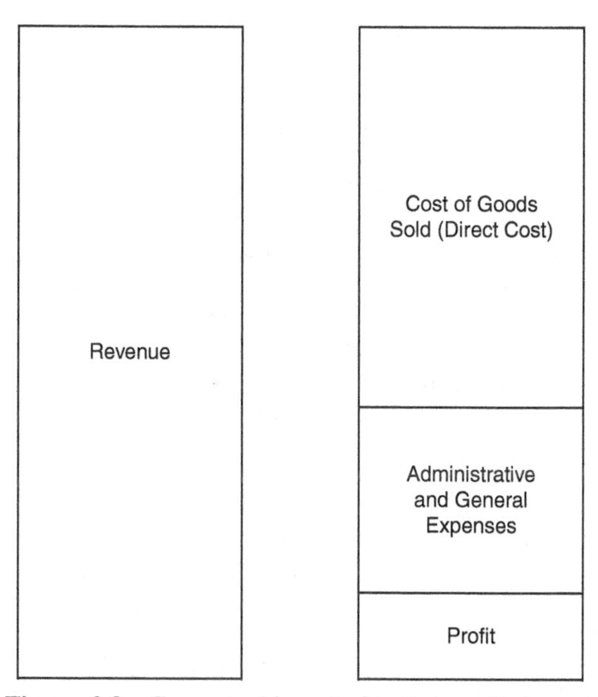

Figure 2.3 Conceptual layout of an Income Statement.

used, its cost is gradually expensed out. The concept and meaning of expense is directly tied to the generation or revenue.

Revisiting the Airport Screen

The airport screen analogy that we made for the Balance Sheet is not applicable to the Income Statement without introducing important distinctions. The Balance Sheet, as we discussed, has been compared to a picture of a company at an instant in time. The Income Statement is closer, in this analogy, to making a video tape of the airport screen and tallying specific data on a sheet of paper divided into two sections, Arrivals and Departures.

As before, our screen will be flickering with the information of incoming and departing flights. We will add a checkmark each time a flight arrives and another each time one departs. The difference between arrivals and departures can tell us how many flights arrived, departed, and were on the ground over a given period of time. We would have many spaces in each sheet, with the names of the airlines and the number of embarking or arriving passengers. We could even track how many flights were ahead of schedule or delayed during the time of the video taping.

COMPONENTS OF AN INCOME STATEMENT – MORE DETAILS

An Income Statement is useful to assess the financial performance of a company. By looking at the amounts, and the proportion of amounts shown in this report, we can see whether the company has grown (i.e., has increased its revenue), whether its profit has increased over the years, whether its net income is reasonable proportionally to its revenue, and many other pieces of information. (Chapter 3 discusses how financial reports are used to make these comparisons).

The Income Statement for Fudd Associates, Inc. is shown in Table 2.2. We will use it to discuss in more detail the accounts found in this report.

Revenue. This item, also referred to as Net Sales or Turnover, is the value earned by a company during the period covered by an Income Statement. Most of the revenue generated by a construction company is attributable to work on the company's projects and is referred to as operating revenue. Some revenue may occur through non-project related activities. A building contractor may sell a truck during the period covered by an Income Statement. The proceedings would be considered as revenue. However, this money is called non-operating revenue, and is usually reported in a separate line under the heading of Revenue. Operating revenue refers to the money earned by a company by performing its main line of business, and non-operating revenue results from incidental, one-of-a-kind financial transactions.

TABLE 2.2 Income Statement for Year Ending December 31, 20XX for Fudd Associates, Inc.

Fudd Associates, Inc. Income Statement For Year Ending December 31, 20XX		
Total revenue		45,226,102
Contract direct cost		
Labor	(4,974,871)	
Materials	(4,563,314)	
Subcontracts	(31,106,513)	
Other direct costs	(827,638)	
Total contract direct cost		(41,476,336)
Gross profit		3,749,766
Selling, general, and admin. expenses		
Variable overhead costs	(1,103,517)	
Fixed overhead costs	(1,248,241)	
Total selling, general, and admin. expense		(2,351,758)
Net income (loss) before income taxes		1,398,008
Income tax		(203,517)
Net Income		1,194,491

Cost of Goods Sold (COGS). In general, this component refers to the money spent to purchase the objects that are in turn sold to the client, or paid for operations directly related to this purchase. For example, Wal-Mart may purchase 1,000 tubes of toothpaste from a manufacturer for \$950 and pay \$50 in transportation. The cost of goods sold for this particular transaction would be \$1,000 or \$1.00 per tube. The total cost of goods for Wal-Mart will be the sum of the cost of all the purchases products for sale.

Often, Cost of Goods Sold is called *Direct Costs* in a construction company's Income Statement. This heading typically reports the cost of labor, material, equipment, and subcontracts. This nomenclature is better aligned with terms of the construction industry for estimating and procurement.

Gross Profit. This is the difference between the total revenue and total cost of goods sold (i.e., direct costs). In the case of the toothpaste transaction, Wal-Mart may sell each tube for \$1.50, resulting in a revenue of \$1,500. The gross profit in this case would be \$1,500 − \$1,000 = \$500. If a contractor receives \$1,000,000 for completing a project, and its direct costs were \$850,000, then the gross profit for this project would be \$1,000,000 − \$850,000 = \$150,000.

Gross profit is not the same as net profit. There is a big component separating gross profit from net profit, namely all overhead costs. If a company could eliminate all overhead costs, then its gross profit would be the same as its net profit. This is impossible: salaries for management must be paid, along with the office rent, electricity, and other normal operation items. Notice in the preceding examples that Wal-Mart enjoys a much larger (fictitious) gross profit than our construction contractor, \$500/\$1,500 = 33% in the former and \$150,000/\$1,000,000 = 15% in the latter. These are reasonable proportions. Retailers have much higher overhead expenses than construction companies.

General and Administrative (G&A) Expenses. These are the administrative costs incurred in the normal operation of any business, such as office salaries, rent, electricity, and so forth. Frequently, these expenses are called *general overhead* in the construction industry.

To contrast the concepts of direct costs and overhead, consider a contractor with several ongoing projects. Each project will have its own separate tally of the cost of labor, materials, equipment and subcontracts. As discussed before, these four components are the direct costs of a project. However, the salaries of the company owner and secretaries cannot be associated with a particular project. Contractors usually obtain the funds to pay these latter costs by allocating a fraction of their total to each project. However, these allocations are intrinsically subjective. When an expense cannot be associated with a particular project or product, it is considered to be overhead.

Construction companies have small overhead proportions compared to other industries, rarely more than 15 cents of each contracted dollar. In the manufacturing sector, the plant (housing) and equipment required for its operations constitute a significant proportion of total costs. In retailing, inventories, salesperson salaries and store space can make up to 40% of the total cost of each dollar of sales. This disparity in the proportion of indirect costs has very significant implications for management priorities in the construction industry. While in most industries managers are judged largely by their ability to control overhead costs, construction managers concentrate on the control of direct costs.

Interest Expenses. Sometimes the cost of borrowing money, that is, the cost of interest during the reported period for all loans taken by the company, is reported in a separate line. Managers may be unfairly penalized for poor performance when interest is included along with other indirect costs. They may have efficiently controlled the company's overhead costs, and the decision to borrow money may be outside their responsibility. When interest expenses are reported separately, a subtotal called *Net Income Before Interest and Taxes* is included. It is computed by subtracting G&A expenses from gross profit. Fudd's Income Statement does not have a separate line for this item.

Taxes and NEBIT. Following the same rationale discussed in relation to interest expenses, it is very common to have a separate line for the cost of taxes during the period. This line is usually called "Provision for Income Taxes" because its amount is based on what the company's accountants estimate that it will pay as a result of the profit (income / earnings) made during the period. Again here, notice the effect of accrual basis accounting. It is immaterial whether or not the tax amount has been paid; what matters is that it is owed.

The line immediately before tax expenses is called *Net Earnings before Income Taxes (NEBIT)* or *Operating Profit*. When interest expenses are reported separately from other overhead expenses, NEBIT is placed immediately below the interest line. Even when interest is merged with other expenses in the overhead section, NEBIT is always included right before the tax expenses line. Management performance is usually measured based on NEBIT and not on net income after taxes.

Net Income. The company's bottom line, also known as net profit and net earnings, is computed as NEBIT minus tax expenses. Let's review how we got here. All revenue (operating and non-operating) is the positive amount from which costs are subtracted. We discussed that, universally, cost of goods sold (i.e., direct costs) are subtracted first, arriving at the company's gross profit. All other costs are then subtracted, first overhead, then interest, then taxes. Subtotals are shown after subtracting overhead and after interest expenses. This journey takes us back to the original point of this section: Net Income = Revenue – Expenses.

The Accounting Tower of Babel

The many synonyms used to refer to the same accounting elements can be daunting. A commonly used financial term in an industry or in part of the world may be rarely used in another. Think of the words *revenue*, *sales*, and *turnover*, which are essentially synonymous. In American construction, the preferred term is *revenue*. In Asia and to some extend the UK, the most common way to address the same account is *turnover*. Whereas American accountants like to refer to the bottom line as *Net Income*, their British colleagues frequently use *Net Profit* instead.

THE STATEMENT OF CASH FLOWS

The Balance Sheet and the Income Statement are the main financial reports discussed in this book, because they play such a critical role in the everyday financial management and operation of a company. An accountant, however, deals with additional financial reports, which also contribute to the understanding of a company's financial situation. *The Statement of Cash Flows* is the most important of these additional reports. It provides a summary of the changes in the amount of available cash during the period covered by an Income Statement. Its last line is the amount of cash that the company has available at the end of the reported period. This report is especially important from an investor's perspective. The disastrous collapse of Enron in the early 2000's could have been averted if its auditors had noticed that the company's available cash was decreasing year after year, despite the excellent (but untrue) Net Income shown in its Income Statement in the same period.

Table 2.3 shows the Statement of Cash Flows for Fudd Associates at the end of 20XX.[9] It shows the changes in cash for three types of company activities, followed by a summary that ends with the amount of cash available at the end of the period covered by the underlying Income Statement. Notice that many lines refer to changes in the balance of Balance Sheet accounts. To develop a Statement of Cash Flows, we need the Balance Sheets at the beginning and the end of the reported period, as well as the Income Statement for the period.

Notice that the balance for the Cash and Cash Equivalents account is exactly the same as the line for Cash and Cash Equivalents at the End of the Period in the Balance Sheet shown in Table 2.1 The two must be the same—otherwise, the company would claim two different amounts when referring to the same physical item, namely Fudd's cash in hand at the end of 20XX. Why is the same information shown in two separate reports? The Statement of Cash Flows summarizes the sources and uses of the company's cash, and provides a perspective that the Balance Sheet and the Income Statement miss altogether.

Let us examine in detail Fudd Associates' Statement of Cash Flows. *Operating Activities* analyzes the cash flow resulting from the company's business operations. This section begins with the company's Net Income for the period, and then adds or subtracts all non-cash items.

The change in Accounts Receivable is subtracted from the starting Net Income because this amount does count in the computation of the Net Income,

[9] The format shown is called the Indirect Method. The alternative presentation is the Direct Method, which differs in the way of presenting the cash from operating activities. Instead of beginning this section with the Net Income and then adding and subtracting all non-cash items to get to the Net Cash Flow from operations, the Direct Method has two basic lines in this section: Cash Inflows and Cash Outflows. The difference between these two lines provides the same Net Cash Flow from operations as the Indirect Method. The vast majority of companies use the Indirect Method.

TABLE 2.3 Statement of Cash Flows for Fudd Associates, Inc.

Fudd Associates, Inc. STATEMENT OF CASH FLOWS For the Year ending December 31, 20XX	
Operating Activities	
Net Income (From the Income Statement)	1,194,491
Adjustments	
Depreciation	234,410
Change in Accounts Receivable	(426,795)
Change in Inventories	(11,105)
Change in Accounts Payable	303,918
Change in other current expenses	91,399
Net cash flow from operating activities	1,386,318
Investing Activities	
Expenditures for fixed assets	(359,344)
Other: Sale of company vehicles	45,034
Net cash flow from investing activities	(314,310)
Financing activities	
Borrowings	750,000
Cash distribution to owners	(236,000)
Payments on Borrowings	(473,294)
Net cash flow from financing activities	40,707
Net change of cash and cash equivalents in period	(81,777)
Cash and cash equivalents:	
At the beginning of period	2,576,075
At the end of period	2,494,298

but does not involve any cash. The subtraction must account for the algebraic sign of the change. If the change is negative (a decrease in the balance from one year to the next), then its subtraction increases Net Income.

The change in Accounts Payable is added to the initial Net Income, since it was subtracted when the Net Income was computed, but did not result in any change in cash.

Other accounts affecting cash are treated similarly, and their treatment is shown with a "+" when their change is added to the initial Net Income, and with a "−" when it is subtracted. Numbers in parentheses are subtracted.

Depreciation for the period is of special interest for construction. This account is an allowance for the decline in value over time of equipment and other capital assets. Companies consider their asset depreciation as a cost, and it is deducted from Net Income. However, since it does not involve any cash, depreciation is added back to the original Net Income in these cash flow adjustments

The second major section of the Statement of Cash Flows is entitled *Investing Activities*, and includes all changes in non-current assets. For a construction company, this section typically would include the purchase and selling of equipment and real estate. A company's net income is not affected when a company buys a new piece of equipment, since it is the even exchange of one asset called cash for another asset called equipment (or the same increase in the asset called equipment and the liability called Accounts Payable, which does not change Net Income, either). However, the amount of available cash is reduced by the purchase. The cost of all non-current assets purchased in the analyzed period is included as a negative amount. The revenue from selling any non-current asset, say a pickup truck, is added to this cash flow subtotal.

The third and last section shown in Table 2.3 is *Financing Activities*. Any cash received from new loans is included here. This money increases the cash flow for the period. Conversely, loan payments are considered negative. This section also includes any issued company stock, which results in the infusion of new money and therefore, increases the cash flow. Cash distributions to the company owners reduce the availability of cash, and are shown as negative amounts.

The algebraic addition of the three section subtotals is the *Net Change in Cash and Cash Equivalents* for the company during the period covered by the Statement of Cash Flows. The *Ending Cash Balance* is found by adding the Beginning cash balance to the computed Net change in cash.

CONTRACT BACKLOG

You may have noticed a problem with the financial statements that we have discussed so far. If revenue includes only the value that the company has already generated during a specific period of time, then where do companies place the value of their contracts pending to be earned? This amount is reported as the contract backlog of the company. This is an important report for understanding the condition of a construction company going forward. A contractor may appear to be doing perfectly well so far, but if it has no backlog, it will be in trouble in the immediate future.

This was the case of many homebuilders at the end of 2006. They had experienced an excellent market for years. However homebuilding construction suffered a significant decline beginning in 2007. The lack of incoming contracts should have been a red flag to investors that there might be trouble ahead. The backlog can also indicate the opposite situation. A company's resources may be overstretched if too many contracts are in the backlog. Adequate resources may not be available to support a rapidly increasing portfolio of projects. In this case, an over abundance of projects in the company backlog would alert investors and lenders to the potential risk reflected by this abnormal situation.

PUBLIC CORPORATIONS

A *public corporation* offers its shares for sale to the general public in a stock market such as the New York Stock Market or NASDAQ. Alternatively, a company can organize as a *privately held corporation*. Stock in such companies is not offered for sale or traded publicly.

Relatively few construction companies choose to organize as public corporations. The Federal Trade Commission (FTC) imposes strict reporting requirements, so that potential investors have a clear picture of each company's financial position (and other management issues) before investing in it. Among the many requirements for financial transparency, public companies must formally report their financial position each quarter (i.e., every three months) by submitting the FTC's form 8-K, and filing an audited annual detailed report (10-K) with the FTC. Every major management event must be reported. These requirements are cumbersome by any standard.

Almost always, public corporations begin as private entities which at one point decide to expand and open their stock to the general public in the stock market. The money from the sale of shares injects new capital into the corporation, and consequently, enhances its ability to undertake expanded operations.

The original shareholders of a private corporation which decides to "go public" have good financial incentives for the move. The first batch of publicly offered shares is called a corporation's *initial public offering (IPO)*. Investors may be willing to pay more than the face value (par value) of each share. This extra income is part of the owners' equity, specifically under the account called *Additional Paid-In Capital*. The difference between the par value and what investors pay to the company for each share can be huge. For example, the par value of the first 19,605,052 shares of Google's IPO was $0.001 (0.1 cent per share), and its selling value in the stock market was $85 per share.

In addition to improving the cash position of the corporation, the original shareholders can pocket extra money, sometimes becoming "instant millionaires," when there is a difference in the price that they paid for their shares and the price that the shares can command if they are sold in the stock market. Anyone who acquired Google stock at its IPO value made a very good deal, even if they paid $85 additional dollars per share. That person could have sold each share for $747 at the end of 2007. The down side is, of course, that this profit of $747 – $85 = $662 per share would be taxed as a *capital gain* for the seller.

Public corporations must report their contract backlog as part of their filings to the FTC. A section of the 10-K is entitled *Management's Discussion and Analysis of Financial Condition and Results of Operation*. This section provides a table listing the reporting company's projects under contract and their earned and pending values. This table is obviously of interest to potential investors.

Public companies also publish an annual report geared to investors. The annual report presents a more attractive version of much of the information given in the 10-K filing. It includes financial information regarding the past year's performance with additional narrative highlighting the company's accomplishments. An annual report is an excellent source of information and it is designed to be easier to read than the austere format of Form 10-K. It must be kept in mind, however, that an annual report is designed to present the company in the best possible light. Negative factors must be reported (by statute) but generally are not printed in big letters.

Privately held corporations do not have the reporting requirements of public corporations. Every company knows its backlog, but it is private information not available to the general public.

REVIEW QUESTIONS AND EXERCISES

1. At the end of the fiscal year 20X2, Rainer Construction Co., Inc. has the following financial profile (all figures are in dollars):

Total assets	1,200,000
Total liabilities	850,000
Total revenue before taxes	300,000
Total expenses	265,000

   (a) What is the total net worth of this company?
   (b) Did the net worth increase, decrease or remain the same compared to FY 20X1? Why?
   (c) What is the income after taxes? (Assume a 50% tax rate.)

2. Define in your words the difference between a Balance Sheet and an Income Statement.

3. The following Balance Sheet figures are available on the Cougar Construction Company:

	Current	Long-Term	Total
Assets	$100,000	$100,000	$200,000
Liabilities	80,000	70,000	150,000

   (a) Find the total net worth of the company.

4. The following data regarding Atlas Construction Company are available:

Current assets	$300,000
Current liabilities	200,000
Long-term liabilities	500,000
Total net worth	200,000

What is the value of the company's fixed assets?

5. Name three accounts typical of the liability side of a Balance Sheet. In what side would you enter "discounts earned?"
6. How are the Balance Sheet and Income Statement related? Discuss the pattern of information flow between these two documents.
7. Given the data shown in Tables P2.7a and P2.7b for Pegasus International, Inc., what method of accounting is being used? How has the working capital of Pegasus International changed from 20X4 to 20X5? Interpret this development considering the increase in revenue from 20X4 to 20X5.

TABLE P2.7a Pegasus International, Inc., Summary of Consolidated Operations (Dollar Amounts and Shares Are in Thousands, Except Per Share Amounts)

	20X5	*20X4*
Revenue	$1,325,423	$801,322
Costs and expenses		
Cost of revenue	1,211,402	726,937
Corporate administrative and general expenses	19,392	14,881
Other (income) and expenses		
Interest on indebtedness	1,354	1,671
Interest income	−6,565	−7,089
Provision for estimated losses on planned disposition of assets	—	400
Total cost and expenses	1,225,583	736,800
Earnings before income taxes	99,480	64,522
Total income taxes	52,429	31,277
Net earnings	47,411	33,245
Preferred dividend requirements	1,803	1,558
Earnings applicable to common stock	$45,608	$31,687

TABLE P2.7b Pegasus International, Inc., Consolidated Balance Sheet

Assets	*31 December, 20X5 and 20X4*	*20X5*	*20X4*
Current assets	Cash	$10,138,000	$13,063,000
	Short-term investments at cost, which approximates market	101,310,000	49,298,000
	Accounts and notes receivable	90,526,000	83,442,000
	Costs and earnings in excess of billings on uncompleted contracts	65,745,000	44,728,000
	Inventories at lower of cost (average and FIFO methods) or market	30,011,000	24,210,000
	Other current assets	7,751,000	9,005,000
	Total current assets	305,481,000	223,746,000
Property, plant, and equipment at cost	Land	11,162,000	11,744,000
	Buildings and improvements	39,622,000	26,909,000
	Machinery and equipment	50,650,000	51,597,000
	Drilling and marine equipment	107,617,000	71,300,000
	Construction in process	42,834,000	44,230,000
	Subtotal	251,885,000	205,780,000
	Less accumulated depreciation and amortization	75,441,000	76,732,000
	Net total property, plant, and equipment	176,444,000	129,048,000
Oil and gas properties at cost	Subtotal	59,814,000	50,683,000
	Less accumulated depletion and depreciation	30,455,000	20,105,000
	Net total oil and gas properties	29,359,000	30,578,000
Other assets	Excess of cost over net assets of acquired companies, less accumulated amortization	15,105,000	16,351,000
	Other	12,058,000	12,542,000
	Total other assets	27,163,000	28,893,000
Total		$538,447,000	$412,265,000

(*continued*)

TABLE P2.7b Pegasus International, Inc., Consolidated Balance Sheet (*Continued*)

Liabilities and Shareholders' Equity		*20X5*	*20X4*
Current liabilities	Accounts payable	$66,050,000	$54,124,000
	Billings in excess of costs and earnings on uncompleted contracts	69,082,000	53,375,000
	Accrued liabilities	58,089,000	37,393,000
	Current portion of long-term debt	1,395,000	1,385,000
	Income taxes currently payable	24,895,000	8,209.00
	Deferred income taxes	49,599,000	31,989,000
	Total current liabilities	269,101,000	186,385,000
Long-term debt due after one year	Total	4,242,000	7,311,000
Other noncurrent liabilities	Deferred income taxes	11,623,000	11,792,000
	Other	9,162,000	4,400,000
Total		$20,785,000	$16,192,000
Contingencies and Commitments			
Shareholders' equity	Capital stock		
	Series B preferred	$567,000	$663,000
	Common	9,368,000	9,203,000
	Additional capital	116,636,000	
		115,184,000	
	Retained earnings	117,748,000	77,327,000
	Total shareholders' equity	244,319,000	202,377,000
Total		$538,447,000	$412,265,000

CHAPTER 3

ANALYZING COMPANY FINANCIAL DATA

INTRODUCTION

Every financial report tells something about a company's financial health. If the company reports a large profit, it may be a better investment than another that reports a smaller profit. But which one is better also depends on other factors. If a company makes one million dollars in annual profit by performing 100 million dollars of work, and another company can make a half a million dollar profit on 25 million dollars of work, it would seem that the latter is a more attractive investment, since the first gets one cent in profit from each dollar of work performed, while the second gets a five-cent profit on each dollar of work performed.

The rationale above seems solid, but in fact it wouldn't be wise to act based solely in the information provided so far. What if each company had the same amount of owners' equity, say five million dollars? From our perspective as investors, the first company now appears to be better: for each dollar invested, the first company gives a profit of 20 cents to the company's owners, but the second company gives only 10 cents on the dollar to the owners. The decision can get even more complicated. What if the second company has been a quite consistent money maker over time, while the first one has had an erratic record, losing money in some years? Or if the second company has a 50 million dollar bank note due in two years? Or if either company is making less profit than its competitors? It is virtually impossible to judge the financial status of a company just by looking at its Balance Sheet, Income Statement, and other primary financial reports without some further point of reference.

This chapter addresses standard metrics, or indicators, for analyzing a set of financial documents. These indicators greatly simplify the understanding of a company's financial status, since they consist of simple ratios and percentages instead of the many and large dollar amounts found in a typical financial statements.

Vertical and Horizontal Analyses

There are many reasons for being interested in the financial status of a company. Lenders and investors are particularly interested in determining the risk of losing the money that they are lending or investing, respectively, in the company. Sureties, labor unions, and even state and federal tax agencies also need to understand a company's financial health. All these parties have developed ways to get a meaningful picture of a company's finances using the primary reports discussed in the previous chapters. These analysis procedures can be grouped into two main categories, namely:

1. ***Vertical Analysis.*** A vertical analysis simplifies the interpretation of financial information by developing a relatively small number of key indicators from the numbers presented in a company's Balance Sheet and Income Statement. This type of analysis takes its name from the fact that since it works with line items from financial statements, it usually requires reading pages "from top to bottom." The bulk of this chapter concerns vertical analysis.
2. ***Horizontal Analysis.*** The purpose of a horizontal analysis is detecting the direction and trends of key financial information. We can make much more sense of vertical analysis if we compare its results over a period of time. Since time is normally shown graphically on the horizontal axis, this type of time-based analysis is called "horizontal." As you will see, we can express these trends using their absolute values, or in proportion to an initial point in time. Notice that although different, vertical and horizontal analyses do not exclude each other. Any good exploration of the significance of a company's financial reports should include both types of analyses if at all possible.

VERTICAL ANALYSIS: FINANCIAL RATIOS

Many financial indicators consist of ratios of items taken from the Balance Sheet and the Income Statement. These ratios provide clues and symptoms of undesirable conditions or highlight issues requiring further investigation. The analysis of an indicator can disclose relationships as well as bases for comparison that reveal conditions and trends that cannot be properly detected by reading the individual items in its formula.

Four main categories of ratios are commonly identified: Liquidity, Profitability, Efficiency and Capital Structure. Each one of these categories, in turn, comprises several ratios. No company needs all possible indicators, since each industry has its own needs and preferences. For example, a common indicator for a retailing business such as its inventory turnover may be near useless in construction. The reverse is also true; some common ratios for construction companies are rarely used in retailing. We will examine only the most common and useful ratios for a typical construction company. The formula and explanation of many others are widely available and follow the same rationale discussed here.

The numerical value of any ratio is in itself of limited value, since it does not furnish a context to understand the financial position of the company. The ratio becomes much more meaningful when it is combined with a horizontal analysis, which shows its range and its trend over time. Furthermore, ratios should be compared with industry averages, which are published by the Construction Financial Management Association (CFMA), Standard and Poor's, Robert Morris Associates and independent consultants. Most public libraries carry at least one of these sources.

LIQUIDITY INDICATORS: CAN THIS COMPANY GET CASH IN A HURRY?

Liquidity refers to the ability of a company to maintain a sound financial position over the short term. In practical terms, good liquidity means that the company can pay its liabilities due in the relatively near future, usually within a year. In turn, paying off a liability requires cash. From a lender's perspective, a company should have more cash and assets that can be easily converted to cash than the amount of liabilities that it needs to pay in the near future. Easily sellable assets are grouped as Current Assets; short-term liabilities are found under Current Liabilities. It follows that a company should have more current assets than current liabilities.

We will look at three financial indicators that deal with the relative liquidity of a company, and are collectively known as Liquidity Indicators. The three of them, namely the Current Ratio, Quick Ratio and Working Capital, address a company's liquidity by relating its Current Assets to its Current Liabilities.

Current Ratio

The Current Ratio can be visualized as the "safety factor" for a creditor's money loaned to a company, computed as the ratio between the total amount of Current Assets and Current Liabilities. Its formula is:

$$\text{Current Ratio} = \text{Current Assets}/\text{Current Liabilities}$$

When a company's Current Ratio is greater than one, it means that the company has more cash, or assets easily convertible to cash, than the amount

TABLE 3.1 Current Ratio Computations for Fudd Associates, Inc

From the Balance Sheet	
Total current assets	$12,880,550
Total current liabilities	$9,774,261
Current Ratio =	
Current Assets / Current Liabilities	
= $12,880,550 / $9,774,261	= 1.32

of money that it will have to pay in the near future. A lender wants this ratio to be as large as possible.

The typical desirable range for a construction company's Current Ratio is between 1.5 and 3.0. A value of less than 1.5 means that the company may struggle to find enough cash or easily convertible assets to pay its current liabilities. The reason for an upper limit for this and other financial indicators is more subtle. An unusually large Current Ratio may indicate an excessively conservative management—which is not desirable.

Table 3.1 shows the current ratio computations for Fudd Associates, Inc. The Balance Sheet and Income Statement used for these computations, as well as for the others in the remainder of this chapter, are taken from Tables 2.1 and 2.2.

The average Current Ratio of most construction companies, even those soundly managed is low compared to the usually acceptable range for other industries. In most industries, a Current Ratio of less than 2.0 raises a red flag with lenders. Construction is a risky line of business, and its relatively low industry averages reflect this fact. Moreover, most contractors have minimal amounts of inventory, relying instead on a pretty tight supply chain from their subcontractors and material suppliers. The larger Current Ratio averages for other industries derive in part from their larger inventories and a much higher proportion of their business done with in-house resources.

Quick Ratio

The Quick Ratio's rationale and computation are very similar to the Current Ratio's, with the difference that the Quick Ratio is more stringent than the Current Ratio. Its formula is:

$$\text{Quick Ratio} = (\text{Near-cash Current Assets})^{*}/\text{Current Liabilities}$$

The Quick Ratio has exactly the same denominator of the Current Ratio, namely the company's Current Liabilities. In the numerator, it also has

* Near-Cash = Easily liquidable

TABLE 3.2 Quick Ratio Computations for Fudd Associates, Inc.

From the Balance Sheet	
Current near-cash assets (above Inventories)	
Cash and cash equivalents	$2,494,298
Account receivables	$8,535,901
Costs & recog'd earnings in excess of billings	$933,165
Total current near-cash assets	$11,963,364
Total current liabilities	$9,774,261
Quick Ratio =	
Current near-cash assets / Current Liabilities	
= $11,963,364 / $9,774,261	= 1.22

accounts from Current Assets, with the difference that it uses only the most liquid (easily convertible to cash) of the Current Assets accounts as opposed to the Current Ratio, which includes all Current Assets in its formula.

The near-cash current assets considered by the Quick Ratio are, more specifically, all those listed above the *Inventories* account (not including Inventories). The accounts in the Current Assets section of a Balance Sheet are listed in order of liquidity, beginning with Cash itself. Although *Inventories* is a Current Assets account, it is not as easily liquidable as Cash (the account) and near-cash accounts such as *Accounts Receivable*. Should a company declare bankruptcy, its inventories could be difficult to sell, and even harder to price.

Since the Quick Ratio includes only the most liquid Current Assets in the numerator, its value will always be less than the Current Ratio of the same company. The Quick Ratio is also called the Acid Ratio because like acid, it is "corrosive" (more demanding) than the Current Ratio. Table 3.2 shows the Quick Ratio computations for Fudd Associates, Inc.

The standard recommended range for Quick Ratio is between 1.0 and 2.0. Construction companies tend to have quite good Quick Ratios. The small amount of inventory carried by most construction contractors makes their Current and Quick Ratios numerically close, and consequently, although their Current Ratio is low as previously discussed, their Quick Ratio is high compared to typical retailers' and manufacturers'.

Working Capital

The Working Capital indicator and the Current Ratio use exactly the same information in their formulas, but instead of dividing Current Assets by Current Liabilities, Working Capital subtracts the balances of these accounts:

$$\text{Working Capital} = \text{Current Assets} - \text{Current Liabilities}$$

This indicator shows the dollars that a company could use to increase its business. Any leftover of Current Assets minus Current Liabilities is in

TABLE 3.3 Working Capital Computations for Fudd Associates, Inc.

From the Balance Sheet	
Total current assets	\$12,880,550
Total current liabilities	\$9,774,261
Working Capital =	
Current Assets – Current Liabilities	
= \$12,880,550 – \$9,774,261	= \$3,106,289

theory a wasted opportunity: If these two amounts were equal (and therefore there was no Working Capital left), theoretically the company would have exactly enough money during the coming year to pay its liabilities for the year. However, this would be true only if the company's performance for the coming year is exactly equal to that of the previous year. The natural variations over time of all account balances make it absolutely necessary to have a safety factor between the two—which is why the Current Ratio should have a value of more than 1.0.

The term "working capital" can mean many things in lay language. Even financial management students are prone to use this term improperly (many times mistakenly considering it synonymous with "owners' equity"). Make sure that you understand the meaning of this term as used for financial purposes. Table 3.3 shows the computations for this indicator for Fudd Associates, Inc.

PROFITABILITY INDICATORS: IS THIS COMPANY MAKING ENOUGH PROFIT?

The ratios considered in this section are called Profitability indicators and provide useful information about a company's ability to make money. As discussed in the introduction to this chapter, knowing a company's profit in itself is not very helpful for a person considering investing in a company or lending money for its operation. It is also not enough information to help a manager assessing the financial success or failure of a particular project. A more important issue is how much money the company generates relative to some key quantity such as its owners' equity or its volume of work performed. Various industry profitability averages are shown in Table 3.4.

Return on Equity

Return on Equity (ROE) is the ratio between a company's Profit (net income) divided by its Owners' Equity, expressed as a percentage. Profit is generally

TABLE 3.4 Sample Industry Profitability Ratios — Based on IRS Return Information

Industry	Construction of Buildings	Computer and Peripheral Equipment	Motor Vehicles and Parts Manufacturers	Gasoline Stations	New and Used Car Dealers	Accommodation
NAICS (Industry Identifier)	236115	334110	336105	447100	441115	721115
Return on Total Assets	8.1%	8.1%	2.5%	5.8%	5.1%	3.5%
Return on Equity before Income Taxes	26.2%	16.8%	1.0%	11.1%	16.0%	0.3%
Return on Equity after Income Taxes	21.7%	10.3%	0.2%	9.5%	14.2%	—
Return on Revenue before Income Taxes	3.7%	7.6%	0.4%	0.7%	0.9%	0.2%
Return on Revenue after Income Taxes	3.0%	4.7%	0.1%	0.6%	0.8%	—

TABLE 3.5 Return on Equity Computations for Fudd Associates, Inc.

(Net Worth = Owners' Equity)	
From the Balance Sheet	
Total net worth	$4,732,935
From the Income Statement	
Net Income	$1,194,491
Return on Equity =	
Net Income / Total Owner's Equity	
= $1,194,491 / $4,732,935	= 0.2524 or 25.24%

considered here as the profit remaining after paying taxes but before paying dividends to owners. The formula for this indicator is:

$$\text{Return on Equity} = \text{Profit/Owners' Equity}$$

For an investor considering whether a particular company is a good investment, its ROE is usually the deal maker or breaker. ROE can be visualized as the interest that the investor gets on his money. If the company shows a 10% ROE, it means that the investor gets a 10 cents profit in one year for each dollar invested. If a totally similar company has an ROE of, say, 12%, the investor would probably opt for the latter, unless noneconomic factors play a major role in the decision.

Table 3.5 shows the ROE computations for Fudd Associates, Inc.

There are many factors to consider other than ROE when investing opportunities do not refer to comparable business schemes. If a bank offers a 10% return on certificates of deposit, most investors would choose this option over investing in a construction company with a 10% ROE, since the bank's offer is much less risky than the option of investing in a construction company. The higher risk of investing in construction companies may be well rewarded. The average ROE for construction compares favorably to many other industries. The management quality of an individual company heavily influences its ROE. For example, Robert Morris Associates (RMA) Annual Studies (2007) shows that Industrial Building Construction contractors have an average ROE of 25.5%. However, the top quartile earns 46.4%, and the bottom quartile earns 5.9%, a quite significant difference.

Return on Revenue

Return on Revenue (ROR) indicates how many cents of profit a contractor is getting from each dollar of revenue. The formula for this ratio is, therefore:

$$\text{Return on Revenue} = \text{Profit/Revenue}$$

TABLE 3.6 Return on Revenue Computations for Fudd Associates, Inc.

From the Income Statement	
Net Income	$1,194,491
Total Revenue	$45,226,102
Return on Revenue =	
Net Income / Total Revenue	
= $1,194,491 / $45,226,102	= 0.0264 or 2.64%

The ROR computations for Fudd Associates, Inc. are shown in Table 3.6.

ROR is most interesting to internal managers, while ROE is crucial for investors and creditors. Since profit and revenue can be tracked for each individual project, ROR can tell the profitability of each one. For individual project managers, ROR can have a great personal impact. A manager whose projects have a high ROR is likely to get a higher bonus than another manager with lower project ROR, if their projects are comparable.

Construction companies have a relatively low ROR compared to other industries, as you can see in Table 3.6. However, construction contractors can perform very large projects with little capital of their own. This makes the denominator in the ROR formula comparatively large and, therefore, the ROR comparatively low. The low ROR construction company average does indicate an intrinsic problem of the industry—just a few cents' difference between the estimated and actual project unit costs makes the difference between a profit or a loss. The business of construction is, indeed, risky.

Return on Assets

A company's Return on Assets (ROA) is computed similarly to its ROE and ROR:

$$\text{Return on Assets} = \text{Profit}/\text{Total Assets}$$

If a company is considering borrowing money, its ROA provides a very useful control value. If the money will be borrowed at an interest rate below the company's ROA, then the company will make a profit from taking the loan. If the loan requires paying a greater interest than the company's ROA, then the loan should not be taken, because the company will pay more in total interest than the profit it can realize. The ROA for Fudd Associates, Inc. is shown in Table 3.7.

Earnings Per Share

A corporation keeps track of its Earnings per Share (EPS). Its formula is:

$$\text{Earnings per Share} = \text{Profit}/\text{Company stock shares}$$

TABLE 3.7 Return on Assets Computations for Fudd Associates, Inc.

From the Income Statement	
Net Income	$1,194,491
From the Balance Sheet	
Total Assets	$15,978,849
Return on Assets =	
Net Income / Total Assets	
= $1,194,491 / $15,978,849	= 0.0748 or 7.48%

The profit, as in the previous cases, normally refers to the net income after taxes. The company stock shares refers, more precisely, to the average number of shares of common stock outstanding.[1]

If a company makes a profit of $50 million for the year and has 40 million shares of common stock in circulation, then its EPS will be 50/40 = $1.25 per share. A shareholder owning 10,000 shares of common stock can easily figure that her investment in this company resulted in an individual profit of $12,500 for the year.

Public corporations report their EPS quarterly. This indicator is the most closely watched by public corporation shareholders. Few construction companies are incorporated as public corporations. The exceptions are real estate developers, which need large amounts of money for their operation and frequently take advantage of the influx of cash resulting from offering their stock in stock markets such as the New York Stock Exchange (NYSE).

EFFICIENCY INDICATORS: HOW LONG DOES IT TAKE A COMPANY TO TURN OVER ITS MONEY?

The indicators under the category of Efficiency consider the time that a company needs to circulate the money it has, owes or is owed. Since this money circulation is the essential activity of any company (a company is a "money pump"), these statistical measures are also called Activity Indicators. The unit for the indicators discussed here is time, usually expressed in days, or cycle times per year.

---

[1] The number of shares can vary during the period reported by the EPS, and, therefore, the number of shares for the period must be averaged. Moreover, there can be several types of shares, such as preferred stock and common stock. Outstanding stock is the number of shares actually held by shareholders.

The rationale for the formulas of ratios in this category is pretty straightforward, as the following two examples show.

A) Suppose that a car has traveled a total of 24,000 miles last year, and the average trip was 2,000 miles long. How long did it take on average to complete each trip? The answer is:

    2,000 miles / 24,000 miles per year = 0.08333 years,

    or in months, 0.08333 × 12 = 1.0 months,

    or in days, 0.08333 × 365 = 30.4 days.

B) Suppose that you have spent a total of $24,000 in materials last year, and the average inventory of materials was $2,000. How long did it take on average to deplete your inventory? The answer is:

    $2,000 / $24,000 per year = 0.08333 years,

    or in months, 0.08333 × 12 = 1.0 months,

    or in days, 0.08333 × 365 = 30.4 days.

As you can see, the two examples above are virtually identical; the only difference being that the first one deals with distance and the second one with money. You can also approach the second example from a different angle. The inventory was depleted 12 times in the year ($24,000 per year / $2,000). Therefore, it took 1/12 of a year, or one month, or 30.4 days, to deplete or turn-over the inventory.

Table 3.8 shows the average ages for several construction specialties for items discussed below.

Average Age of Inventory

This indicator was introduced in the preceding example. The formula that follows formalizes the concept.

$$\text{Average Age of Inventory} = (\text{Average Inventory Balance}/\text{Total Cost of Inventory}) \times 365$$

TABLE 3.8 Important Average Ages for the Construction Industry (in days)

Specialty	All Companies	Industrial and Nonresidential	Heavy and Highway	Specialty Trade
Days in Acct. Receivable	51.4	48.4	45.5	67.7
Days in Inventory	3.2	0.2	3.8	2.6
Days in Acct. Payable	41	46.8	34.7	29.2
Operating Cycle	31.5	22.1	37.2	49.6

© 2007 Construction Financial Management Association in RMA Annual Statement Studies

The Average Inventory Balance is found by averaging the balance for the Balance Sheet account called Inventory as many times as possible over the period covered by the Income Statement. If this period is one year, the Average Inventory Balance could be found by averaging out the balances of the Inventory account at the end of each quarter: March 31, June 30, September 30, and December 31.

The denominator of the formula comes from the Income Statement, which contains an account called Cost of Inventories. This account shows the cost of the entire inventory purchased in the year. It is the equivalent of the total mileage in the first example above. Finally, the 365 in the formula comes from the days in a year. If the Income Statement covered a semester, then instead of 365 we would use 180, the days usually assumed for one half year. These steps can be seen in Table 3.9, which consists of the computations of the Average Age of Inventory for Fudd Associates, Inc.

Materials and parts required for a project are kept in inventory, on average, for the number of days shown by this indicator. Having parts sitting idle in the warehouse is not necessarily bad, since the more parts are available, the less likely it is that the project will stop for lack of materials or parts. However, a company's inventory is not generating any money. On the contrary, the company may be paying interest if it was purchased on credit.

As previously mentioned, most construction contractors don't keep large inventories, and this indicator is not as critical in the construction industry as most of the others discussed here. Some specialties do keep sizable inventories and should pay close attention to their Average Age of Inventory. Large homebuilders maintain large inventories (since they tend to provide materials to their subcontractors). Specialty contractors such as electrical contracting companies also maintain large inventories.

TABLE 3.9 Average Age of Inventory Computations for Fudd Associates, Inc.

From the four most recent quarterly Balance Sheets	
Inventory at 3/31/XX	\$139,788
Inventory at 6/30/XX	\$91,899
Inventory at 9/30/XX	\$115,550
Inventory at 12/31/XX	\$122,106
Average Inventory	\$117,336
From the Income Statement	
(Assuming Cost of Inventory = Cost of Materials)	
Cost of Inventory	\$9,563,314
Average Age of Inventory =	
(Average Inventory / Cost of Inventory) × 365	
= (\$117,336 / \$9,563,314) × 365	= 4.5 days

The average age of inventory is important to manufacturing companies that fabricate products and hold them in inventory prior to sale. This is not the way the construction of a project proceeds. We do not say that the work completed to date on a project is held in inventory. Therefore, inventory-based ratios are not as directly applicable in construction as in the manufacturing sector.

Average Age of Accounts Receivable (Collection Period)

The rationale for the Average Age of Accounts Receivable is conceptually similar to the previous examples. Revenue is our "total mileage" for this indicator because for practical purposes, in construction all the revenue is first billed and then collected.

If the Income Statement covers one year, the formula for this indicator is:

Average Age of Accounts Receivable

$$= (\text{Average balance of Accounts Receivable}/\text{Revenue}) \times 365$$

As in the previous case, if the Income Statement does not summarize a year of operations, the 365 would be replaced by the number of days of the period covered by the Income Statement.

The computations for Fudd Associates, Inc. shown in Table 3.10 can help in understanding these computations.

Keeping a short collection period is of utmost importance to any construction company. A company can be rich in assets, and yet go bankrupt if it has large balances of Accounts Receivable but no cash to pay its immediate obligations. Construction companies are at constant risk of exhausting their cash reserves due to the standard contract payment terms and the large amount of subcontracted work on a typical job.

TABLE 3.10 Average Age of Accounts Receivable Computations for Fudd Associates, Inc

From the four most recent quarterly Balance Sheets	
A/R at 3/31/XX	$6,588,333
A/R at 6/30/XX	$5,375,100
A/R at 9/30/XX	$6,590,000
A/R at 12/31/XX	$8,635,901
Average A/R	$6,797,334
From the Income Statement	
Total Revenue	$45,226,102
Average Age of Accounts Receivable =	
(Average A/R / Total revenue) × 365	
= $(6,797,334 / $45,226,102) × 365	= 54.9 days

Average Age of Accounts Payable

The same approach used for the average ages of Inventory and Accounts Receivable is applicable here once more. The formula for this indicator is:

$$\text{Average Age of Accounts Payable} = [\text{Average balance of Accounts Payable}/(\text{Direct Costs} - \text{Cost of Labor})] \times 365$$

As before, an average balance for the Accounts Payable must be computed for the analyzed period. The 365 is replaced by the number of days covered by the Income Statement if the period is not one year.

The denominator in the above formula is a bit different from the previous indicators because the equivalent of "total miles" discussed on the example at the beginning of this section is not as straightforward. The majority of purchases for any project are initially posted as Accounts Payable, since they are purchased on credit and later paid off. The big exception to this rule is the cost of labor, which normally doesn't pass through Accounts Payable since it is paid virtually as soon as it is incurred. To compare apples to apples, then, the cost of labor must be deducted from the total Direct Cost. If a project has other accounts paid on a cash (non-credit) basis, they should be subtracted as well. Table 3.11 consists of the Average Age of Accounts Payable computations for Fudd Associates, Inc.

The practical effect of the Average Age of Accounts Payable is the opposite to its Accounts Receivable counterpart. The more you delay paying a debt, the more money you have to operate your company.

TABLE 3.11 Average Age of Accounts Payable for Fudd Associates, Inc.

From the four most recent quarterly Balance Sheets	
A/P at 3/31/XX	\$4,546,338
A/P at 6/30/XX	\$2,786,200
A/P at 9/30/XX	\$4,300,290
A/P at 12/31/XX	\$3,659,156
Average A/P	\$3,822,996
From the Income Statement	
Direct costs	\$41,476,336
Labor Costs	\$4,974,871

Average Age of Accounts Payable =
(Average A/P / (Direct costs − labor costs)) × 365
= (\$3,822,996 / (\$41,476,336 − \$4,974,871)) × 365 = 38.2 days

TABLE 3.12 Average Age of Underbillings for Fudd Associates, Inc.

From previous Balance Sheets (not shown)	
Average Underbillings	\$937,486
From the Income Statement	
Total Revenue	\$45,226,102
Average age of Underbillings =	
(Average Underbillings / Total revenue) × 365	
= (\$937,486 / \$45,226,102) × 365	= 7.6 days

Other Average Ages

The computation of average ages for Costs and Estimated Profit in Excess of Billings (Underbillings) and Billings in Excess of Costs and Estimated Profit (Overbillings) follow a similar logic, resulting in similar formulas, as you can see in Table 3.12 and Table 3.13.

Similar to previous examples, the average amounts for Fudd's Underbillings and Overbillings were computed as the simple average of the respective balances for these Balance Sheet accounts at the end of each quarter. The amount of Revenue, as in previous examples, comes from Fudd's Income Statement.

Operating Cycle

What happens to a dollar invested in a construction project? As shown in Figure 3.1, the dollar sits on each of the arrows for the number of days found for each indicator in the Efficiency category, either increasing the time that it takes for the dollar to come back to us (Inventory, Underbillings, and Accounts Receivable) or decreasing this time (Underbillings and Accounts Payable). This total time is called the company's *operating cycle*, usually expressed in days.

TABLE 3.13 Average Age of Overbillings for Fudd Associates, Inc.

From previous Balance Sheets (not shown)	
Average Overbillings	\$742,820
From the Income Statement	
Total Revenue	\$45,226,102
Average age of Overbillings =	
(Average Overbillings / Total revenue) × 365	
= (\$742,820 / \$45,226,102) × 365	= 6.0 days

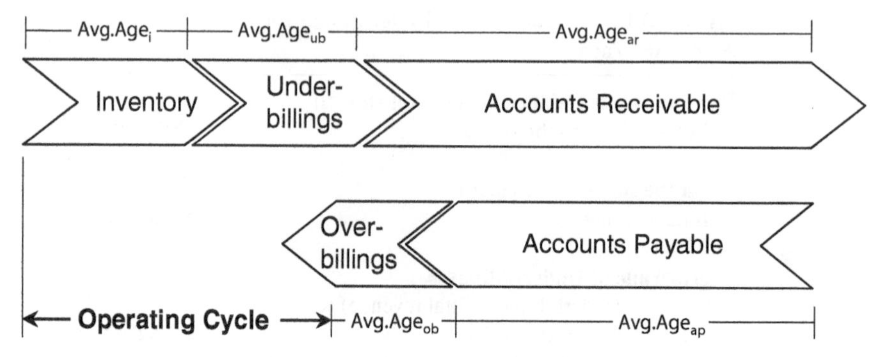

Figure 3.1 Operating cycle.

Turnover Ratios

Although the use of *average ages* is very common in construction, other industries tend to use the inverse of the average ages computed in this section. These inverse amounts are called Turnover Ratios, and show how many times in a year a company cycles its inventories, accounts receivable, or any other of the items discussed in this section. Accounts Receivable Turnover, for example, is the inverse of the Average Age of Accounts Receivable. If this age is expressed in years (say 0.10 years), then its turnover is simply its numerical inverse (1/0.10 = 10.0 times per year). If the age is in days, then a conversion is needed, as shown in the following example.

A company has an average age of inventory of 30.4 days. The inventory turnover for this company is:

$$(365 \text{ days in a year})/(30.4 \text{ days}) = 12.0 \text{ times/year}$$

An inventory turnover of 12.0 times per year means that the company's inventory was used up and replenished 12 times.

Some ratios are customarily computed as turnovers and virtually never as average ages. They have the same conceptual framework as the previously discussed turnovers. They show how many times an item has been leveraged (i.e., multiplied) compared to another during the period covered by the Income Statement.

Revenue to Assets Turnover

One ratio that is frequently expressed as a turnover instead of as an average age is the Revenue to Total Assets Turnover. It indicates how much annual volume of work a company did compared to its Total Assets. A large value

TABLE 3.14 Revenue to Assets Turnover for Fudd Associates, Inc.

From the Balance Sheet	
Total Assets	$15,978,849
From the Income Statement	
Total Revenue	$45,226,102
Revenue to Assets turnover =	
Total Revenue / Total Assets	
= $45,226,102 / $15,978,849	= 2.83

for this ratio is desirable to the company owners but implies added risk to potential creditors. Its formula is:

$$\text{Revenue to Assets Turnover} = \text{Revenue/Total Assets}$$

Table 3.14 shows the computations of this indicator for Fudd Associates, Inc.

CAPITAL STRUCTURE INDICATORS: HOW COMMITTED ARE THE OWNERS?

The final indicator category that we will address here is called Capital Structure. It addresses the relative size of a company's assets, liabilities, and owners' equity. The indicators in this category are useful for internal managers because they show whether the company is borrowing too much money compared to the amount of its total assets and owner's equity. That is, is the company overinvested?

Investors and creditors may stay away from a construction company that has borrowed, for example, 10 million dollars and whose owners have a total equity of, say, half a million dollars. To understand the rationale here, compare such a company with another that has borrowed the same dollar amount but whose owners have five million dollars of equity in the company. Should a financial crisis arise, two very negative things can happen in the case of the first company. First, since all creditors must be paid off by liquidating the company's assets, there would be a real possibility that the proceedings of the sale plus the company's owners' equity would not be enough to pay all liabilities. Second, owners would have little incentive to remain in business during the crisis, and could just "cut and run" with a small personal loss compared to the amount of trouble they might incur. In the case of the second company, creditors would have to count less on the money generated by the sale of assets to get paid, and the company owners would think twice before

TABLE 3.15 Debt to Equity for Construction Specialties

Specialty	All Companies	Industrial and Nonresidential	Heavy and Highway	Specialty Trade
Debt to Equity	2.5	3.8	1.4	1.7

leaving behind the company and its creditors. They would stand to lose more of their $5 million equity stake.

Debt to Equity

The term *debt* as used here is synonymous with Total Liabilities (which is admittedly confusing). The Debt to Equity ratio, therefore, shows the relation of a company's Total Liabilities compared to its Owners' Equity. The introductory discussion for this ratio category in the previous paragraphs implicitly pointed out the importance of this indicator, since it establishes how committed the owners are to their company. The formula for this ratio is:

$$\text{Debt to Equity} = \text{Total Liabilities/Total Owners' Equity}$$

The higher the value of a company's Debt to Equity, the riskier the company is in the eyes of potential creditors. Table 3.15 compares this ratio across several construction specialties.

Assets to Equity (Leverage)

Instead of using Debt to Equity, some companies use the Assets to Equity ratio, which is also called Leverage or Gearing. This formula is:

$$\text{Assets to Equity (Leverage)} = \text{Total Assets/Total Owners' Equity Leverage}$$

Debt to Equity and Assets to Equity are related as follows:

$$\text{Assets to Equity (Leverage)} = \text{Debt to Equity} + 1$$

Table 3.16 gives a brief proof of this relationship.

Table 3.17 shows the computation of this ratio for Fudd Associates, Inc., which numerically checks with the above equation.

OTHER INDICATORS

As mentioned in the introduction to this chapter, there are dozens of financial indicators, each one being of more relevance to some industries than to others.

TABLE 3.16 Proof of Assets to Equity = Debt to Equity + 1

Assets to Equity = Total Assets / Owners' Equity	①
But, Total Assets = Total Liabilities + Owners' Equity	②
Replacing Total Assets in ①,	
Assets to Equity = (Total Liabilities + Owners' Equity) / Owners' Equity	③
Distributing the denominator,	
Assets to Equity = (Total Liabilities / Owners' Equity)	
+ (Owners' Equity / Owners' Equity)	④
Since Debt to Equity = Total Liabilities / Owners' Equity	⑤
and Owners' Equity / Owners' Equity = 1	⑥
Assets to Equity = (Debt to Equity) + 1	⑦

One indicator of particular importance for heavy/highway contractors relates to the quality of assets.

This indicator provides a rough idea of how new or depleted a company equipment fleet is, by dividing the depreciated value of the equipment by its purchase value. A value close to one indicates a newer (nondepreciated) fleet. The formula for this indicator is:

$$\text{Asset Newness} = (\text{Purchase Value} - \text{Accumulated Depreciation})/\text{Purchase Value}$$

As the formula implies, this indicator varies from 1.0 for a totally new fleet to 0.0 to a completely depreciated one. Although values in between these two extremes provide an idea of a fleet's age, it is not intended to accurately reflect its maintenance and condition. Accounting practices do not consider the market value of any asset, and, therefore, the book value of any piece of equipment (purchase value minus accumulated depreciation) is only a crude approximation of its practical worth. Table 3.18 shows the computations for the asset newness of Fudd Associates, Inc.

TABLE 3.17 Assets to Equity Computations for Fudd Associates, Inc.

From the Balance Sheet	
Total Assets	$15,978,849
Total Owners' Equity	$4,732,935
Assets to Equity =	
Total Assets / Total Owners' Equity	
= $15,978,849 / $4,732,935	= 3.38

TABLE 3.18 Asset Newness Computations for Fudd Associates, Inc.

From the Balance Sheet
(PP&E = Property, Plant and Equipment)

PP&E purchase cost	$4,261,559
PP&E Accum. Depreciation	$(2,344,097)
	$1,917,462

Asset Newness =

$$\frac{\text{(PP\&E purchase cost)} - \text{(PP\&E Accum. Depreciation)}}{\text{PP\&E purchase cost}}$$

$$= \frac{\$1{,}917{,}462}{\$4{,}261{,}559} = 0.45$$

HORIZONTAL ANALYSIS: TRACKING FINANCIAL TRENDS

A critical factor revealed by comparative financial statements is trend. Tracking how the numerical value of a financial item such as Return on Equity or Accounts Receivable has changed over time provides information regarding the performance of a company. As previously mentioned, this type of analysis is called "horizontal" analysis since the parameters of interest are plotted against time and time is usually shown on the horizontal axis. We will consider two types of graphs used in a horizontal financial analysis: Time Series and Index-Number Trend Series.

TIME SERIES GRAPHS

A comparison of the value of a financial item over a number of years can reveal much about its direction (improving or worsening), its velocity (fast or slow), and its amplitude (large or small). This graph is more revealing and easier to understand than any narrative, and shows this information in an extremely compact way. Figure 3.2 shows the trend of the Return on Equity

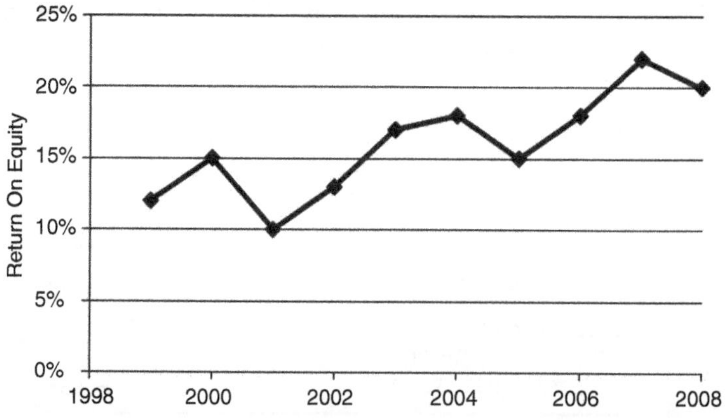

Figure 3.2 Fudd, Inc. Return on Equity for the past 10 years.

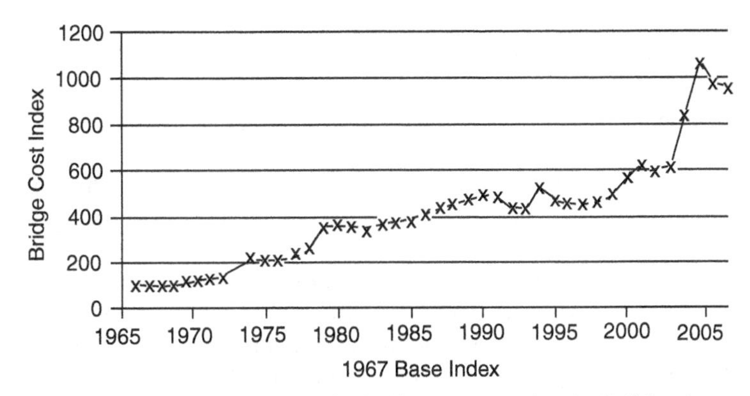

Figure 3.3 Index series for bridge construction in California.

for Fudd Associates, Inc. You can readily appreciate not only that Uncle Fudd has been successful in keeping an acceptable ROE, but that it has improved over time.

Index-Number Trend Series

When a comparison of financial statements is made over a longer multiyear period, a year-to-year numerical assessment such as the side-by-side comparison becomes cumbersome. The best way to perform such longer-term comparisons is by means of index numbers. A year's index number is simply the ratio between the value analyzed for that year and a reference year called the base year.

Figure 3.3 consists of the indexes for the cost of bridge construction in California from 1967 to 2007. In this particular series, 1967 is assigned a value of 100, and the number in each following years is computed by comparing the cost of building the same bridge in the year with its cost in 1967. The index for 2007 is 960. This means that the standard bridge used for the index costs $960/100 = 9.6$ times more in 2007 than in 1967. The index for 1987 is 441, which means that it would cost $960/441 = 2.18$ times to build the same bridge in 2007 compared to 1987.

CONCLUSION

Financial indicators and all techniques developed to gain insight into a company's financial health are empirical by nature. They are the result of distilling from a company's Balance Sheet and Income Statement a compact representation of the company's performance and value. Anyone can invent a new indicator, as is the case of private financial analysts (e.g., the Fails Management Institute), who keep track of unconventional statistics that they have found to be useful for their practice. For example, a company could calculate the ratio between its Total Revenue and its number of superintendents. This metric could be used to assess whether the company has too few superintendents for its current revenue, or vice versa, if it has too many.

We have considered a number of financial indicators in this chapter, and yet they are but a fraction of the metrics available to a company. For example, some companies compute their revenue to fixed assets ratio and their book value per share (total owners' equity/common shares outstanding). These ratios are important for manufacturing plants and public corporations, respectively, and could be included in a construction financial management text. But, inevitably, a selection must be made of the most useful indicators. The reader is encouraged to investigate and even create additional metrics that could be valuable for their particular case.

REVIEW QUESTIONS AND EXERCISES

1. Name and discuss four principal tools of financial analysis.
2. Give two examples of an activity ratio. What do these ratios reflect?
3. Analyze the operations of the Jumbo Construction Company across a five-year period based on the data in Table P3.1.
   (a) Plotting significant times and ratios of your choice aid in your analysis. What is your assessment of the financial health of Jumbo Construction?
4. Given the balance sheet in Table P3.2, calculate the following ratios:
   (a) Current ratio
   (b) Quick ratio
   (c) Debt-equity ratio
5. How does trend analysis supplement the basic financial ratio calculations and their interpretation?
6. Given a current ratio of 1.7, to which percentage of book value would you have to liquidate current assets to be able to pay off current liabilities?
7. Given that the current ratio (i.e., current assets/current liabilities) for a company over the years is as shown, what can you say about the operation of this company?

20X1	20X2	20X3	20X4	20X5
1.50	1.51	1.42	1.20	0.95

TABLE P3.1 Jumbo Construction Company Five-Year Consolidated Summary of Operations

	20X5	20X4	20X3	20X2	20X1
Revenues	$832,420,000	$391,236,000	$182,684,000	$214,973,000	$462,822,000
Interest and dividend income	2,526,000	3,802,000	2,627,000	1,304,000	1,005,000
Cost of contracts	775,552,000	347,425,000	148,846,000	183,862,000	434,120,000
G & A expenses	36,129,000	35,100,000	29,207,000	26,676,000	23,220,000
Interest expense	2,199,000	1,089,000	—	—	—
Income before taxes	21,066,000	11,424,000	7,258,000	5,739,000	6,487,000
Provision for income taxes	10,700,000	5,700,000	3,600,000	2,800,000	3,500,000
Net income	12,666,000	5,724,000	3,658,000	2,939,000	2,987,000
Net income per share	5.51	2.51	1.62	1.31	1.34
Dividends per share	.70	0.45	0.35	0.33	
Average number shares	2,298,000	2,281,000	2,252,000	2,236,000	2,230,000

TABLE P3.2 Balance Sheet, Cyclone Construction Co. (31 December 2012)

Assets		Liabilities	
(a) Current assets		(f) Current liabilities	
Cash on hand and on deposit	$389,927.04	Accounts payable	$306,820.29
Notes receivable, current -	16,629.39	Due subcontractors	713,991.66
Accounts receivable, including		Accrued expenses and taxes	50,559.69
retainage of $265,686.39	1,222,346.26	Equipment contracts, current	2,838.60
Deposits and miscellaneous receivables	15,867.80	Provision for income taxes	97,816.66
Inventory-construction material	26,530.14	Total	1,171,826.90
Prepaid expenses	8,490.68	(g) Deferred credits	
Total	1,679,791.31	Income billed on jobs in	
		progress at 31 December 19X4	2,728,331.36
(b) Notes receivable, non-current	12,777.97	Costs incurred to 31 December	
		19X4 on uncompleted jobs	2,718,738.01
(c) Property			9,593.35
Buildings	5,244.50	Total current liabilities	1,181,420.25
Construction equipment	188,289.80	Equipment contracts, non-current	7,477.72
Motor vehicles	37,576.04	(h) Total liabilities	1,188,897.97
Office furniture and equipment	13,596.18		
Total	244,706.52	Net worth	
(d) Less accumulated depreciation	102,722.51	(i) Common stock, 4,610 shares	461,000.00
Net property	141,984.01	Retained earnings	184,655.32
		(0) Total net worth	645,655.32
(e) Total assets	1,834,553.29	(k) Total liabilities and net worth	1,834,553.29

TABLE P3.3 Balance Sheet, Peachtree Construction (31 December 2016)

Assets	
Current assets	
Cash	$243,146
Accounts receivable	
Trade accounts	201,573
Retainage	42,147
Total accounts receivable	243,720
Material inventory	1,873
Work in process (costs and estimated earnings in excess of billings)	76,142
Prepaid expenses	6,148
Other current assets	782
Total current assets	571,811
Fixed assets	
Machinery and equipment	542,173
Cars and trucks	49,214
Furniture and fixtures	5,812
Total depreciable assets	597,199
Less accumulated depreciation	291,314
Net fixed assets	305,885
Total assets	877,696
Liabilities	
Current liabilities	
Accounts payable	236,144
Retainage payable	21,312
Notes payable	84,211
Accrued liabilities	15,827
Work in process (billings in excess of costs and estimated earnings)	5,163
Other current liabilities	1,421
Total current liabilities	364,078
Long-term liabilities	138,481
Total liabilities	502,559
Net worth	
Capital stock	15,000
Retained earnings	360,137
Total net worth	375,137
Total liabilities and net worth	$877,696

TABLE P3.4 Income Statement, Peachtree Construction (Year ended 31 December 31, 2015)

	$	*% of Sales*
Net sales-total revenue costs	2,143,761	100
Materials	1,013,913	47.3
Labor (includes all payroll taxes and union fringes)	548,271	25.57
Subcontracts	201,798	9.41
Other direct costs	51,450	2.4
Total direct costs	1,815,432	84.68
Gross profit	328,329	15.32
Operating expenses		
Variable overhead		
Auto and truck	12,089	0.57
Communications	5,575	0.26
Interest	1,275	0.05
Miscellaneous	213	0.01
Office supplies	1,943	0.09
Travel and entertainment	871	0.04
Total variable overhead	21,966	1.02
Fixed overhead		
Contributions	675	0.03
Depreciation	8,238	0.38
Dues and subscriptions	514	0.02
Insurance	12,475	0.58
Legal and audit	2,315	0.1
Licenses and taxes	175	0.01
Payroll taxes (office only)	51,790	2.42
Rent	2,506	0.12
Repairs and maintenance	375	0.02
Salaries-office	23,418	1.1
Salaries-officers	38,410	1.79
Total fixed overhead	140,891	6.57
Total overhead	162,857	7.59
Net profit before taxes	165,472	7.73
State and federal taxes	77,076	3.6
Net profit	88,396	4.13

8. Given the financial statement shown here as Table P3.3 calculate the:
   (a) Current ratio
   (b) Quick ratio
9. Based on the Income Statement for Peachtree Construction, shown in Table P3.4, what is the ratio of operating profit to completed contract sales. Is this ratio acceptable? Explain.

TABLE P3.5 Key Financial Data, Clairmont Constructors

(000's)	2011	2012	2013	2014	2015
Sales	2,460	2,560	2,830	2,835	2,850
% Sales growth		4	10.5	1	1
Labor	800	820	780	810	900
Material	925	900	1,010	1,010	1,040
Cubcontr.	500	550	720	765	680
Gross margin	235	290	320	250	230
Gross margin %	9.5	11.3	11.3	8.8	8
G & A	133	172	202	218	220
G & A % sales	5.4	6.7	7.1	7.7	7.7
Income before tax	102	118	118	32	10
Income before tax%	4.1	4.6	4.2	1.1	0.4
Admin. salary	24	27	32	38	39
Assets	800	812	830	850	830
Sales to assets	3.07	3.15	3.4	3.33	3.43
Equity	350	287	285	280	205
Sales to equity	7.03	8.92	9.93	10.12	13.9
Current Assets	520	532	475.8	510	511.5
Investments	100	110	120	110	120
Investments	100	110	120	110	120
Property and equipment	280	280	354.2	360	338.5
Accu. depreciation	100	110	120	130	140
Current liabilities	325	380	390	425	465
Long term liabilities	125	145	155	145	160
Current ratio	1.6	1.4	1.22	1.2	1.1
Debt to equity	1.28	1.83	1.91	2.04	3.04
Income to equity %	29.1	41.1	41.4	11.4	4.8
Income to assets %	12.7	14.5	14.2	3.8	1.2
Sales to current assets	4.73	4.81	5.94	5.56	5.57

Source: J. J. Adrian, *Construction Accounting* (Reston, Va.: Reston, 1978).

10. A five-year summary of key financial data for a construction firm is shown in Table P3.5. Evaluate and analyze the performance of this company during the period. What actions, if any, should be taken to improve the company's position?

11. Discuss why a firm's current ratio may differ from another in the same industry.

12. Illustrate how a decrease in working capital can be accompanied by an increase in the current ratio.

CHAPTER 4

ACCOUNTING BASICS

INTRODUCTION

In order to understand how revenue and expenses are accounted for on a day-to-day basis, a basic knowledge of the bookkeeping process is necessary. Most construction students have very little contact with courses related to accounting and business management. On the other hand, business students take many courses addressing the minutiae of bookkeeping and financial accounting. The unique nature of construction, as discussed in Chapter 1, makes construction accounting quite different from that encountered in the manufacturing and service sectors. Therefore, students who have been exposed to standard bookkeeping protocols can still profit from this brief introduction to, what we might call, construction accounting procedures.

This chapter introduces some of the fundamental concepts that govern the acquisition and flow of financial data through the accounting system of a construction-oriented firm. This introductory treatment is amplified using simple flow diagrams. Appendix B uses this flow-graphing approach to illustrate the processing of various types of transactions typical of a construction company.

TRANSACTION PROCESSING

A simplified diagram reflecting the sequence of actions that occurs during a typical bookkeeping operation is shown in Figure 4.1.

The flow unit that is processed in the accounting sequence is called a transaction. A transaction occurs when a source document such as a billing from

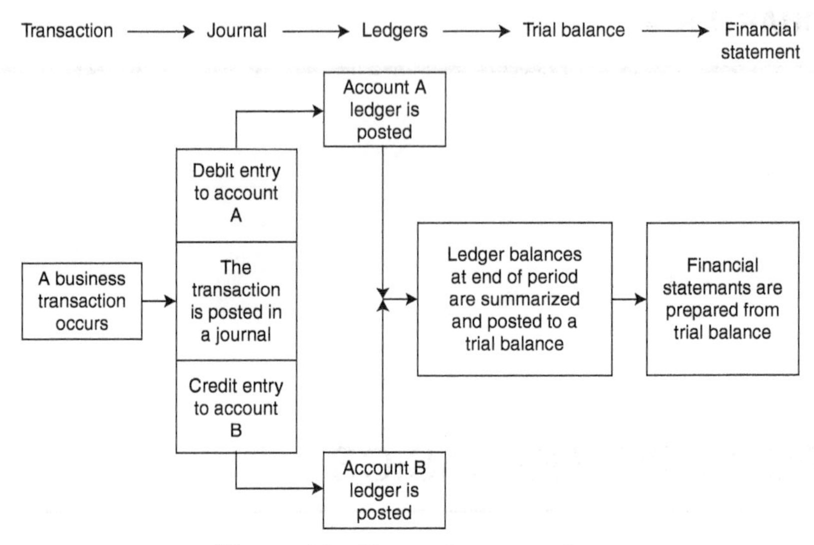

Figure 4.1 Transaction processing.

a subcontractor or a time based obligation (e.g., payroll payment deadline) occur. Transactions are also generated in the course of closing accounts to prepare Balance Sheets, Income Statements, and associated reports. Some typical types of source documents that lead to the generation of a transaction are listed in Table 4.1. By monitoring the flow of transactions through the accounting system, the manager can derive information regarding the company's balance of revenue versus expense as well as the financial status of the company.

Transactions are processed to reflect increase or decrease in the balances in accounts maintained by the company. As discussed in Chapter 2, an account can be viewed as a record used to accumulate and monitor the balance fluctuations generated by some source of financial activity. The initial documentation of this activity is accomplished through entries made to the accounting system.

Accounts are categorized in accordance with what they pertain to as previously discussed in Chapter 2: Assets, Liabilities, Net Worth, Revenues, and Expenses. These categories are, of course, the main components of a Balance

TABLE 4.1 Source Documents

Examples	Provide Information About
(A) Check stubs or registers	Cash disbursements
(B) Purchase invoices received from vendors	Purchases of materials or services
(C) Invoices from subcontractors for payment	Progress payment requests
(D) Equipment records	Rates of depreciation

Sheet and an Income Statement. Appendix A contains the typical accounts in a construction company's Chart of Accounts.

Accounts may be maintained manually by using a standard form in a loose-leaf binder or similar physical document. Presently, all construction companies of any size use computerized accounting systems. This notwithstanding, it is important to understand how data (transactions) are acquired and processed either manually or using a computer.

JOURNALIZING THE TRANSACTION

As shown in Figure 4.1, a new transaction is initially recorded or journalized. Conceptually, journalizing is a matter of chronologically recording transactions as they occur (or are recognized) on a day-by-day basis. Journals establish the time at which transactions are recognized and enter the accounting system. That is, the journal lists the point in time at which transactions are acquired and enter the accounting system.

The most widely used protocol for transaction processing is referred to as the "double-entry" system. In this protocol each transaction causes an increase or decrease in two different accounts. In effect, the two entries balance each other. Using this approach, all accounts can be checked at any time to ensure that all accounts are "in balance." The "balance" here is the balance given by the basic equation:

$$\text{Assets} = \text{Liabilities} + \text{Net Worth} + (\text{Revenues} - \text{Expenses})$$

At the time a transaction is journalized, a decision must be made as to which two accounts will be affected, and, whether the accounts involved will increase or decrease in value.

Accountants utilize the terms debit and credit to refer to the way in which the transaction will ultimately affect the individual account balances. Debit and credit refer to different effects, depending on what type of account (e.g., Asset, Revenue, etc.) is being addressed. Debits cause an increase in the balance of Asset and Expense accounts, while causing a decrease in the balance of Liability, Net worth, and Revenue accounts. Conversely, credits decrease the balance of Asset and Expense accounts, while increasing the balance of Liability, Net Worth, and Revenue accounts. This relationship is summarized as shown in Table 4.2.

TABLE 4.2 Debits and Credits by Account Type

Asset		Liability		Net Worth		Revenue		Expense	
Debit	Credit	Debit	Credit	Debit	Credit	Debit	Credit	Debit	Credit
+	–	–	+	–	+	–	+	+	–

Each transaction causes one account to be debited and another account to be credited. The effect on the balance of the accounts affected varies in accordance with the types of accounts involved. That is, a transaction may cause the balance in one account to decrease, while increasing the balance of the second account. However, some transactions cause the balances of both accounts to increase. Other transactions cause the balances of both accounts to decrease. For instance, if a transaction causes an asset to be debited (e.g., cash) and a revenue account to be credited (e.g., revenue Project 101), the effect is to increase the balances of both accounts. If another transaction causes a liability to be debited and an expense account to be credited, the effect is to decrease the balance in both accounts. Transactions affecting accounts of the same type cause an increase in one and a decrease in the other.

As noted above, the double-entry method of bookkeeping is a method of maintaining balances between accounts. These balances are ultimately reflected in the Balance Sheet. A result of using the double-entry method of bookkeeping is that pairs of entries are recorded chronologically in the company journals.[1] We will now introduce some typical transactions and describe how they are journalized.

A Transaction to Enter Initializing Capital

In order to gain an overview of the construction accounting flows, consider the schematic diagram shown in Figure 4.2. Let us assume that Cousin Elmer and Uncle Fudd have decided to form Fudd Associates, Inc., and each has contributed $10,000 cash as initial capital. This leads to entries in the Asset account and the Net Worth accounts, as shown in Figure 4.2. The $20,000 contributed to form the company results in debits to the Asset account called cash and balancing credit entries in the Net Worth account called Capital Stock. Here is a case where the transactions increase the balances in both accounts affected. The entry of these two transactions in the company journal would appear as Shown in Table 4.3.

A Vendor Billing Transaction

Let us assume now that the first billing to arrive at Fudd Associates, Inc. is an invoice (request for payment) from Ace Electric Company for work done on wiring in the job site trailer on Fudd Associates Project 101. The charge of $265 must be booked again the "Work-in-progress (expense)" account for this mobilizations cost on Project 101.[2] That is, it is a project expense and must be recorded as such. Therefore, a debit (increase in balance) to the work-in-progress (expense) account must be made. The company now has an

[1] Larger companies have multiple profit centers and many journals are maintained.

[2] Mobilization is actually a subaccount or subsidiary account of the general ledger account "Work-in-progress expense – Project 101."

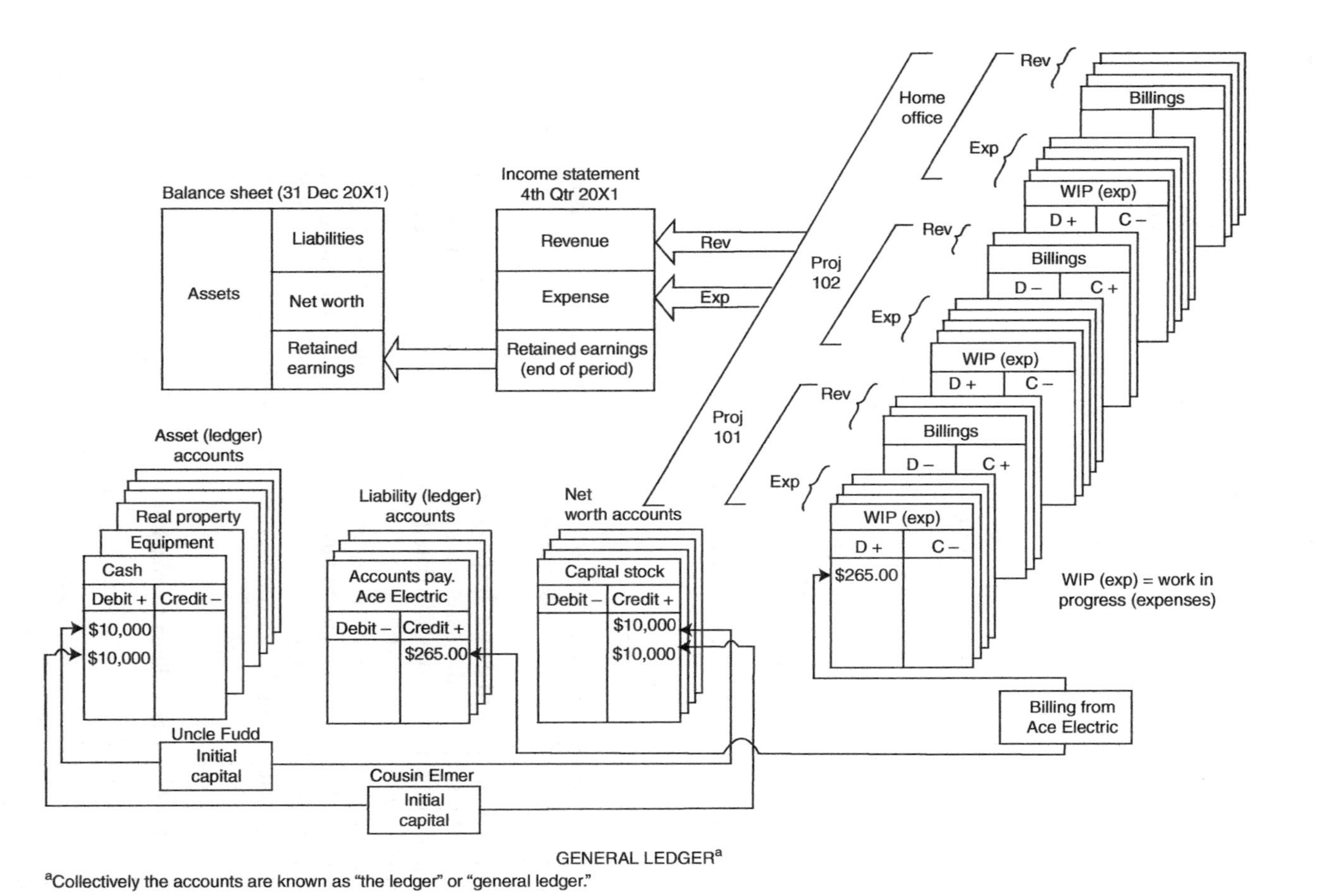

[a]Collectively the accounts are known as "the ledger" or "general ledger."

Figure 4.2 Accounting flows for simple transactions.

TABLE 4.3 Initial Capital Journal Entries

Journal				
	Date	Description	Debit	Credit
1	Dec. 1 20XX	Cash	$10,000	
		Elmer T. Fudd capital		$10,000
2	Dec. 1 20XX	Cash	$10,000	
		Ferdinand Fudd, capital		$10,000

obligation to pay Ace Electric for the work performed. This obligation must also be reflected by the accounting system. Therefore, a balancing entry—a credit to the outstanding balance payable to Ace Electric—must also be made. Obligations to pay vendors, subcontractors, and other suppliers are reflected in the Accounts Payable area of the Liability general ledger accounts. As noted in Figure 4.2, the total of all active accounts is referred to in accounting parlance as the general ledger. The entries to the company journal generated by the billing from Ace Electric are shown in Table 4.4. Convention in recording transactions requires that the debit be entered first and then the credit. This leads to a debit upper left, credit lower right appearance in the journal.

A Billing to the Client

Now let us assume that there is a mobilization payment clause on Contract 101 and Fudd Associates sends a billing to the client, Donut Factory, Inc., requesting payment for the mobilization amount of $5,000. As shown in Figure 4.3, this will cause the project billings account (Billings – Project 101) to be credited and the asset account—Accounts Receivable—to be debited. The request for payment is dated 7 December 20XX, and the journal entry for this transaction is shown in Table 4.5.

In the last two transactions, we have made entries based on the fact that obligations to make payment have occurred. In journalizing the invoice from Ace, Electric Co., Fudd Associates, Inc. was recognizing an obligation to pay for something. However, payment has not been made; cash has not been

TABLE 4.4 Vendor Billing Journal Entries

Journal				
	Date	Description	Debit	Credit
1	5 Dec 20XX	Work-in-progress expense-Project 101	$265	
2		Ace Electric Invoice #101-1		$265

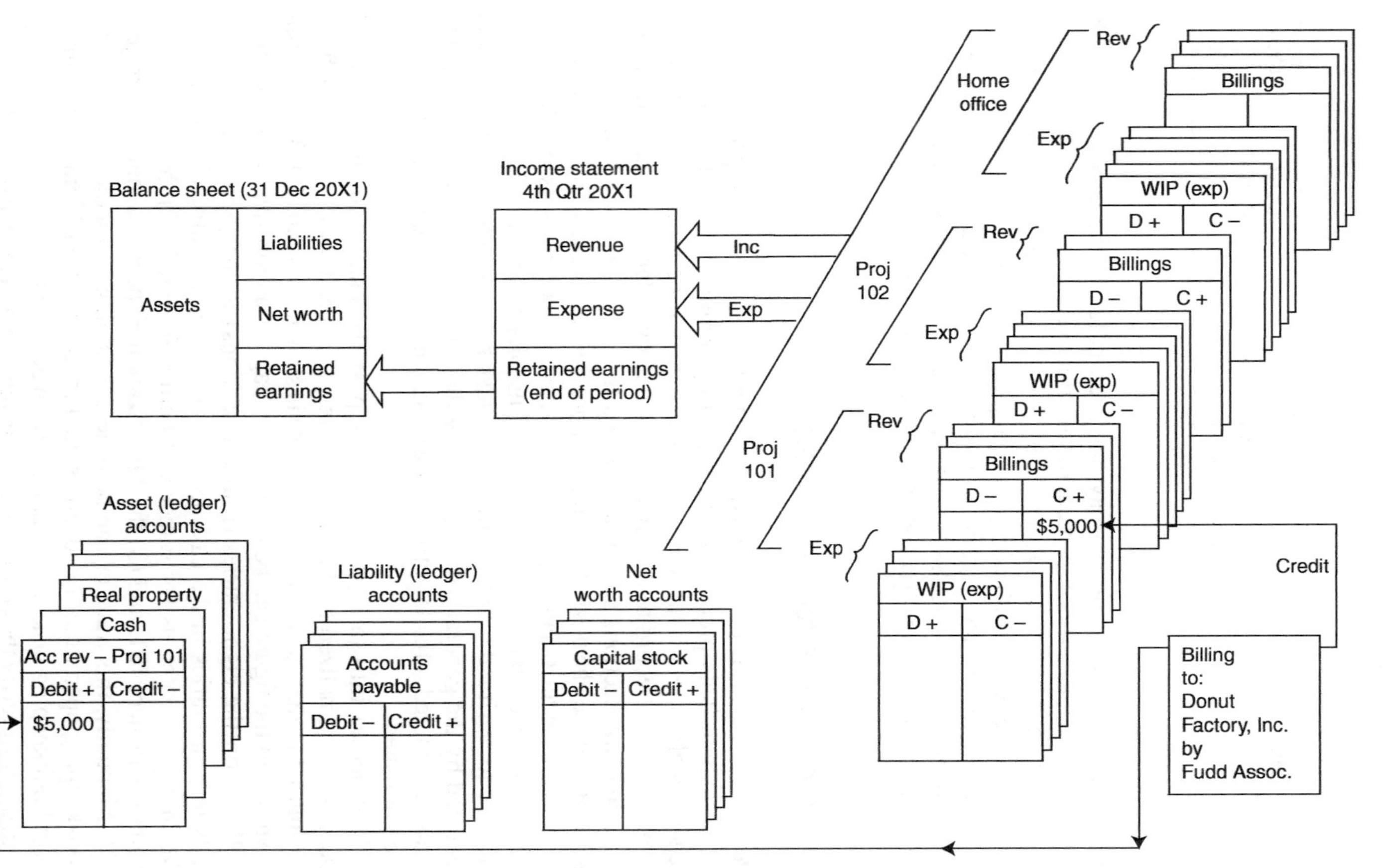

Figure 4.3 Billing to client.

TABLE 4.5 Billing to Client Journal Entries

Journal				
	Date	Description	Debit	Credit
1	7 Dec 20XX	Accounts Receivable – Job 101	$5,000	
2		Billings – Job 101		$5,000

transferred. The obligation to pay has been noted in the accounts payable designated Ace Electric. It is recognized effective 5 December 20XX.

In sending a billing to Donut Factory, Inc., for a mobilization payment, Fudd Associates is establishing an obligation on the part of the client to pay generating revenue. Again, no monies have been transferred. However, for accounting purposes, the impact on the various account balances of this transaction has been recognized effective 7 December 20XX.

POSTING ENTRIES TO THE LEDGER

Periodically, data are transferred from the journal(s) to the appropriate account in the ledger. This process is called posting. Figures 4.2 and 4.3 illustrate schematically the accounts in the ledger that will be posted. The objective of posting is to record to the appropriate general ledger accounts the impact of various transactions so that at selected times the balance of entries in all accounts can be checked. As mentioned previously, accounting is the process of maintaining the balance implied by the basic equation assets = liabilities + net worth + (revenue – expense). This is usually done following the posting process to verify the entries that have just been transferred. Information reflected by the general ledger regarding the financial status of the company is not considered reliable until posting of all entries from the journal has been accomplished.

In order to illustrate posting, let us post the entries for the month of December journalized for Fudd Associates, Inc. The flow of the entries from the journal to the various ledger accounts is shown in Figure 4.4. The account as shown in the figure can be thought of a page in a loose-leaf ledger book.

Based on the eight entries (four transactions) in the journal, posting results in four entries in the cash ledger and individual entries in (1) the Net Worth capital accounts for Elmer and Ferdinand Fudd, (2) the Billings account for Project 101, and (3) the Mobilization Expense account for Project 101. Manually, this transfer can be time-consuming and tedious. Therefore, bookkeeping operations are ideally suited for automation. Computers handle most bookkeeping tasks once data has been acquired.

In order to verify the entries made up to any point in time, a trial balance can be made by summarizing the entries in all active accounts and checking to see

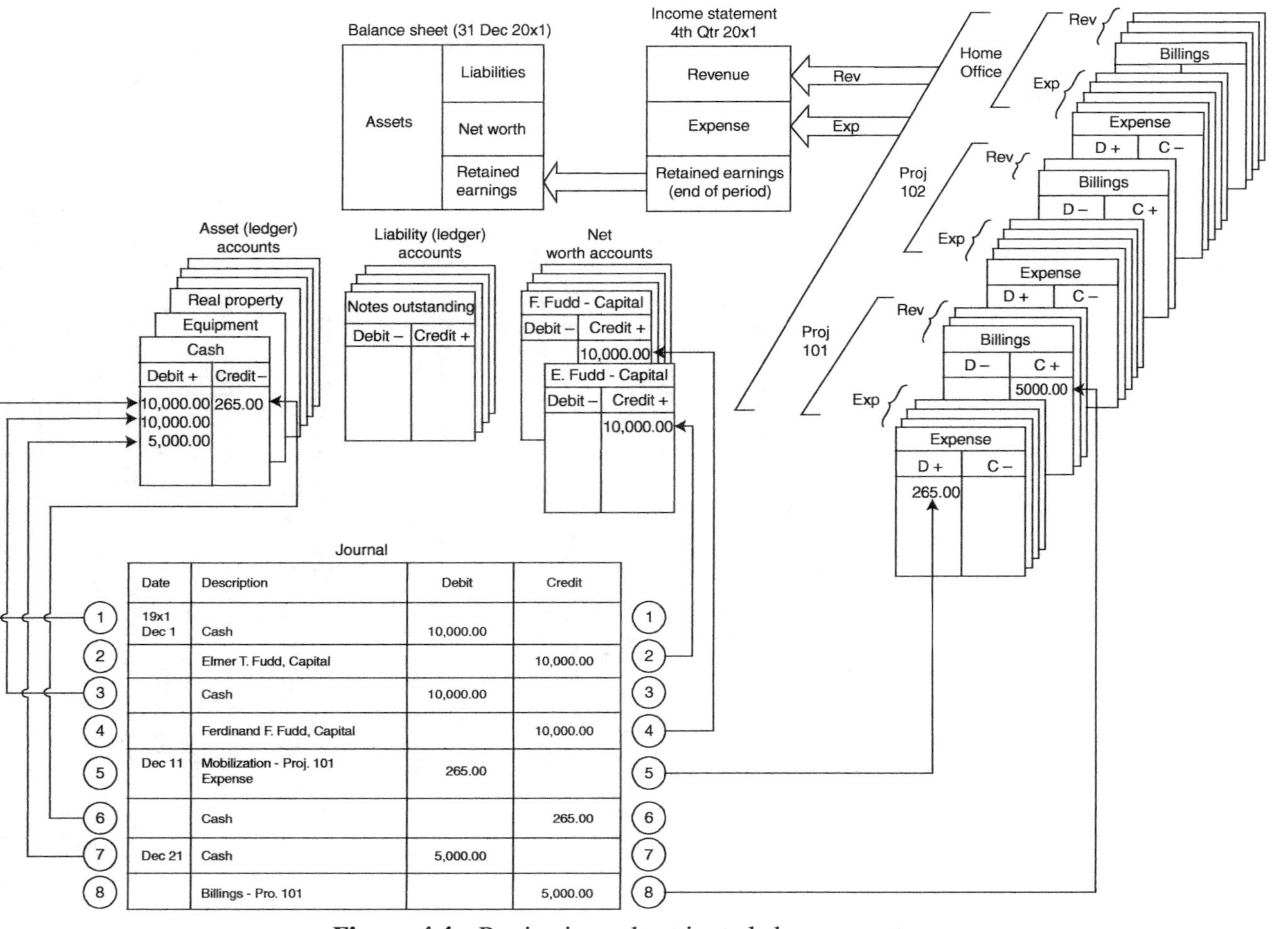

Figure 4.4 Posting journal entries to ledger accounts.

TABLE 4.6 Trial Balances

Fudd Associates, Inc. (31 December 20XX)			
Account	Account No.	Debit Balance	Credit Balance
Cash		$24,735.00	
E. Fudd-capital			$10,000.00
F. Fudd –capital			10,000.00
Billings – Project 101			5,000.00
Expense – Project 101		265.00	265.00
		$25,000.00	$25,000.00

that the sum of debits and credits in all accounts balance. Assume that a trial balance for the posting activity in Figure 4.4 is accomplished in connection with closing[3] the books as of 31 December 20XX. The trial Balance Sheet is shown in Table 4.6.

As can be seen, as of 31 December, Fudd Associates has five active accounts. The account balances have been summarized and entered on the trial Balance Sheet. Since four accounts have single entries, these are entered directly. The cash account balance is based on four entries, and the account has a debit balance of $24,735.00. The balance of accounts and of the basic accounting equation is verified, since the debit and credit balances both equal $25,000. It is reassuring that the accounts balance. However, this does not ensure that the entries have been made to the correct accounts. Whether or not the debits and credits are posted to the proper accounts requires further review.

The operations required to organize the data contained in the general ledger for financial statement production are referred to as closing activities. Closing activities are discussed in the following sections.

RELATIONSHIP OF WORK-IN-PROGRESS AND REVENUE/EXPENSE ACCOUNTS

Terminology regarding project revenues and expenses varies from firm to firm and project to project, depending upon the method of accounting used. However, the term "work-in-progress" is used broadly in accounting for billings and costs associated with projects under accrual based systems of accounting. In general, billings (revenue-generating requests for payment) and costs are accounted for as work-in-progress (WIP) transactions until such time as revenue (based on billings) and expenses (based on invoices, etc.) are recognized in a formal sense. This formal recognition of revenue and expense

[3] The process and objective of closing accounts will be discussed in detail later.

typically occurs at the point in time at which accounts are closed in order to generate the Income Statement.

When income recognition occurs in the course of closing accounts, the WIP billings become revenue and the WIP costs become expenses. In this sense, the WIP accounts can be viewed as holding accounts in which revenue and expense information is gathered so that at closing the amount of revenue and expenses can be compared to calculate income. As you will see later, this calculation is the basis for recognizing income and calculating the associated taxes which must be paid. This progression of collecting billings and cost information and, at closing, transforming it into revenue and expenses is shown schematically in Figure 4.5.

The terminology "work-in-progress" (as a prefix) may appear in a company's books as construction in progress (CIP), contracts in process, or some similar phrase. The general idea is, however, that both billings and costs are "in progress" until the moment at which they are declared revenues and expenses for purposes of income recognition. Therefore, at the time billings and costs enter the bookkeeping system, they are typically carried in work-in-progress accounts until the point at which the closing of accounts occurs. This is a subtle point that has not been emphasized previously in discussing the flow of project-related transactions. A better understanding can be gained by studying transactions 2, 3, 6, and 8 in Appendix B.

One additional point deserves comment. At the time of closing accounts (i.e., the end of an accounting period—month, quarter, etc.), some revenue earned has not been billed to the client, and some expenses incurred may not have been documented in terms of an invoice from the vendor or a billing from the appropriate subcontractor. Although these transactions have

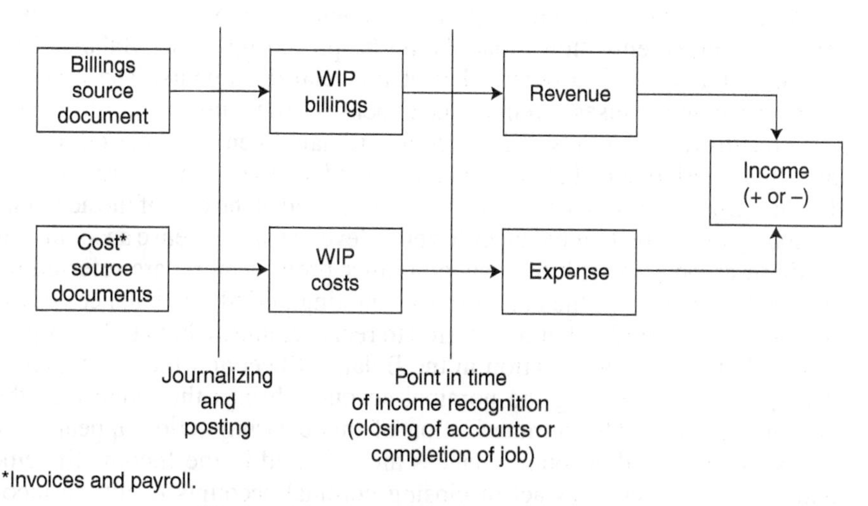

Figure 4.5 Concept of work-in-progress.

accrued, they have not been documented in terms of source documents. Since they have occurred, it is important include them in the financial statements. Payroll is a good example. Assume that closing to prepare financial reports occurs on the 15th of the month. If the office staff is paid monthly, half of their pay will have accrued as an obligation by the 15th of the month. Similarly, half of a month's rent will accrue on the company's building and must be reflected in any financial statement prepared as of the 15th of the month.

The key idea is that, even though a transaction is not documented in terms of a source document, if, in fact, a revenue or expense has accrued, this transaction should be accounted for at the time of closing accounts. Therefore, revenue earned but not billed should be recognized at closing. Expenses incurred at accrued for which invoices or billings have not been received should also be recognized.

CLOSING THE ACCOUNTING CYCLE

At certain times, normally at the end of accounting periods, it is necessary to check on the financial health of the firm and its operations. Accounting periods are determined on the basis of management needs and the requirements of outside agencies for information concerning the financial status of the firm. Accounting periods may be as short as one week or as long as one quarter. Government regulations impacting most firms require that reports be submitted for tax purposes on a quarterly basis. The so-called accounting cycle is terminated at the end of the accounting period by closing accounts and generating financial statements.

At the end of the accounting period, accounts are closed so as to recognize revenue and expense flows that should be picked up in the Balance Sheet, Income Statement, and other relevant financial documents. The process of closing accounts causes revenues and expenses to be transferred from nominal accounts to real accounts. Real accounts remain open from period to period and their period end balances are reflected in the financial reports (e.g., Balance Sheet, etc.). Nominal accounts are closed at the end of the accounting period by bringing their balances to zero. Revenue and expense accounts, such as those associated with projects and other profit centers, are referred to as nominal accounts. At the end of the accounting period, the balances of these accounts are cleared out or transferred to real accounts, which link to reported items of summary information in the Balance Sheet and Income Statement. This process of clearing the nominal accounts brings the balance in these accounts to zero. The summarization of this clearing action appears in the real accounts. Real account balances are reflected in the Income Statement and Balance Sheet. This act of closing nominal accounts results in income recognition (based on earned value).

RECOGNITION OF INCOME

The assessment of income earned on a construction project is complicated since the "sale" is, in effect, occurring across the life of the project. That is, the project itself is a large end item that will be paid for in increments throughout the period of construction. If the project extends over several accounting periods, the question is "How much income should be recognized in a given period?"

If the accrual basis of account is being used, it is possible to sum billings to the client during a given period and view this as revenue for income recognition purposes. However, on longer-term contracts use of the billings method can lead to deviation from the actual revenues earned base on work performed. That is, the amount billed may not be the amount earned.

The billings that are submitted by the construction contractor are called progress payment requests. These partial payments for a given period (usually a month) of the total amount bid are calculated by estimating the amount of work performed in a given pay period. The client then pays the corresponding portion of the total work for the period just completed. Many factors (including the accuracy of the estimate of work completed) can cause a condition referred to as overbilling or underbilling. When revenue is based on progress payments billed to a particular point in the project, the revenue (and associated income) billed may be either great than or less than the revenue earned. If the cumulative amount billed is greater than the revenue earned, an overbilling has occurred. If the cumulative amount billed is less than the revenue earned, an underbilling condition exists.

Therefore, the billings method does not reflect the true financial position of the project (and the firm) because the income earned has been misrepresented. This problem develops on contracts that run over many progress payment periods. The problem of over and underbilling is typical of the long-term (multiperiod) construction projects.

Percentage-of-Completion Method of Income Recognition

In order to better reflect the actual revenue earned on a particular project at a given time, the percentage-of completion (POC) method can be used. In the POC method, gross revenue is recognized on each contract in proportion to the progress made on that contract. The revenue (and income) generated based on the progress is recognized whether or not it is billed to the client. The key criterion for recognizing revenue is that it has been earned. This gives a better indication of the firm's position in the financial statement and is, therefore, the recommended method for income recognition and statement preparation.

Several methods are used, in practice, to establish the percentage completed in terms of value earned at a particular point in the project's life cycle. These are:

Cost-to-cost method
Effort-expended method
Units-of-work-performed methods

The cost-to-cost or cost-completion method calculates the percentage complete on the basis of the following ratio:

Cost incurred to date on project/Estimated total cost of project

Another way of calculating this ratio is:

Cost incurred to date/Cost incurred to date + estimated cost to complete

In the effort-expended method, the percentage is based on the ratio of units of resource effort (e.g., man-hours or similar measure) to the total effort estimated to complete. This ratio can be expressed in terms of labor hours or labor cost. Based on man-hours, the calculation could be made as follows:

Man-hours to date expended on project/Total estimated man-hours on project

or, in terms of labor cost, as follows:

Labor cost incurred/Labor cost incurred + estimated labor cost to complete

This method of calculation tends to offset distortions in the cost-to-cost method, which occur when materials are delivered to the site (and therefore cost is incurred) but are not installed (i.e., the revenue is not yet earned).

The problems inherent in both the cost-to-cost and effort-expended methods relate to the fact that inefficient work during the early phases of the project (and the consequent higher expenses and/or resource expenditure) generates apparently higher revenues by increasing the numerator of the ratios above. This tends to inflate the earned revenue amount.

To correct for inflated earned revenue, the percentage of completion can be calculated by using physical units of production as the basis for calculation. Using the units-of-work-performed method, we find that the POC ratio becomes:

Units of work performed to date/Estimate total units of work in project

On a heavy construction earthwork job, this ratio might be:

> Cubic yards of material excavated to date/Estimated total cubic yards in project

This physical completion method of determining the POC is appropriate on projects where a single unit of work (e.g., cubic yard, lineal foot) can be identified as representative of the work to be accomplished on the total project. It does, however, assume that revenue is generated and earned in proportion to the extent of physical completion of the selected work item. This method neglects earned revenue associated with mobilization and demobilization of the job. For instance, if extensive haul and access roads must be constructed before the first lineal foot of pipeline can be constructed, using the footage of pipeline as the work unit would result in no revenues earned during the extended period of access road construction.

Clearly, all of the methods used to determine the percentage of completion are based on estimates. The American Institute of Certified Public Accountants (AICPA) *Construction Contractors: AICPA Audit and Accounting Guide* indicates that the use of the percentage-of-completion method depends on the ability to make reasonably dependable estimates.

In general, the POC method is the preferred method of income recognition and should be used for all projects unless the unique and risk-associated aspects of a particular project are so significant as to preclude development of reliable estimates.

Completed-Contract Method of Income Recognition

The basic idea underlying the completed-contract method is that the recognition of income is deferred until the end of the project. In general, for all projects of any size and continuing over more than one quarter (e.g., three month period), use of this method is not allowed and, if used, is carefully scrutinized by the federal and state tax agencies.

The advantage from the contractor's point of view is that, at the end of the project, actual as-built information regarding the costs incurred and the revenues generated is available. Therefore, a "settling up" can occur, which gives the most accurate evaluation of income generated by the project. The determination of project income is no longer dependent upon estimates. Income recognition and tax liability does not occur until the contract is completed.

In general, this deferment of tax payment is not viewed favorably by federal and state tax agencies. In fact, this method has been misused to defer tax payments over multiyear periods on billion dollar contracts. Litigation has established that this use of the completed contract method is not appropriate and tax law has been modified to disallow the use of this method in such cases. As a practical matter, the method can only be used by small contractors (e.g., small electrical and plumbing firms) who operate primarily on a work order basis.

If the method is used for a project in progress during the generation of financial reports, revenues and expenses are reported and maintained in "holding accounts" (e.g., Construction in progress – Job 103, etc.) until the project is complete. In such cases, until the job is completed, progress payments received are considered as "advances" against the ultimate successful completion of the work. These accumulated advances are shown as liabilities and are ultimately cleared as the end of the standard accounting period (e.g., quarter).

The project expenses are considered to be contractor-owned assets until the project is accepted as complete by the owner. The accumulation of expenses is captured in appropriate accounts (e.g., Construction in progress – expenses – job 103). At completion, progress payments accumulated (i.e., all progress payments received) are compared with the total cost accumulation to determine the income recognized.

TRANSACTIONS DURING A PERIOD

In order to understand the process of closing and statement preparation, it is necessary to consider the activity during an accounting period. During a given period revenues and expenses occur randomly or on an "as scheduled" basis (e.g., submission of a progress payment request at the end of the month). These revenues and expenses are journalized and ultimately transferred to the general ledger accounts. The Chart of Accounts (see Appendix A) being used by a firm is a listing of all of the ledger accounts. In the following discussion, the general ledger accounts and the Chart of Accounts can be viewed as synonymous. As stated previously, the Chart of Accounts is organized into the following:

Revenue accounts
Expense accounts
Asset accounts
Liability accounts
Net worth (or equity) accounts

In other words, the Chart of Accounts is organized so as to provide feeder information to the financial statements. By examining the accounts in Appendix A, it is possible to recognize this breakdown of accounts into the five categories listed above. The movement of transactions to the general ledger accounts is shown in Figure 4.6.

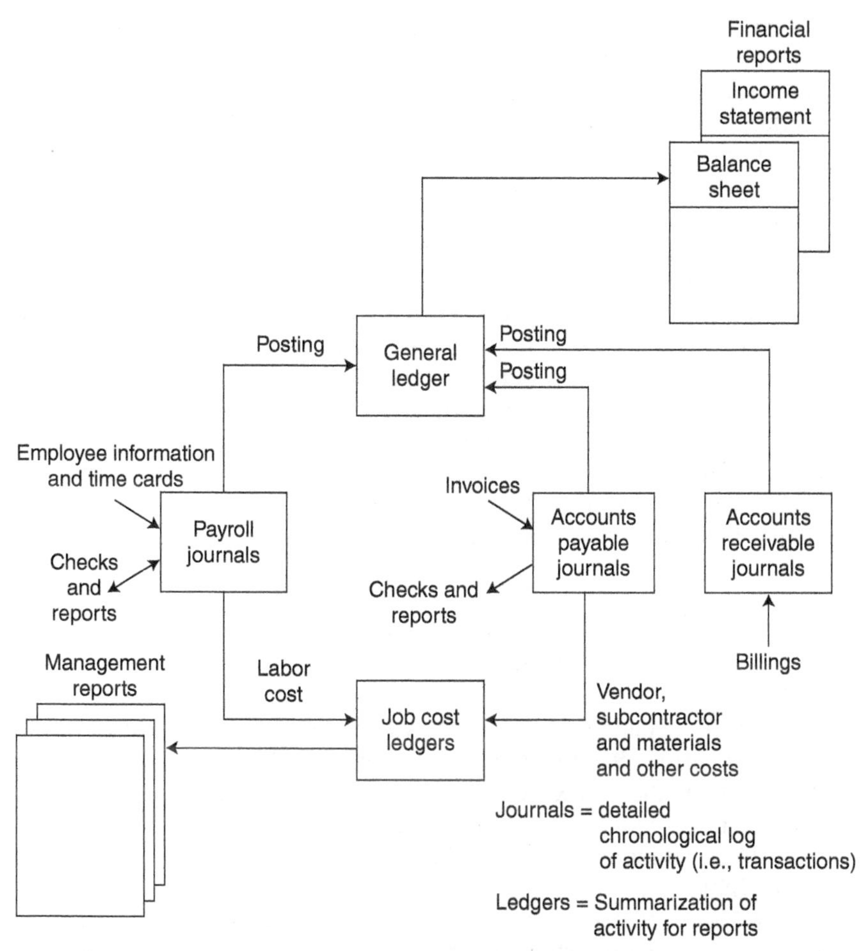

Figure 4.6 General accounting flows.

Since in construction the focus of production is job- or project-oriented, the generation of revenue and expense transactions is generally associated with a particular job or project. As can be seen in Figure 4.6, most of the expenses associated with a project are generated in the form of payroll transactions or from the Accounts Payable area. These expense transactions are journalized in the payroll and accounts payable journals and ultimately posted to the appropriate general ledger accounts in the Chart of Accounts. Similarly, billings to the client are journalized in the accounts receivable journal and posted to the appropriate general ledger accounts.

Posting of these transactions to the general ledger occurs as a batch processing operation, which lumps different expenses together without trying to sort them for purposes of identification. That is, the weekly expenses of Job 101 will be consolidated and batch posted to the single GL Work-in-progress account for a given project (i.e., Work-in-progress – expense – Job 101). This procedure means that no attempt is made to relate the particular expense to a physical or cost component.

Alternatively, a set of job cost accounts can be maintained to allocate costs to various cost centers (e.g., concrete—slab on grade, hand excavation, etc.). These job cost accounts can be thought of as Work-in-progress subsidiary accounts established to provide detailed cost reporting for each project. The specific purpose of these project accounts is to capture information linking various costs to the detailed cost centers of the project with which they are associated. Project cost accounts will be discussed in detail in Chapter 5. A typical listing of project cost accounts is shown in Figure 5.4.

The simultaneous flow of information to the Chart of Accounts (GL accounts) and to the Job Cost accounts is shown schematically in Figure 4.6. The cost accounts are established to obtain detailed information relating expenses to particular cost centers within the job.

POSTING TO THE GENERAL LEDGER DURING THE ACCOUNTING PERIOD

In order to understand the operation of the general ledger during a given accounting period, consider the situation presented in Figure 4.7.

It is assumed that Apex Construction has three projects in progress during the period under consideration. These three projects are numbered from 101 to 103. The terminology used for Revenue- and Expense-related accounts is given in Table 4.7.

For purposes of demonstration, projects will use both the Percentage-of-Completion and Completed-Contract methods of income recognition. The general ledger account names used to record revenue and expense accumulation when using each method are given in Table 4.7. Projects 101 and 102 will use the POC method. Project 103 will use the completed contract or CC method. This is summarized in Table 4.8.

As can be seen from Figure 4.7, billings associated with Projects 101 and 102 are posted to revenue accounts within the Chart of Accounts (i.e., billings -101 and billings-102). Similarly, expenses associated with these jobs are posted to the appropriate expense accounts (i.e., WIP – expense-101 and WIP-expense-102). The accumulation of these billings and expenses will be used at the end of the period to calculate the incremental amount of income to be recognized.

The billings on the completed-contract job are not recognized at this time as revenues. Recognition is deferred until the project is completed. Therefore, the billings for the CC project are posted to the appropriate advanced billings

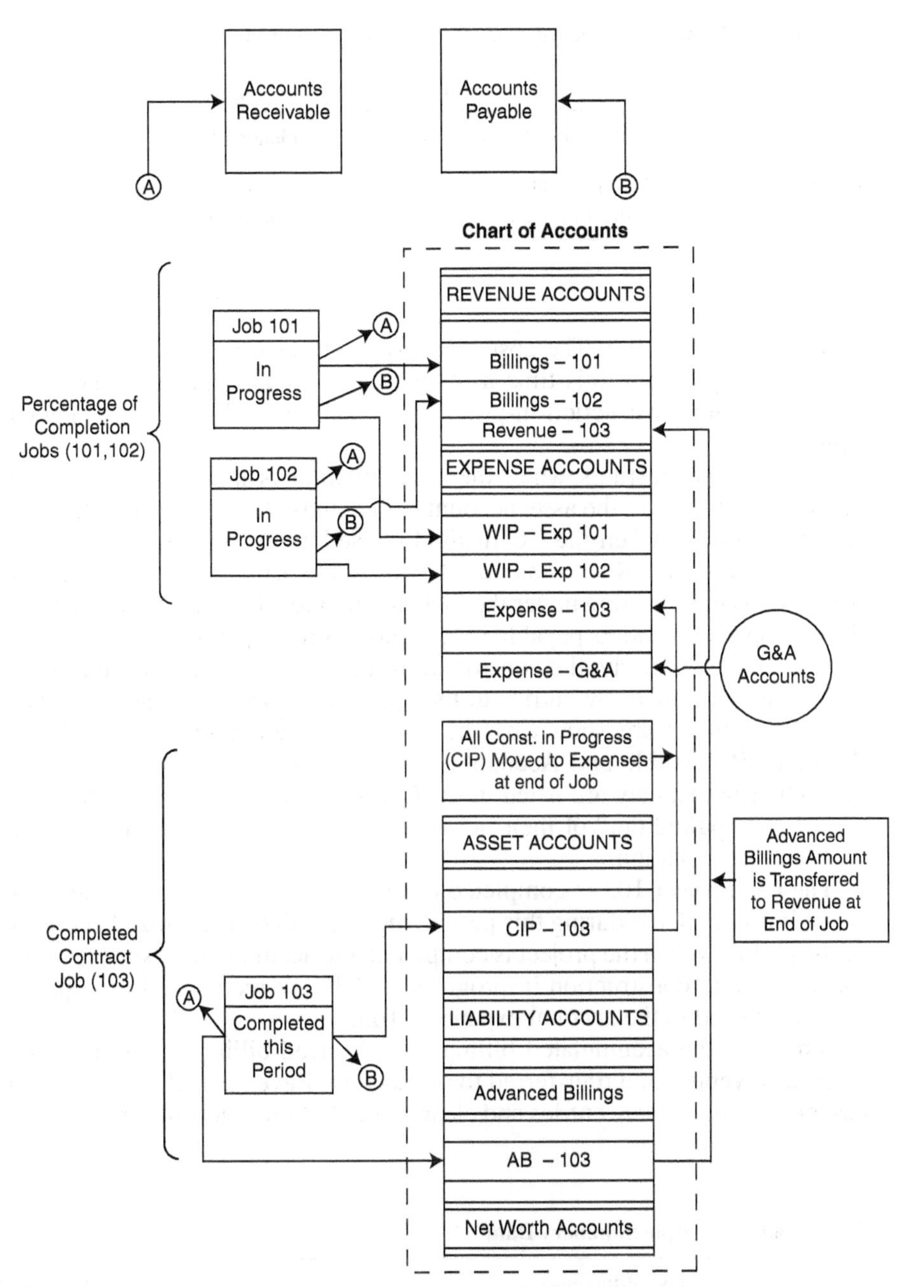

Figure 4.7 During-the-period transactions.

TABLE 4.7 General Ledger Accounts Posted When Using POC and CC

	Method of income recognition	
Type of transaction	General Ledger under POC	General Ledger under CC
Revenue	Billings - Job X	Advanced Billings - Job Y
Cost	Work-in-progress expense - Job X	Construction in Progress - Job Y

account (i.e., AB 103). The costs and expenditures associated with this job are not recognized at this time as expenses. They are, therefore, posted to the appropriate construction-in-progress account (e.g., CIP-103), as shown in Figure 4.7.

The revenue postings when the POC method is used are balanced by appropriate entries to the asset account—Accounts Receivable. The Advance Billings postings when the CC method is used are also offset by balancing entries to Accounts Receivable, as shown by the arrow labeled A in Figure 4.7. The expense posting under the POC method result in balancing entries to the appropriate accounts payable ledge accounts (e.g., payroll, Jones Lumber, and the like). Similarly, the construction-in-progress amounts posted on the CC project are offset by entries to the appropriate accounts payable. These balancing entries are shown by the arrows labeled B in Figure 4.7. The Accounts Receivable and Accounts Payable themselves are general ledger accounts (that is, they are listed in the Chart of Accounts) and are shown in Figure 4.7 "pulled out" of their positions as asset and liability accounts for clarity of presentation.

The CC Project 103 is completed during the period under consideration. Since it is completed during this period, income will be recognized at the end of the period. When the project is completed, the accumulated expenses in the asset account "construction-in-progress – 103" are recognized as expenses and transferred to the account expense – 103.

Similarly, the accumulated billings in advanced billings -103 are recognized as revenues and transferred to the account Revenue 103. These actions essentially consolidate entries and clear these "holding" accounts preparatory

TABLE 4.8 Example Summary Data

Project	Method of Income Recognition	Amount Billed	Bid Price	Estimated Expense
101	POC	$20,000	$100,000	$75,000
102	POC	$110,000	$200,000	$160,000
103	CC	Completed	$250,000	$225,000

to income recognition at the end of the period. These flows are shown by arrows connecting the appropriate accounts in Figure 4.7.

General and Administrative costs are assumed to recognized on a period-by-period basis. Postings of these overhead-type costs are shown as entries to the expense account G&A expenses.

CLOSING ACTIONS AT THE END OF THE PERIOD

At the end of the period, actions must be taken to "settle up" and recognize the amount of income to be reported for the period. These are the closing actions that clear the revenue and expense accounts and lead to calculations of income. In the case of Jobs 101 and 102, income is being recognized period by period (i.e., on percentage-of-completion basis.) When recognizing income with the Completed-Contract method, income is recognized only in the period in which the job is completed. In this case, job 103 in completed in the period under consideration. Therefore, the profit achieved on the entire project will be recognized in this period. Actions occurring at closing for these three projects are shown schematically in Figure 4.8.

Income calculation for Projects 101 and 102 requires the calculation of the percentage complete. The cost-to-cost basis is used in this example. The period expenses for Project 101 amounts to $25,000 versus the total estimated expense of $75,000. This means that the percentage complete is 33.3 percent and the earned revenue is $33,333. The amount of income recognized is (33,333 minus 25,000) $8,333. This is recognized in the account Income Summary and ultimately in the Income Statement. Comparing the amount billed, $20,000 with the earned revenue, $33,333, indicates that an underbilling has occurred. This underbilling is posted to the asset account *Costs and estimated earnings in excess of billings*, and amounts to $13,333. Similar actions relating to Project 102 result in the recognition of $20,000 in income. The total estimated cost of $160,000 versus $80,000 expenses incurred indicates that the percentage complete is 50%. The earned revenue is $100,000 and a profit of $20,000 is recognized in the Income Summary. On this project, an overbilling has occurred, since the amount billed ($110,000) is greater than the amount earned ($100,000). This overbilling of $10,000 is reflected in the liability account *Billings in excess of costs and estimated earnings.*

Project 103, which was completed during the period and on which the deferred income will now be recognized is handled in a manner similar to the two percentage-of-completion contracts. The exception is that the expenses are compared to the total project bid price (i.e., the project is 100% complete). This is, of course, the advantage of the of Completed-Contract method. We know the expenses so the calculation of income recognized is not based on an estimate, but rather on final as-built amounts. Restrictions have been placed on the use of the CC method, however, and it can be used only in very special situations. As a practical matter, for projects that are completed over several

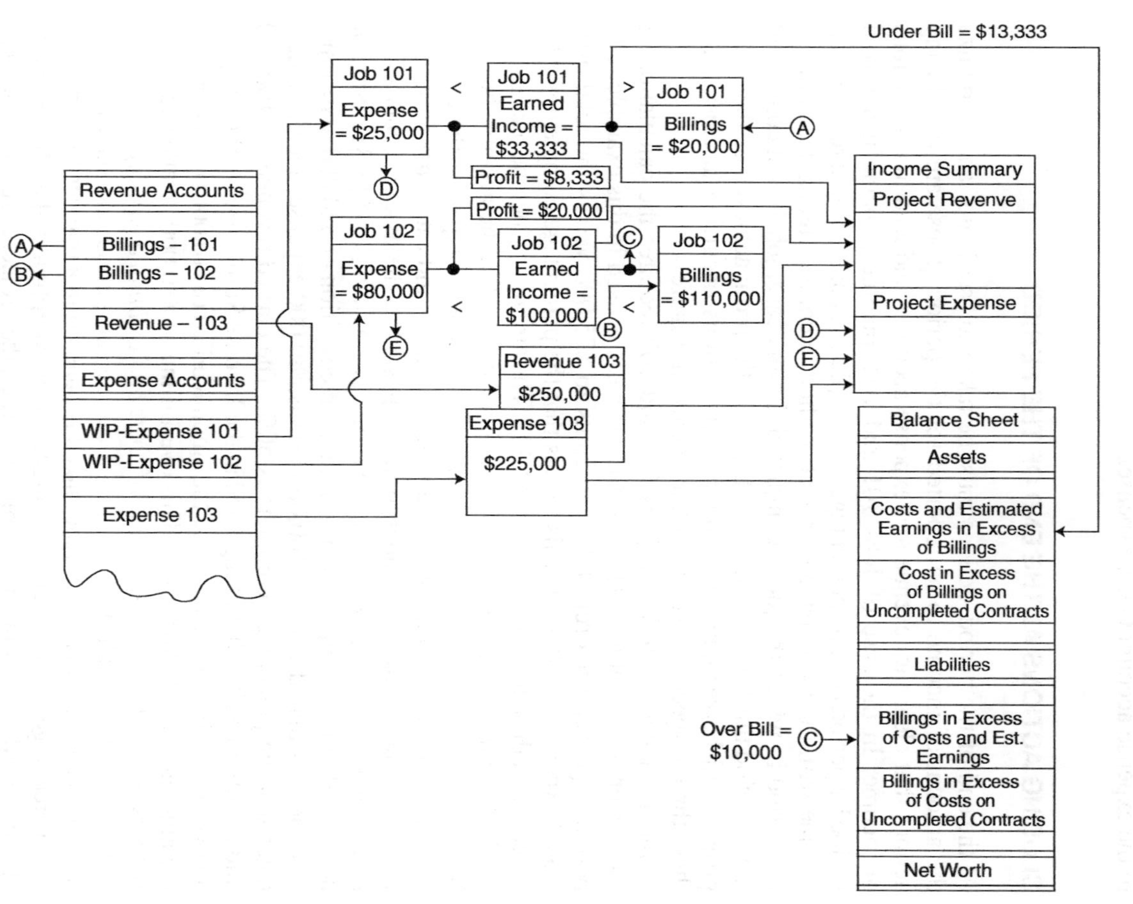

Figure 4.8 Closing actions at end of period.

accounting periods the POC method is the method of income recognition required by the Internal Revenue Service (IRS).

In this case, the bid price (modified by any change order amounts) is compared to the actual expenses of $225,000. The earnings of $25,000 are transferred at this time to the income summary. Earnings recognized on Projects 101,102, and 103 for this period amount to $53,333.

REVIEW QUESTIONS AND EXERCISES

1. What is the difference between journalizing and posting?
2. Is it possible to post into an asset account and a revenue account? Give two examples with debiting and crediting an asset account.
3. Today is 3 February 20XX. You are using the accrual method of accounting. You have received an invoice from Halco Industries in the amount of $565 for work performed on Project 101. You also mailed an invoice to Ajax International for payment of work performed by your company in the amount of $6,238 on Project 101. On 10 February you send a check to Halco to cover their billing and receive a check from Ajax on 12 February in the amount of your billing to them.
   Make the proper journal entries on a sheet captioned as in Table P4.1.

TABLE P4.1 Company Journal

Transaction#	Date	Description	Debit	Credit

4. Journalize the transactions shown in Table P4.2 for Tip Top Associates, Inc., incurred during the month of July 20XX (assume accrual basis).

TABLE P4.2 Transactions for Tip Top Associates

6-Jul	Initializing capital deposited in bank	$12,000
10-Jul	Office supplies purchased	250
12-Jul	Purchase computer (paying $1000 with balance on account)	3,500
15-Jul	Office rent	425
20-Jul	Billing to Jones County Board of Education	1,255
25-Jul	Salaries—employee payroll	655
28-Jul	Paid on account for computer	800
30-Jul	Cost of expended supplies	145

5. Transactions during the period 1 December 2015 to 30 November 2016 are given below for ABC Construction.

(1) Total income (fees) from construction contracts	$400,000
(2) Material inventory purchased on account	80,000
(3) Total office rent paid	20,000
(4) Salaries paid to engineers	90,000
(5) Accounts receivable collected	60,000
(6) Account payable paid	30,000
(7) Subcontractor accounts paid	70,000
(8) Bidding expenses	20,000
(9) Building depreciation	40,000
(10) Construction equipment depreciation	30,000

Using the transactions presented in Appendix B as a guide:

- Journalize all of the above transactions.
- Post journal entries to appropriate accounts.
- Develop the Income Statement for period 1 December 2015 to 30 November 2016.
- Develop the Balance Sheet for ABC Construction as of 30 November 2016.

6. What is a "real" account? How does it differ from a nominal account?
7. What is the reason for closing accounts?

TABLE P4.3 Balance Sheet, CPM Construction Company (31 December, 2013)

Assets		Liabilities	
Cash	75,000	Accounts payable	85,000
Accounts receivable	110,000	Notes payable	50,000
Buildings	300,000	Long-term loans	60,000
Less accumulated depreciation on the buildings	(150,000)	Total liabilities	195,000
Construction equipment	240,000	Net worth	
Less accumulated depreciation on equipment	(80,000)	Capital stocks	250,000
Other assets	20,000	Retained earnings	70,000
Total assets	515,000	Total net worth	320,000
		Total net worth and liabilities	515,000

TABLE P4.4 Transaction for CPM Construction Company

Transaction Number	Date	Transaction
1	1/2/X4	CPM Co. bought construction equipment for $130,000 for which the company paid $15,000 cash and remaining on account.
2	2/4/X4	CPM Co. was billed $20,000 by Smith Material Supplier for cost of material.
3	3/4/X4	CPM paid $20,000 to Smith Material Supplier related to transaction #2.
4	3/8/X4	CPM billed client for $320,000 (bill #1 on Job 101).
5	4/7/X4	CPM was billed $60,000 by National Rental Co. for renting construction equipment.
6	5/8/X4	CPM received $290,000 from client for bill #1 on Job 101.
7	6/7/X4	CPM paid $60,000 to the National Rental Co. for the bill received on 4/7/X4.
8	7/3/X4	CPM paid $70,000 cost of labor.
9	8/16/X4	CPM was billed $45,000 by subcontractor.
10	9/16/X4	CPM paid $45,000 to the subcontractor for the bill received on 8/16/X4.
11	10/1/X4	CPM billed client for $280,000 (bill #2 on Job 101).
12	10/20/X4	Accounts receivable of $20,000 are collected.
13	11/15/X4	CPM received $265,000 from client for bill #2.
14	12/15/X4	Accounts payable of $40,000 are paid.
15	12/25/X4	CPM paid $145,000 in payroll expense.
16	12/30/X4	Dividends paid in the amount of $20,000 to stockholders.
17	12/30/X4	Building depreciation of $30,000 recognized for the year.
18	12/30/X4	CPM depreciates its construction equipment for the total of $65,000 each year (this includes also the depreciation of the equipment bought on 1/2/X4.)

8. Can you think of construction contractors who can use the Completed-Contract method on a regular basis due to the type of contracts that they normally undertake?

9. How are the Income Statement and Balance Sheet linked when closing accounts?

10. The Balance Sheet of CPM Construction Company as of 31 December 2013 is given in Table P4.3. Assume that this company is using the POC method of income recognition. Further, assume that 65% of the projects with total bid price of $850,000 have been completed in 2014.

(a) Journalize the following transactions. Identify each transaction as asset, liability, revenue, expense, and so forth (Table P4.4).

(b) Establish relevant accounts for posting. Divide them into categories as Assets, Liabilities, and Net Worth.

(c) Close accounts as of 31 December 2014.

(d) Develop the Income Statement for the year 2014.

(e) Develop the Balance Sheet as of 31 December 20X4.

CHAPTER 5

PROJECT-LEVEL COST CONTROL

OBJECTIVES OF PROJECT-LEVEL COST CONTROL IN CONSTRUCTION

As had been emphasized in the first four chapters, construction is a project-based activity. In dealing with financial accounting, the focus has been on the company and the generation of revenue and income at the level of the firm. The company income is the aggregate of income generated from the projects in progress (and completed) as well as other revenue-generating activities (e.g., equipment rental, etc.). We now must address the project level to better understand how revenue and the associated income flows from these profit centers are tied into the company-level accounting system.

Within the project format, there are a number of issues that are critical to making a given job profitable:

- Tracking costs to date is one of the primary jobs of a successful project manager.
- On more sophisticated projects, finding the earned value to date is essential to avoiding cost overruns.
- Projection of profit to date, based on revenues and costs to date allows the PM to assess the profitability of the project.
- Actual costs must be compared to budgeted costs in order to determine whether the project is on the cost profile originally predicted.

- A viable cost control system must be developed to ensure the early detection and assessment of financial problems on the project.
- An effective cost control system must be able to detect and reflect both profitable and negative cost trends.
- A good cost control system can be the basis for accurate calculation of unit costs, which can be used for pricing future work.

UNIQUE ASPECTS OF CONSTRUCTION COST CONTROL

Cost control in the manufacturing and service sectors is dealt with prior to the sale of the product or service. The manufacturer of a refrigerator or TV set knows the direct cost of manufacturing a given unit prior to sale. In construction, the project is "sold" first, before the costs of construction are incurred. Since the cost is fixed on a manufactured item or service prior to sale, profit can be influenced by increasing or decreasing the sale price based on the strength of the market.

In the manufacturing and service sectors, pricing may, and typically does, vary as units are sold over time. Both the retail and wholesale prices of a lawn mower, for instance, vary over time based on market conditions. If the market is strong and demand is high, the sales price of the product or service can be increased. In discussing the sale of bicycles in the Chapter 1, it was noted that if the initial sale price leads to low sales volume (customers are not attracted to purchase), the price can be reduced. Simply stated, if the market is weak, the price of the unit of sale may be reduced. Conversely, a strong market usually supports an increase in price.

In construction, the price and revenue from the project are effectively fixed at the time the bid is quoted and accepted. Therefore, decisions, regarding pricing and profit are agreed to by the client and the contractor prior to beginning construction.[1] Since the price and levels of expected revenues are, as a practical matter, fixed when the bid is accepted, construction project managers are primarily involved with management and control of costs. The quoted price cannot be changed or renegotiated unless major deviation from the conditions, as shown in the contract documents, can be established.

Once the bid price is accepted by the client, project managers are focused on cost control. The lower the cost, the greater the profit generated on a given project. Pricing is subject to very little variation. Therefore:

$$\text{Project Profit} = \text{Bid Price as adjusted (Fixed)} - \text{Cost of Work (Variable)}$$

[1] Certain contracts (Unit price, Cost plus, etc.) do provide for small adjustments based on field conditions and quantities installed.

Because "cost of work" is the variable in this equation, management is heavily involved in data collection, reporting, and analysis with the objective of controlling and reducing (where possible) the project cost. Most construction companies have developed cost control systems to capture variations as they occur during construction.

TYPES OF COSTS

Contractor-incurred costs associated with the construction of a given facility relate to:

1. Direct cost consumed in the realization of a physical subelement of the project (e.g., labor and material costs involved in pouring a footer).
2. Production support costs incurred by project-level support resources or required by the contract (e.g., the superintendent's salary, site office costs, builder's risk insurance, etc.).
3. Costs associated with the operation and management of the company as a viable entity (i.e., *General and Administrative "G & A" costs*) such as the costs associated with preparation of payroll in the home office, preparation of the estimate, marketing, and salaries of company officers.

The production support costs are normally referred as *project indirect costs* or *field overhead*. G & A costs are also referred to as *home office* overhead. All of these charges must be recovered before income to the firm is generated.

The term *overhead* is used loosely to refer to all non-direct costs (i.e., 2 and 3 above) on the project. Overhead costs in construction are typically less than 5% of total project costs. Conversely, the direct cost percent of total cost in construction is relatively high. By comparison, overheads in other industries tend be much higher than those in construction. For this reason, careful management of direct costs is critical to success in completing a construction project. (Why would overheads in the fast-food industry be higher than those in construction?)

THE CONSTRUCTION ESTIMATE

Project cost control starts with the estimate prepared to support submittal of a bid for any given project. In order to bid on a job, an estimate of all direct, indirect, and home office overhead costs for the project is prepared. The line items and quantities required to complete the project are estimated ("taken off") from its plans and specifications. The total cost associated with each direct cost line item is the product of its quantity times its unit cost. Ideally, unit costs are based on data from past projects of a similar nature performed by the company. In most cases, however, a project includes new

Jefferson Contractors, Inc.
ESTIMATE SUMMARY
Estimate No. 6692 By: DWH Date: 1 August X4
Owner: NASA Project: VA Building

Code	Description	MH	Labor	Material	Sub	Owner	Total
01	Site improvements						
02	Demolition						
03	Earthwork						
04	Concrete						
05	Structural Steel	1,653	18,768	15,133			33,901
06	Piling						
07	Brick & masonry						
08	Buildings						
09	Major equipment	2,248	26,059	1,794			27,853
10	Piping	2,953	34,518	57,417	1,500	34,541	127,976
11	Instrumentation				33,000		33,000
12	Electrical				126,542		126,542
13	Painting				14,034		14,034
14	Insulation				4,230		4,230
15	Fireproofing			530	1,110		1,640
16	Chemical cleaning						
17	Testing						
18	Const. equipment					35,666	35,666
19	Misc. directs	1,008	10,608	2,050		2,000	14,658
20	Field extra work						
Sub Total Direct Cost		7,862	89,953	76,924	180,416	72,207	419,500
21	Con. tools/sup.			7,361			7,361
22	Field payroll/ burden					16,580	16,580
23	Start-up asst.						
24	Ins. & taxes					5,268	5,268
25	Field sprvsn.	480	7,200			2,038	9,238
26	Home off. exp.					2,454	2,454
27	Field emp. ben.					10,395	10,395
Sub Total Indirect Cost		480	7,200	7,361		36,735	51,296
Adjustment Sheets							
Total Field Cost		8,342	97,153	84,285	180,416	108,942	470,796
28	Escalation						
29	Overhead & profit		8,342	5,057	9,021	10,190	32,610
30	Contingency						18,076
31	Total Project Cost						521,482

Figure 5.1 Typical estimate summary.

items whose unit cost must be obtained from published references or from a detailed analysis of each resource required by the item. A typical summary estimate is shown in Figure 5.1.

The line items in the estimate can be referred to as estimating accounts. The costs developed during the estimating period form the basis of the project budget, which consists of cost accounts establishing target values and cost goals for the project.

When the estimate of costs is distributed across the life of the project by plotting cost amounts during each period of the project against time, the cost estimate is transformed into a time-scaled budget. The budget also reflects corrections and refinements made to the estimate following acceptance of the bid and prior to commencement of the work. There is normally a one-to-one relationship between the estimating accounts in the estimate and the cost accounts in the project budget. The costs developed during the estimating phase are transferred to the budget and represent target values against which field costs and performance can be measured.

COST CONTROL SYSTEM

In construction, each project is unique and requires a customized cost control system. The project cost control system is a management information system that gathers cost data designed to aid the manager in controlling the project. As such, the project cost control system is to the manager what the cockpit dials and instruments are to the pilot of an airplane. It is a monitoring system designed to provide timely feedback to management regarding actual performance on a given project vis-à-vis project performance goals, such as those contained in the project budget. A secondary function of the job cost system is the collection of data to be used as a basis for estimating future jobs. The system collects historical data on actual costs of labor, equipment, and materials, which provide reference information in pricing out future projects. Therefore, the cost control system serves two major functions: (1) cost monitoring and control versus the project budget, and (2) collection of data for estimating future projects.

As construction proceeds, field data are collected in the appropriate job cost accounts. At this point, the project manager is in a position to compare actual field data with the expectations reflected in the budget. Variances are monitored between expected costs and actual field production rates and charges. By examining the variances, the manager can detect which accounts are seriously deviating from planned progress and can take corrective action. This approach is often referred to as management by exception. A conceptual diagram of the relationship of the financial accounting and cost accounting systems is shown in Figure 5.2. Cost management focuses on monitoring and controlling the cash flowing into and through the project cost accounts.

BUILDING A COST CONTROL SYSTEM

The form and design a of a cost control system depends upon the needs of the contractor and the effort that must be expended to implement and maintain the system. As with all management systems, the effort and time expended in operating the system must be weighed against the savings achieved by using the system. To be effective, the cost control system must produce the right amount of information at the right time (e.g., think pilot and aircraft instruments). The extent to which project management supports a project

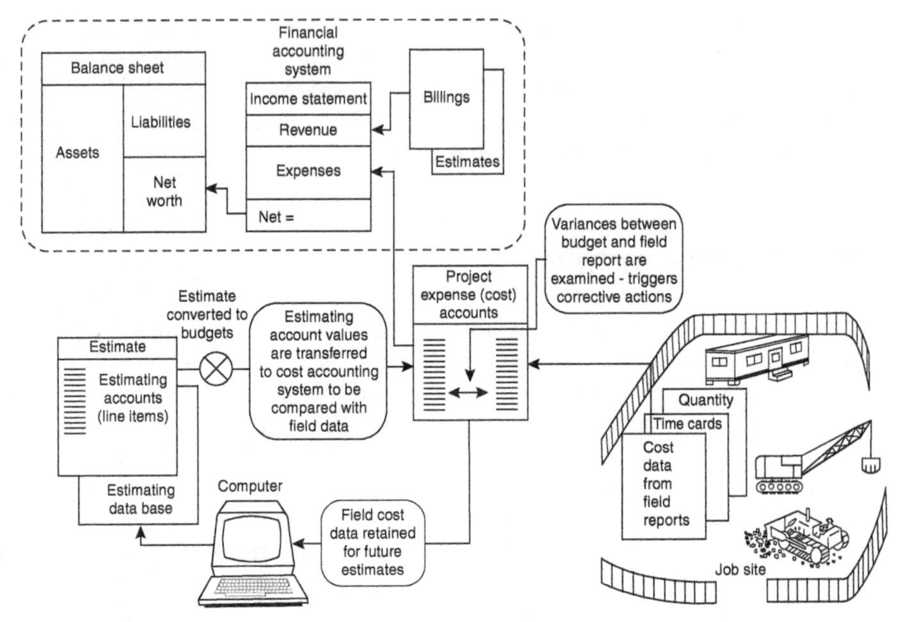

Figure 5.2 Cost control system layout.

control system depends upon management attitude, the financial risks inherent in the form of contract and the size of the profit margin.

The design, implementation, and maintenance of a project cost control system can be considered a multistep process. The five steps shown schematically in Figure 5.3 form the basis for establishing and maintaining a cost control system. The following questions regarding each step in the implementation of the cost control system must be addressed:

1. ***Chart of cost accounts***. What will be the basis adopted for developing estimated project expenditures, and how will this basis be related to the firm's general accounts and accounting functions? What will be the level of detail adopted in defining the project cost accounts, and how will they interface with the general ledger accounts of the company's financial accounting system?
2. ***Project cost plan/budget***. How will the cost accounts be utilized to allow comparisons between the project estimate and budget with actual costs as recorded in the field? How will the project budget estimate be related to the construction plan and schedule in the formation of a project cost control framework?
3. ***Cost data collection***. How will cost data be collected and integrated into the cost-reporting system?

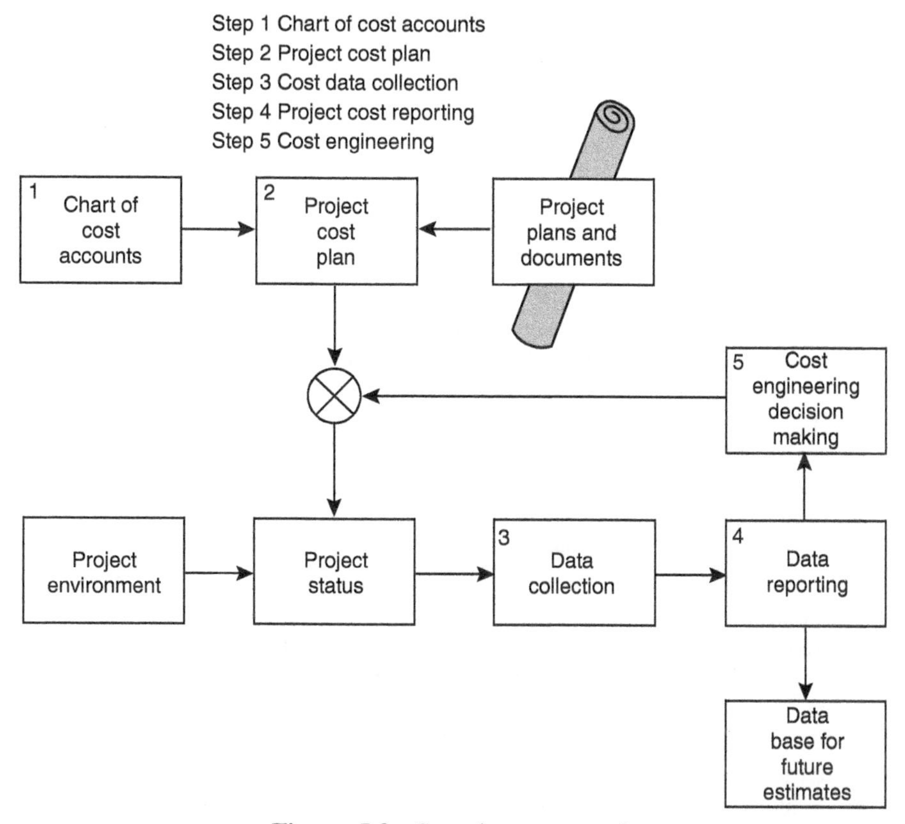

Figure 5.3 Steps in cost control.

4. ***Project cost reporting***. What project cost reports are relevant and required by project management in its cost management of the project?
5. ***Cost engineering***. What cost engineering procedures should project management implement in its efforts to minimize costs?

These are basic questions that management must address in setting up the cost control system.

COST ACCOUNTS

Cost accounts define the level of detail at which job cost information will be maintained. Cost accounts can be very rudimentary or very detailed, depending on the complexity of the project and the needs of management. Cousin

Elmer and Uncle Fudd of Fudd Associates may decide to define just four cost accounts for each project as follows:

1. Labor Cost
2. Equipment Cost
3. Material Cost
4. Other Costs

This amounts to keeping four physical folders into which documents for each of the four types of costs are aggregated. This level of detail is, of course, very coarse and is not particularly helpful in locating potential cost overruns or for estimating future jobs.

The more complex the project, the finer the level of detail must be. The numerical or alphanumeric designations of individual line items defined for control purposes are called cost codes. Therefore, cost accounts are defined in terms of a cost code or numbering system.

The cost accounts can be thought of as a subsection of the overall chart of accounts (which includes revenue and other general ledger accounts). Project cost accounts link to the Project expense accounts in the general ledger (G/L). Depending upon the desires of management and the purpose of the project cost control system, these accounts can be fully integrated into the general ledger.

Many contractors find it desirable to maintain the cost accounts separately from the G/L accounts. The aggregate amounts of the individual cost accounts are posted to summary (work in progress) expense accounts for each project. This approach is shown conceptually in Transaction 3 of Appendix B.

COST ACCOUNT STRUCTURE

The first step in establishing a project cost control system is the definition of project-level cost centers. The primary function of the cost account breakdown is to divide the total project into significant units of work that can be measured in the field (see Figure 5.4).

In most construction firms, cost systems have a structured sequence corresponding to the order of appearance of the various trades or types of construction processes typical of the company's construction activity. Heavy construction contractors are interested in earthwork-related accounts such as grading, ditching, clearing and grubbing, and machine excavation. References and information published by the American Road and Transportation Builders Association (ARTBA) emphasize these type of accounts. The *MasterFormat* published by the Construction Specifications Institute focuses on building or vertical construction accounts. A breakdown of the major classifications

MASTER LIST OF PROJECT COST ACCOUNTS
Subaccounts of General Ledger Account 80.000
PROJECT EXPENSE

		Project Work Accounts 100–699			Project Overhead Accounts 700–999
100		Clearing and grubbing	700		Project administration
101		Demolition		.01	Project manager
102		Underpinning		.02	Office engineer
103		Earth excavation	701		Construction supervision
104		Rock excavation		.01	Superintendent
105		Backfill		.02	Carpenter foreman
115		Wood structural piles		.03	Concrete foreman
116		Steel structural piles	702		Project office
117		Concrete structural piles		.01	Move in and move out
121		Steel sheet piling		.02	Furniture
240		Concrete, poured		.03	Supplies
	.01	Footings	703		Timekeeping and
	.05	Grade beams		.01	security
	.07	Slab on grade		.02	Timekeeper
	.08	Beams		.03	Watchmen
	.10	Slab on forms	705		Guards
	.11	Columns		.01	Utilities and services
	.12	Walls		.02	Water
	.16	Stairs		.03	Gas
	.20	Expansion joint		.04	Electricity
	.40	Screeds	710		Telephone
	.50	Float finish	711		Storage facilities
	.51	Trowel finish	712		Temporary fences
	.60	Rubbing	715		Temporary bulkheads
	.90	Curing	717		Storage area renta!
245		Precast concrete	720		Job sign
260		Concrete forms	721		Drinking water
	.01	Footings	722		Sanitary facilities
	.05	Grade beams	725		First-aid facilities
	.07	Slab on grade	726		Temporary lighting
	.08	Beams	730		Temporary stairs
	.10	Slab	740		Load tests
	.11	Columns	750		Small tools
	.12	Walls	755		Permits and fees
270		Reinforcing steel	756		Concrete tests
	.01	Footings	760		Compaction tests
	.12	Walls	761		Photographs
280		Structural steel	765		Surveys
350		Masonry	770		Cutting and patching
	.01	8-in. block	780		Winter operation
	.02	12-in. block	785		Drayage
	.06	Common brick	790		Parking
	.20	Face brick			Protection of adjoining property
	.60	Glazed tile	795		Drawings
400		Carpentry	796		Engineering
440		Millwork	800		Worker transportation
500		Miscellaneous metals	805		Worker housing
	.01	Metal door frames	810		Worker feeding
	.20	Window sash	880		Genera! clean-up
	.50	Toilet partitions	950		Equipment
560		Finish hardware		.01	Move in
620		Paving		.02	Set up
680		Allowances		.03	Dismantling
685		Fencing		.04	Move out

Figure 5.4 List of typical project expense (cost) accounts.

within the MasterFormat account system is shown in Table 5.1. A portion of the second level of detail for the classifications 0 to 3 is shown in Figure 5.5.

PROJECT COST CODE STRUCTURE

Once project cost accounts are established, each account is then assigned an identifying numeric or alphanumeric code known as a cost code. Once segregated by associated cost centers, all the elements of expense constituting a work unit can be properly recorded by cost code.

A variety of cost-coding systems exists in practice, and standard charts of account are published by organizations such as the National Association of Home Builders (NAHB), Associated General Contractors (AGC), and the Construction Specifications Institute (CSI) These standardized codes are very useful when a contractor needs to use manuals or other types of published information to estimate a unit of price for a line item. Standardization allows a uniform definition of the item, its unit of measure, and even its location in the estimate. Furthermore, sections in project specifications usually follow the structure of a standard code, thus simplifying the cross-referencing of items in the project's cost estimate and its drawings.

The MasterFormat code as used in the R.S. Means Building Construction Cost Data identifies three levels of detail. At the highest level, the major work classification, as given in Table 5.1, is defined. Also at this level, major subdivisions within the work category are established. For instance, 030-level accounts pertain to concrete, while 031 accounts are accounts specifically dealing with concrete forming. In a similar manner, 032 accounts are reserved for cost activity associated with concrete reinforcement.

At the next level down, a designation of the physical component or subelement of the construction is established. This is done by adding three digits to the work classification two-digit code. For instance, the three-digit code for footings is 158. Therefore, the code 031158 indicates an account dealing with concrete forming costs for footings.

At the third and lowest level, digits specifying a more precise definition of the physical subelement are used. For instance an account code of 031 158 5000 can indicate that this account records costs for forming concrete footings of a particular type (see Figure 5.6). This level of refinement is very great, and the account can be made very sensitive to the peculiarities of the construction technology to be used. Further refinement could differentiate between forming different types of footings with different types of material.

At this level, the cost engineer and construction manager have a great deal of flexibility in reflecting unique aspects of the placement technology that lead to cost fluctuations and thus must be considered in defining cost centers.

TABLE 5.1 Major Divisions in CSI's MasterFormat Uniform Construction Index 2004

Procurement ands Contracting subgroup
  00 – Procurement and Contracting Requirements
General Requirements subgroup
  01 – General Requirements
Facilities Construction subgroup
  02 – Existing Conditions
  03 – Concrete
  04 – Masonry
  05 – Metals
  06 – Wood, Plastics, and Composites
  07 – Thermal and Moisture Protection
  08 – Openings
  09 – Finishes
  10 – Specialties
  11 – Equipment
  12 – Furnishings
  13 – Special Construction
  14 – Conveying Equipment
  15 – Reserved for future use in Mechanical
  16 – Reserved for future use in Electrical
Facilities Services subgroup
  21 – Fire Suppression
  22 – Plumbing
  23 – Heating Ventilating and Air Conditioning
  25 – Integrated Automation
  26 – Electrical
  27 – Communications
  28 – Electronic Safety and Security
Site and Infrastructure subgroup
  31 – Earthwork
  32 – Exterior Improvements
  33 – Utilities
  34 – Transportation
  35 – Waterway and Marine Construction
Process Equipment subgroup
  40 – Process Integration
  41 – Material Processing and Handling Equipment
  42 – Process Heating, Cooling, and Drying Equipment
  43 – Process Gas and Liquid Handling, Purification and Storage Equipment
  44 – Pollution Control Equipment
  45 – Industry-Specific Manufacturing Equipment
  48 – Electrical Power Generation

0 Conditions of the Contract		*2 Site Work*		*2 Site Work (continued)*		*3 Concrete*	
0000-0099.	unassigned	200	Alternates	260	Roads and Walks	300	Alternates
1 General Requirements		0210-0209.	unassigned	261	Paving	0301-0309.	unassigned
0.100.	Alternates of Projectscope	120	Clearing of Site	262	Curbs and gutters	310	Concrete Formwork
0.101-0109.	unassigned	211	Declination	263	Walks	0311-0319.	unassigned
110	Schedules and Reports	212	Structures moving	264	Road and parking Appurtenances	320	Concrete Reinforcement
0111-0119.	unassigned	213	Clearing and grubbing	0265-0269.	unassigned	0321-0329.	unassigned
120	Samples and Shop Drawings	0214-0219.	unassigned	270	Site Improvements	330	Cast-in-Place Concrete
0121-0129.	unassigned	220	Earthwork	271	Fences	331	Heavyweight aggregate concrete
130	Temporary Facilities	221	Site grading	272	Playing fields	332	Lightweight aggregate concrete
0131-0139.	unassigned	222	Excavating and backfilling	273	Fountains	333	Post-tensioned concrete
140	Cleaning Up	223	Dewatering	274	Irrigation systems	334	Nailable concrete
0141-0149.	unassigned	224	Subdrainage	275	Yard improvements	335	Specially finished concrete
150	Project closeout	225	Soil poisoning	0276-0279.	unassigned	336	Specially placed concrete
0151-0159.	unassigned	226	Soil compaction control	280	Lawns and Planting	0337-0339.	unassigned
160	Allowances	227	Soil stabilization	281	Soil Preparation	340	Precast Concrete
0161-0169.	unassigned	0228-0229.	unassigned	282	Lawns	341	Precast concrete panel
		230	Piling	283	Ground covers and other plants	342	Precast structural concrete
		0231-0234.	unassigned	284	Trees and shrubs	343	Precast prestressed concrete
		235	Caissons	0285-0289.	unassigned	0344-0349.	unassigned
		0236-0239.	unassigned	290	Railroad Work	350	Clementitious Decks
		240	Shoring and bracing	0291-0294.	unassigned	351	Poured gypsum deck
		241	Sheeting	295	Marine Work	352	Insulating concrete roof decks
		242	Underpinning	296	Boat Facilities	353	Cementitious unit decking
		0243-0249.	unassigned	297	Protective Marine Structures	0354-0399.	unassigned
		250	Site drainage	298	Dredging		
		0251-0254.	unassigned	299	unassigned		
		255	Site utilities				
		0256-0259.	unassigned				

Figure 5.5 Detailed codes for classification with CSI MasterFormat.

031	Concrete Formwork										
031		Struct C.I.P. Formwork	CREW	DAILY OUTPUT	LABOR-HOURS	UNIT	1996 BARE COSTS				TOTAL INCL. O&P
							MAT.	LABOR	EQUIP.	TOTAL	
	150	FORMS IN PLACE, FOOTINGS Continuous wall, 1 use	C-1								
	5000	Spread footings, 1 use		305	.105	SFCA	1.51	2.50	.09	4.10	5.75

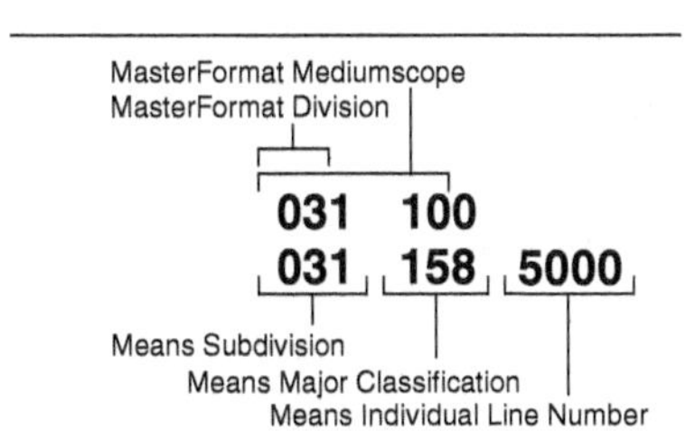

Figure 5.6 Cost (line item) structure in the MasterFormat code.

Large and complex projects in industrial and energy-related construction may require cost codes that reflect additional information, such as the project designation, the year in which the project was started, and the type of project. Long and complex codes in excess of 10 digits can result. An example of such a code is shown in Figure 5.7. This code, consisting of 13 digits, specifically defines the following items:

1. Year in which project was started (2014).
2. Project control number (15).
3. Project type (5 for power station).
4. Area code (16 for boiler house).
5. Functional division (2, indicating foundation area).
6. General work classification (0210, indicating site clearing).
7. Distribution code (6, indicating construction equipment).

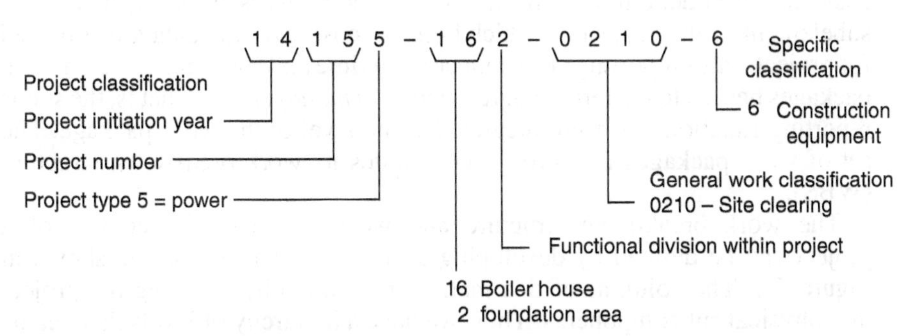

Figure 5.7 Classification of accounts: Typical data structure.

The distribution code establishes what type of resource is being costed to the work process (i.e., clearing), the physical subelement (i.e., foundations) in what area of which project. Typical distribution codes might be as follows:

1. Labor
2. Permanent materials
3. Temporary materials
4. Installed equipment
5. Expendables
6. Construction equipment
7. Supply
8. Subcontract
9. Indirect

Clearly, these codes "slice and dice" the project into a very high level of detail. This makes the four cost folders used by Fudd Associates, Inc. look pretty paltry. An impressive concentration of information can be achieved by proper design of the cost code. Such codes are also ideally suited for data retrieval, sorting, and assembly of reports on the basis of selected parameters (e.g., all construction equipment costs for concrete forming on project 10 started in a given year). The desire to cram too much information into cost codes, however, can make them so large and unwieldy that they are confusing to upper-level management.

COST ACCOUNTS FOR INTEGRATED PROJECT MANAGEMENT

In large and complex projects, it is advantageous to break the project into common building blocks for control both of cost and time. The concept of a common unit within the project that integrates both scheduling and cost control has led to the development of the work breakdown approach. The basic common denominator in this scheme is the work package, which is a subelement of the project on which both the cost and time data are collected for project status reporting. The collection of time and cost data based on work packages has led to the term *integrated project management*. That is, the status reporting function has been integrated at the level of the work package. The set of work packages in a project constitutes its work breakdown structure (WBS).

The work breakdown structure and work packages for control of a project can be defined by developing a matrix similar to the one shown in Figure 5.8. The columns of this matrix are defined by breaking the project into physical subcomponents. Thus, we have a hierarchy of levels that begins with the project as a whole and, at the lowest level, subdivides the project into

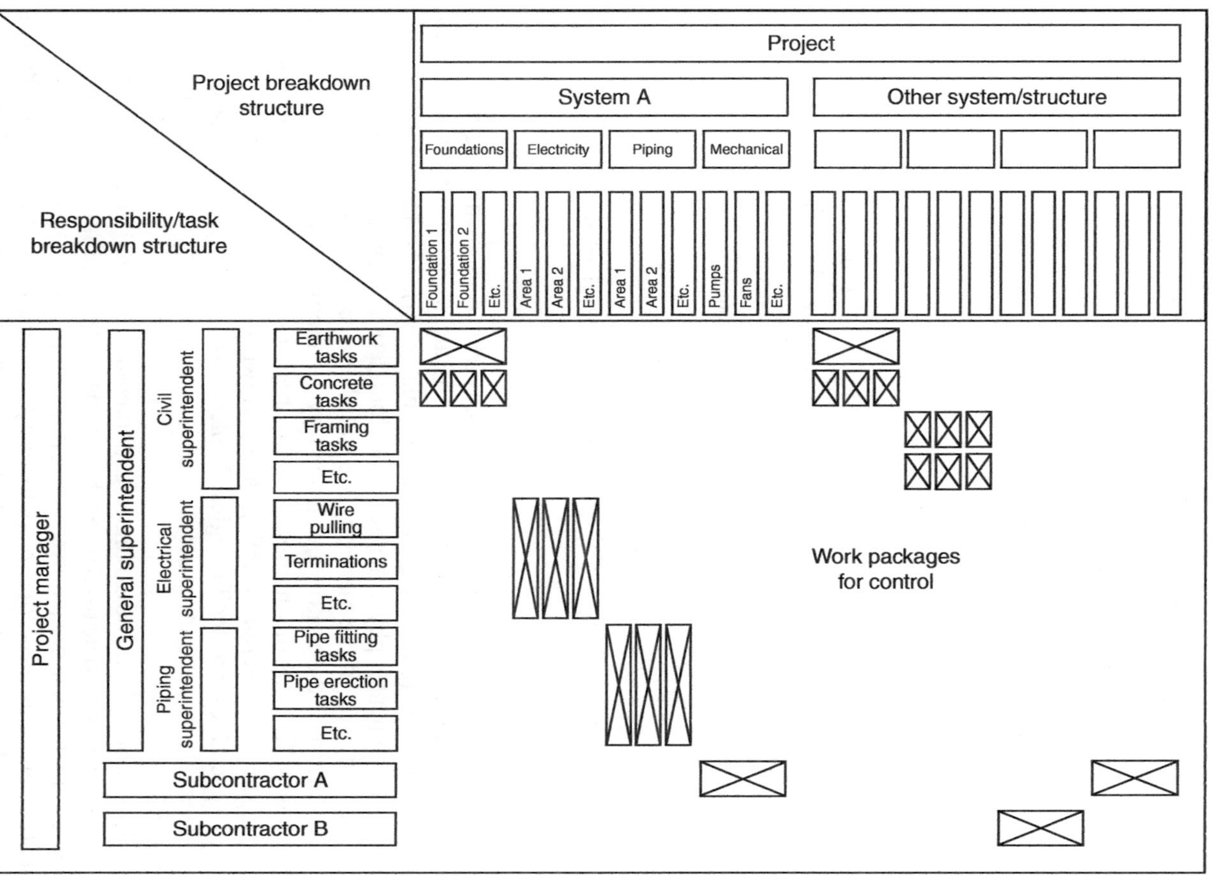

Figure 5.8 Project control matrix.

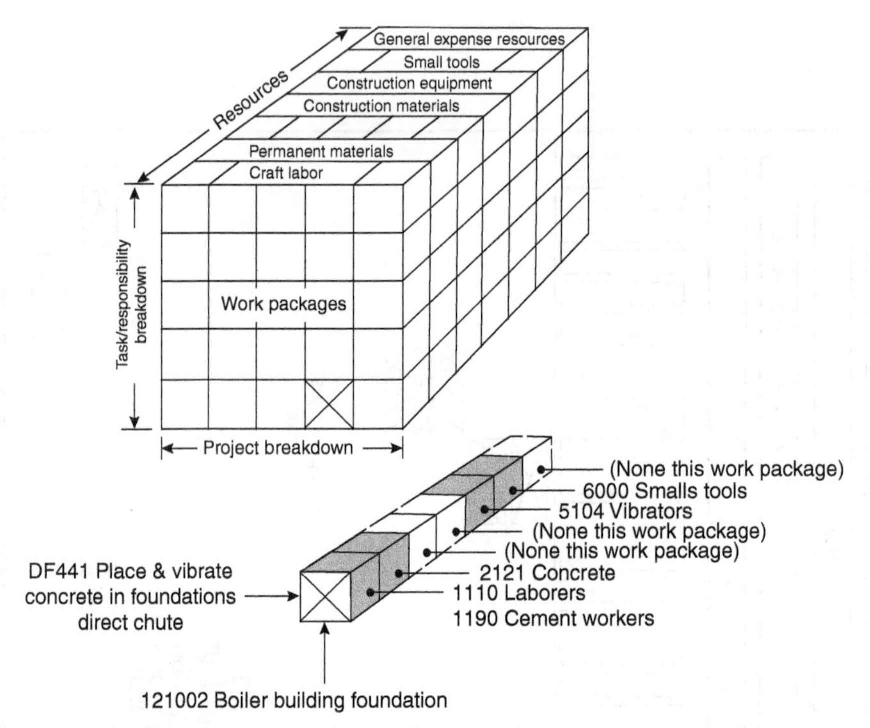

Figure 5.9 Three-dimensional visualization of work-package-oriented cost accounts.

physical end items such as foundations and areas. As shown in Figure 5.8, the project is subdivided into systems. The individual systems are further divided into disciplines (e.g., civil, mechanical, electrical). The lowest level of the hierarchy indicates physical end items (foundation 1, etc.). Work packages at this lowest level of the hierarchy are called control accounts.

The rows of the matrix are defined by technology and responsibility. At the lowest level of this hierarchy, the responsibilities are shown in terms of tasks, such as concrete, framing, and earthwork. These tasks imply various craft specialties and technologies. Typical work packages then are defined, for example, as concrete tasks on foundation 1 and earthwork on foundations 1 and 2.

This approach can be expanded to a three-dimensional matrix by considering the resources to be used on each work package (see Figure 5.9). Using this three-dimensional breakdown, we can develop definition in terms of physical subelement, task, and responsibility, as well as resource commitment. A cost code structure to reflect this matrix structure is given in Figure 5.10 This 15-digit code defines units for collecting information in terms of work package and resource type. Resource usage in terms of monetary units, quantities, man-hours, and equipment-hours for a foundation in the boiler building

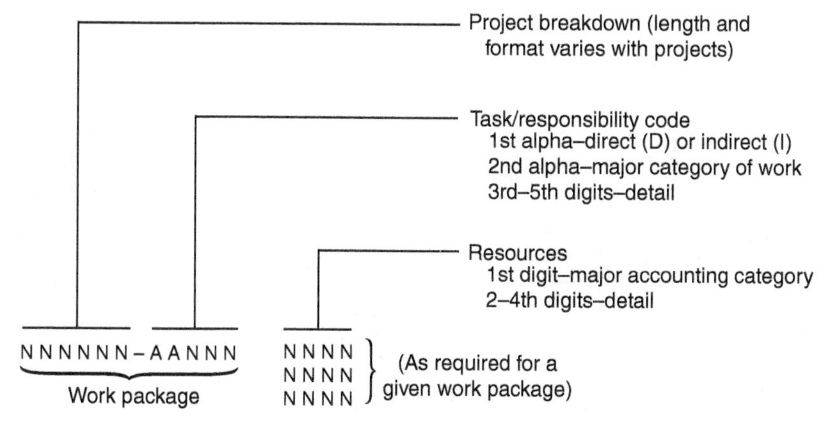

Figure 5.10 Basic cost code structure.

would be collected under work package code 121002. If this work relates to placement and vibration of concrete by using a direct chute, the code is expanded to include the alphanumeric code DF441. The resource code for the concrete is 2121. Therefore, the complete code for concrete in the boiler building foundations placed by using a chute would be 121002-DF441-2121.

This code allows collection of cost data at a very fine level. Scheduling of this work is also referenced to the work package code as shown in Figure 5.11. The schedule activities are shown in this figure as subtasks related to the work *package*.

EARNED VALUE ANALYSIS

Earned Value Analysis (EVA), also known as Earned Value Management (EVM), is the main formal project cost control technique currently used in construction. It also integrates elements of schedule control, using simple measures that can be understood by most construction professionals.

To illustrate the context for EVA, consider a situation where a carpet installing subcontractor has agreed to complete a contract for a six-story building. The owner will furnish all materials and pay the subcontractor by time billed. The contract specifies a duration of three weeks at a cost of $450,000. Two weeks into the job, two stories have been carpeted, but four stories remain untouched. Furthermore, the carpet installed so far has cost half of the total budget for this item. Most people will conclude that it is very likely that this subcontractor will take longer than planned to complete its contract and will end up with a cost overrun. If finishing two floors out of six has taken two weeks, isn't it reasonable to infer that the total job will take six weeks? Or that the total cost will be 150% of the budget, since one third of

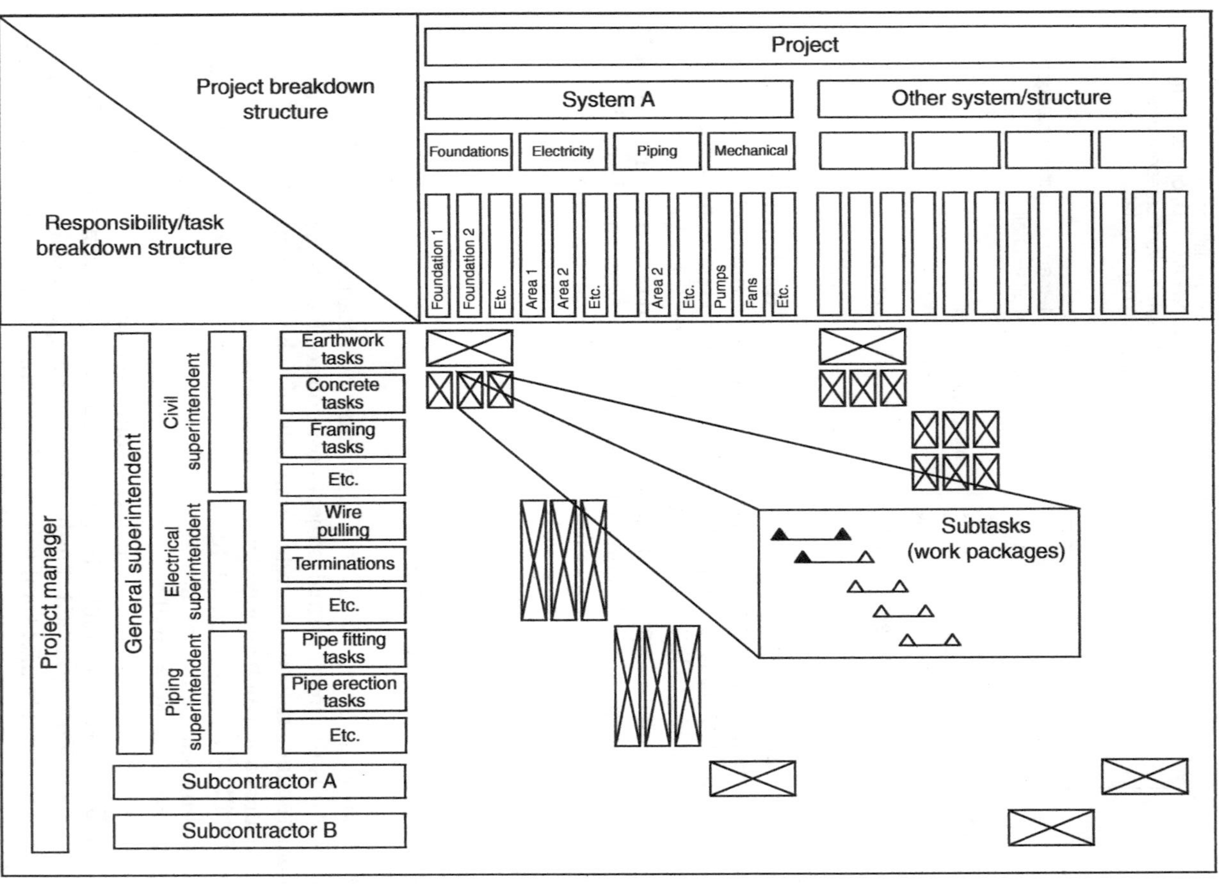

Figure 5.11 Project control matrix with scheduling of subtasks.

the work has taken half of the budget? These figures are simple extrapolations of the performance so far, and reflect the assumption that the cost and time performance rates up to now are representative of the rates for the remainder of this subcontractor's job.

EVA uses exactly the preceding thought process to assess whether a project is ahead or behind schedule, and whether it will cost less or more than its budget. It compares the value of the work performed to date (its "earned" value) with the value of the work planned to have been done to date and with the actual cost of the work to date. If the earned value is less than the planned value, EVA will report that the project is behind schedule. If the earned value is less than the actual cost, it will report that the project is over budget.

EVA's rationale for tracking cost and schedule is as simple as it is powerful. Its validity is confirmed by its use throughout many years and in many projects. It was originally implemented by the Department of Defense in the late 1970s to help better control complex projects. This original system was called the Cost and Schedule Control Systems Criteria or C/SCSC. It proved to be so effective that other government agencies (e.g., Department of Energy, etc.) adopted C/SCSC as a means of maintaining oversight on complex projects such as nuclear and conventional power plants. Private owners such as power companies implemented similar systems, since reporting to various government authorities encouraged or required the use of C/SCSC and earned value concepts. Nowadays, EVA is supported by virtually all scheduling and cost control packages such as Primavera Project Planner® and Microsoft Project®. Its principles permeate the management of virtually all large construction projects, even in cases where their project managers may not be explicitly aware of the technique as discussed in this section.

The idea of earned value is based upon a rigorous development of percent complete of the budgeted costs associated with individual work packages or line items. Each work package has an initial budget or estimate which is defined as the Budgeted Cost at Completion or BCAC. As work proceeds on an individual work package or account, assessment of the percent complete is made at various study dates. The initial schedule establishes an expected level of work completion as of the study date. The level of expected production is often shown as an S-curve plotting the cost or units of production (e.g., units produced, work hours expended, etc.) against time. This cost/production curve is referred to as the baseline. At any given time (study date), the units of cost/production indicated by the baseline are called the Budgeted Cost of Work Scheduled (BCWS).

The tracking system requires that field reports provide information about the Actual Cost of Work Performed (ACWP) and the Actual Quantity of Work Performed (AQWP). The "earned value" is the Budgeted Cost of Work Performed (BCWP). The relative values for a given work package or account at a given point in time (see Figure 5.12) provide information about the status

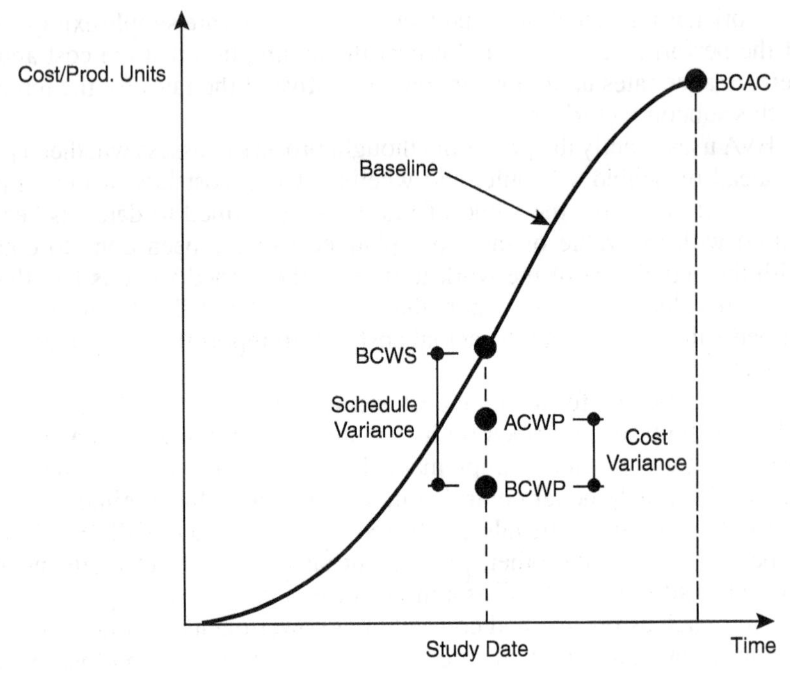

Figure 5.12 Control values for Earned Value Analysis.

in terms of cost and schedule variance. The six parameters that form the foundation of the "earned value" concept are:

1. *BCWS*. Budgeted Cost of Work Scheduled = Value of the baseline at a given time ACWP: Actual Cost of Work Performed; Measured in the field.
2. *ACWP*. Actual Cost of Work Performed – Measured in the field.
3. *BCWP*. Budgeted Cost of Work Performed = [% Complete] × BCAC.
4. *BCAC*. Budgeted Cost At Completion = Contracted Total Cost for the Work Package.
5. *AQWP*. Acutal Quantity of Work Performed – Measured in the field.
6. *BQAC*. Budgeted Quantity at Completion – Value of the Quantity Baseline as Projected at a Given Point.

In order to put these terms into context, consider the small project shown in Figure 5.13 The project consists of two control accounts: A and B. "A" consists of two subaccounts, A.1 and A.2. The study date (e.g., September 1, etc.) information for these work packages is given in TABLE 5.2. In this example, the budget is expressed in worker hours so the baseline for control is in worker hours. The estimated total number of worker hours for this scope

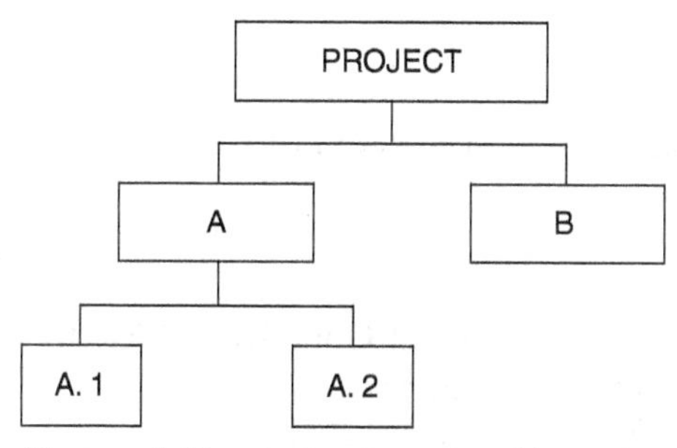

Figure 5.13 A simple project hierarchy.

of work is 215 (the sum of the estimated worker hours for A.1, A.2, and B). The BCWP, or earned value for a given work package, is given as:

$$BCWP_i = PC_i \times BCAC_i$$

Where i is the work package or account label, and PC is the percent complete as of the study date.

The percent complete (PC) for each package is based on the ratio of the Actual Quantities (AQWP) divided by the Budgeted Quantity at Completion (BQAC) based on the latest quantity assessment. If we know that the original quantity estimate is 100 units, but updated information indicates that a total of 120 units will be required to complete the work, completion of 50 units would not indicate 50 percent complete. The correct PC would be 50/120 (e.g., AQWP/BQAC).

Based on the information in Table 5.2, the PC for each work package in the small project would be:

$$PC(A.1)\ 35/105 = 0.333$$

$$PC(A.2)\ 60/77\ 0.780$$

$$PC(B) = 100/125\ 0.800$$

TABLE 5.2 Study Date Data for Simple Project

	BCAC	ACWP	BQAC	AQWP	PC (%)	BCWP	ECAC
A							
A.1	100	40	105	35	33.3	33.3	120
A.2	50	35	77	60	78.0	39.0	45
B	65	50	125	100	80.0	52.0	62.5
Total	215	125	—	—	57.8	124.3	227.5

Project PC (PPC) = Total BCWP = 124.3 ÷ 215 = 57.8%
$ECAC_i$ = Estimated Cost at Completion for Work *Package i* = $ACWP_i/PC_i$

Then

$$\text{BCWP (Project)} = .333\ (100) + .78\ (50) + .8\ (65)$$
$$= 33.3 + 39 + 52 = 124.3$$

Therefore, the Project Percent Complete (PPC) for the small project is: PPC $= [\frac{124.3}{215}] \times 100 = 57.8$ percent.

This simple example illustrates several points:

1. The PC for a given package is based on the ratios of the AQWP/BQAC.
2. The PPC is calculated by relating the total BCWP (i.e., earned value) to the total BCAC for the project scope of work.
3. The total work earned is compared to the work required. The values of units to be earned are based on the originally budgeted units in an account/work package and the percent earned is based on the latest projected quantity of units at completion.

Worker hours are used to here to demonstrate the development of the PPC. However, other cost or control units may be used according to the needs of management.

It is very important to know that schedule and cost objectives are being achieved. Schedule and cost performance can be characterized by cost and schedule variances as well as cost performance and schedule performance indices. These values in C/SCSC are defined as follows:

CV, Cost Variance = BCWP − ACWP

SV, Schedule Variance = BCWP − BCWS

CPI, Cost Performance Index = BCWP/ACWP

SPI, Schedule Performance Index = BCWP/BCWS

Figures 5.14 a, b, and c plot the values of BCWP, ACWP, and BCWS for the small project data given in Table 5.2. At any given study date, management will want to know what are the cost and schedule variance for each work packages. The variances can be calculated as follows:

$$\text{CV(A.1)} = \text{BCWP (A.1)} - \text{ACWP (A.1)} = 33.3 - 40 = -6.7$$
$$\text{CV (A.2)} = \text{BCWP (A.2)} - \text{ACWP (A.2)} = 39 - 35 = +4$$
$$\text{CV (B)} = \text{BCWP (B)} - \text{ACWP (B)} = 52 - 50 = +2$$

Since the CV values for A.2 and B are positive, those accounts are within budget (i.e., the budgeted cost earned is greater than the actual cost). In other words, less is being paid in the field than was originally budgeted. The negative variance for A.1 indicates it is overrunning budget. That is, actual cost is greater than the cost budgeted.

This is confirmed by the values of the CPI for each package:

$CPI\ (A.1) = 33/40 < 1.0$ A value less that 1.0 indicates cost overrun of budget.

$CPI\ (A.2) = 39/35 > 1.0$

$CPI\ (B) = 52/50 > 1.0$ Values greater than 1.0 indicate actual cost less than budgeted cost

The schedule variances for each package are as follows:

$$\text{SV (A.1)} = \text{BCWP (A.1)} - \text{BCWS (A.1)} = 33.3 - 50 - 16.7$$

$$\text{SV (A.2)} = \text{BCWP (A.2)} - \text{BCWS (A.2)} = 39 - 32 = +7$$

$$\text{SV (B)} = \text{BCWP (B)} - \text{BCWS (B)} = 52 - 45 = +7$$

These computations are summarized in Figure 5.14.

The positive values for A.2 and B indicate that these items are ahead of schedule. The negative value for A.1 indicates a scheduling problem. The calculation of the SPI values will confirm this assessment. Overall, it can be stated that A.2 and B are ahead of schedule and below cost, while A.1 is behind schedule and over cost.

Six scenarios for permutations of ACWP, BCWP, and BCWS are possible as established by Singh (Singh, 1991). The various combinations are shown in Figure 5.15 and Figure 5.16.

The reader is encouraged to verify the information in Figure 5.16.

The earned value approach requires a comprehensive knowledge of work packaging, budgeting, and scheduling. It is a data-intensive procedure and requires the acquisition of current data on the ACWP and AQWP for each work package or account. It is a powerful tool, however, when management is confronted with complex projects consisting of hundreds of control accounts. In large projects consisting of thousands of activities and control accounts, it is a necessity. Without it, projects can quickly spiral out of control. A more detailed presentation of this topic is beyond the scope of this chapter. The interested reader should refer to more detailed publications that describe the Earned Value Management System (EVMS) and the inherent procedures associated with its implementation.

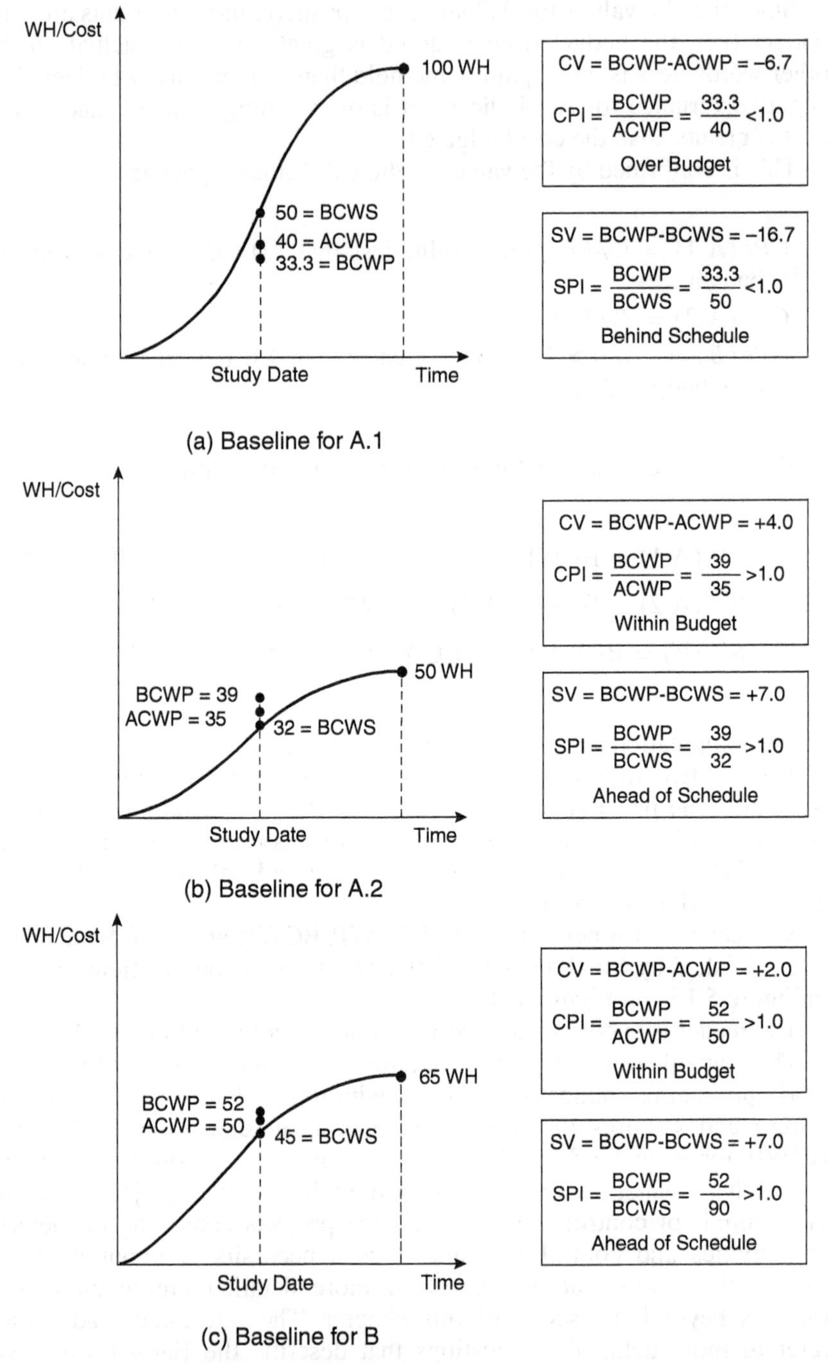

Figure 5.14 States of Control account for single project.

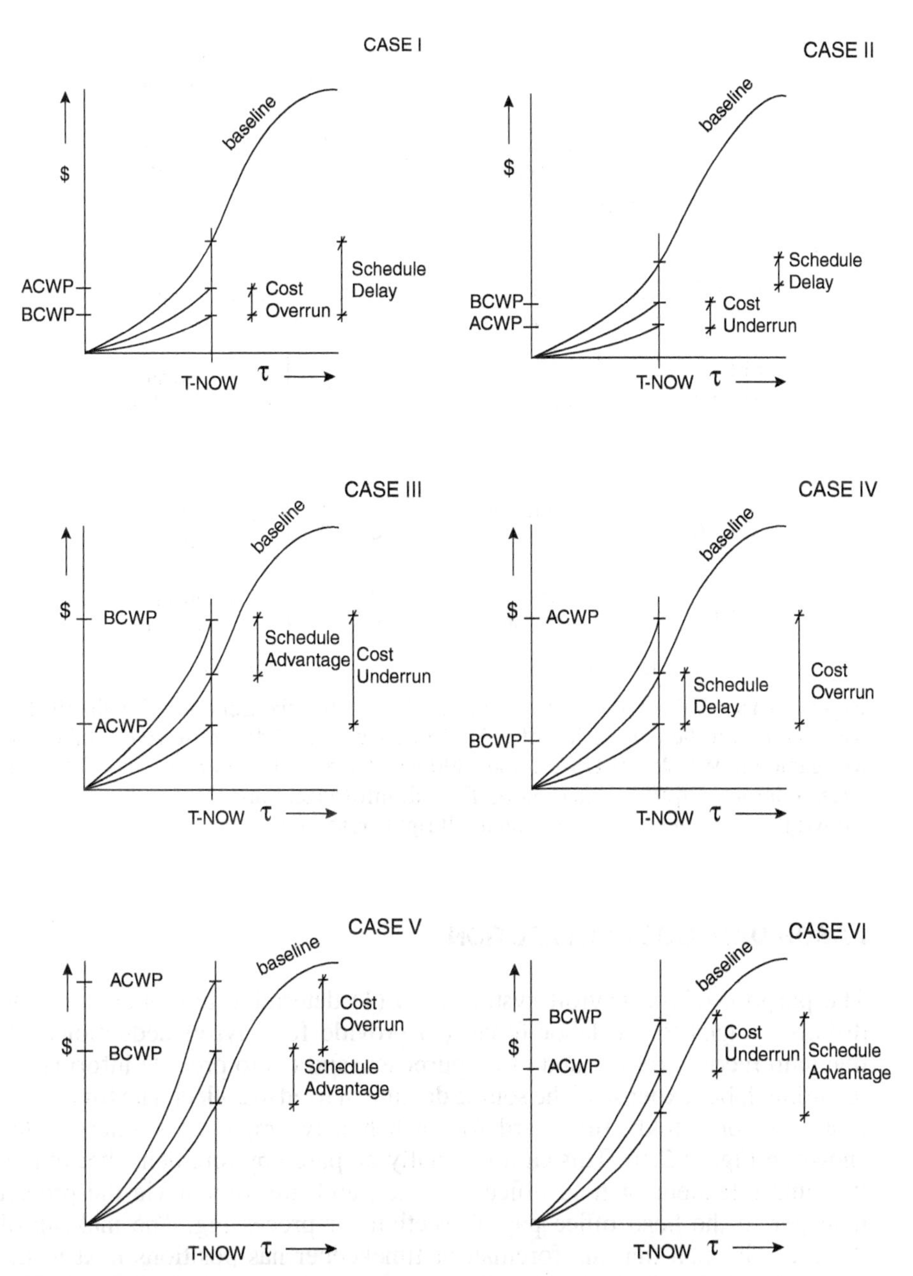

Figure 5.15 Scenarios for permutations between ACWP, BCWP, and BCWS (Singh, 1991).
Reprinted with the permission of AACE International, 209 Prairie Ave., Suite 100, Morgantown, WV 25601 USA. Phone 800-858-COST/304-296-8444. Fax: 304-291-5728. Internet: http://www.aacei.org. E-mail: info@aacei.org.

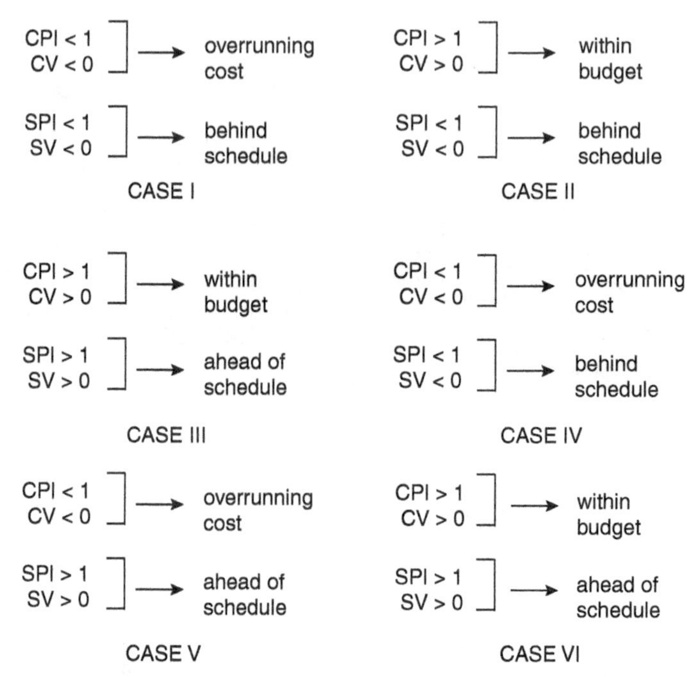

Figure 5.16 Values of CPI, CV, and SPI, SV for the six scenarios (Singh, 1991). Reprinted with the permission of AACE International, 209 Prairie Ave., Suite 100, Morgantown, WV 25601 USA. Phone 800-858-COST/304-296-8444. Fax: 304-291-5728. Internet: http://www.aacei.org. E-mail: info@aacei.org.

LABOR DATA COST COLLECTION

The purpose of the payroll system is to (1) determine the amount of and disburse wages to the labor force, (2) provide for payroll deductions, (3) maintain records for tax and other purposes, and (4) to provide information regarding labor expenses. The source document used to collect data for payroll is a daily or weekly time card for each hourly employee similar to that shown in Figure 5.16. This card is usually prepared by foremen, checked by the superintendent or field office engineer, and transmitted via the project manager to the head office payroll section for processing. The makeup of the cards is such that the foreman or timekeeper has positions next to the name of each employee for the allocation of the time worked on appropriate cost subaccounts. The foreman in the distribution made in Figure 5.17 has charged 4 hours of A. Apple's time to an earth excavation account and 4 hours to rock excavation. Apple is a code 15 craft, indicating that he is an operating engineer (equipment operator). As noted, this distribution of time allows the generation of management information aligning work effort with cost center. If no allocation is made, these management data are lost.

Dewey, Cheatum, and Howe
Company

Report No. *16*

DAILY LABOR DISTRIBUTION REPORT

Date *12 September 04*

Job No. *101*

Location *Peachtree Corners Shopping Mall*

Foreman's signature

Description of work performed and cost account number: *80,103* | *80,104* | *80,260.01* | *80,260.07*

Employee or badge number	Name	Code	Craft or union	Rate	Hours (*80,103*)	Hours (*80,104*)	Hours (*80,260.01*)	Hours (*80,260.07*)	Hours	Hours	Total hours ST	Total hours PT
65	*Adam Apple*	ST	*15*	*16.50*	*4*	*4*					*8*	*0*
		PT										
14	*Ella Del Fabbro*	ST	*10*	*12.50*			*8*				*8*	*0*
		PT										
22	*Charles Hoarse*	ST	*10*	*12.50*			*6*	*2*			*8*	*0*
		PT										
		ST										
		PT										
		ST										
		PT										
		ST										
		PT										
		ST										
		PT										
		ST										
		PT										
		ST										
		PT										
				Totals	*4*	*4*	*14*	*2*			*24*	*0*

Approved by

Figure 5.17 Foreman's daily labor distribution report.

The flow of data from the field through preparation and generation of checks to cost accounts and earnings accumulation records is shown in Figure 5.18.

This data structure establishes the flow of raw data or information from the field to management. Raw data enter the system as field entries and are processed to service both payroll and cost accounting functions. Temporary files are generated to calculate and produce checks and check register information. Simultaneously, information is derived from the field entries to update project cost accounts. These quantity data are not required by the financial accounting system and can be thought of as management data only.

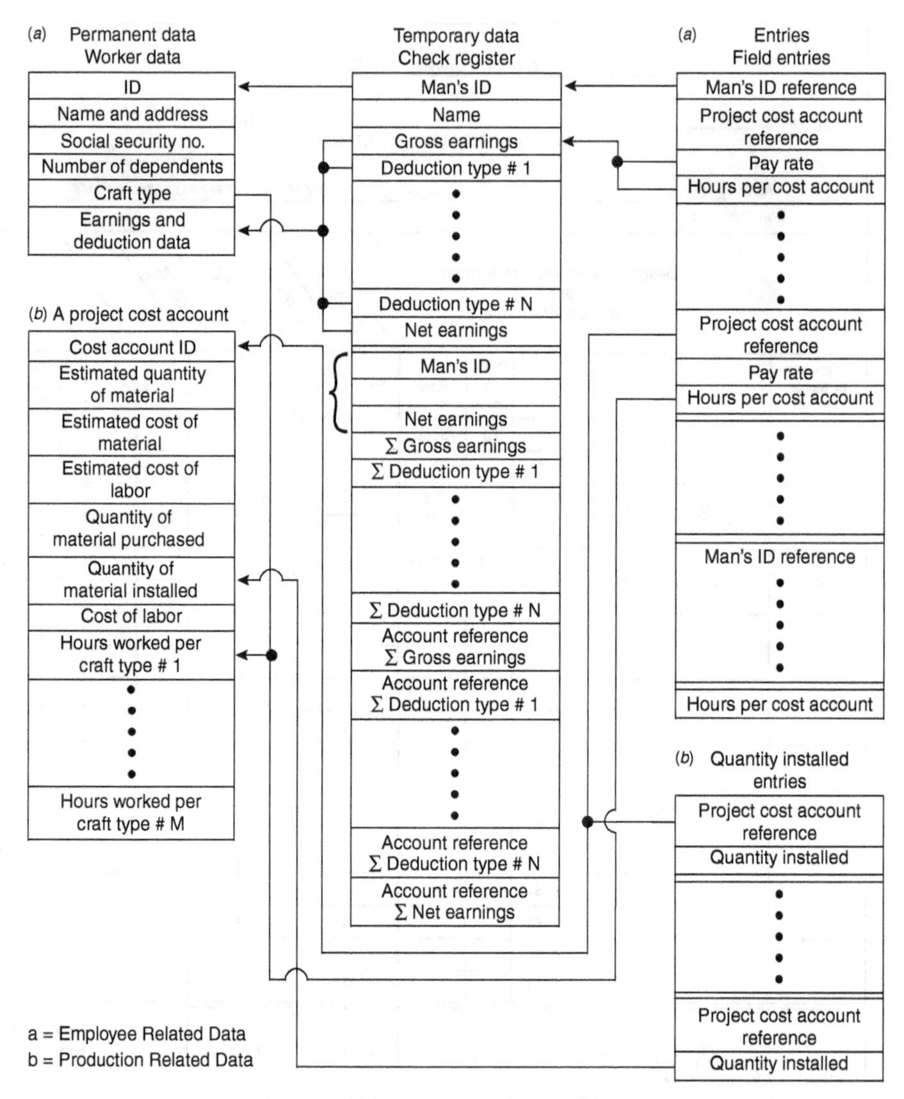

Figure 5.18 Payroll data structure.

From the time card, the worker's ID (badge number), pay rate, and hours in each cost account are fed to processing routines that cross check them against the worker data (permanent) file and use them to calculate gross earnings, deductions, and net earnings. Summations of gross earnings, deductions, and net earnings are carried to service the legal reporting requirements placed on the contractor by insurance carriers (Public Liability and Property Damage, workmen's compensation), the unions, and government agencies (e.g., Social Security and Unemployment).

REVIEW QUESTIONS AND EXERCISES

1. As a construction project manager, what general categories of information would you want to have on a cost control report to properly evaluate what you think is a developing overrun on an operation, "place foundation concrete," that is now under way and has at least 5 weeks to go before it is completed?
2. What are the major functions of a project coding system?
3. List advantages and disadvantages of the MasterFormat coding system.
4. Assume that you are the cost engineer on a new $12 million commercial building project. Starting with your company's standard cost code, explain how you would develop a project cost code for this job. Be sure that the differences in purpose and content between these two types of cost codes are clear in your explanation. Specify any additional information that may be needed to draw up the project cost code.
5. Develop a cost code system that gives information regarding:
   - When project started
   - Project number
   - Physical area on project where cost accrued
   - Division in Uniform Construction Index
   - Subdivision
   - Resource classification (labor, equipment)
6. The following planned figures for a trenching job are available: The amount of excavation is 100,000 cubic yards. The resource hours and cost are shown in Table P5.1.

   TABLE P5.1 Project Effort and Costs

Resources (hours)		Cost
Machines	1000	$100,000
Labor	5000	$100,000
Trucks	2000	$62,500

   At a particular time during the construction, the site manager realizes that the actual excavation will be in the range of 110,000 cu yd. Based on the new quantity, he figures that he will have 30,000 cu yd left.

   TABLE P5.2 Actual Effort and Costs

Resources		Cost
Machines	895 hours	$85,000
Labor	6011 man-hours	$79,000
Trucks	1684 hours	$50,140

From the main office, the following job information is available (Table P5.2):

What would concern you as manager of this job?

7. Categorize the following costs as (a) direct, (b) project indirect, or (c) fixed overhead:
   - Labor
   - Materials
   - Main office rental
   - Tools and minor equipment, field office
   - Performance bond
   - Sales tax
   - Main office utilities
   - Salaries of managers, clerical personnel, and estimators
8. Calculate the cost and scheduling variances for each of the work packages shown in Table P5.3. What is the percent complete for the entire package?

TABLE P5.3 Data for Problem 8

	Work Hours			Quantities		
	EST	ACT	Forecast	EST	ACT	Forecast
A	15000	8940	15500	1000	600	1100
B	2000	1246	1960	200	93	195
C	500	356	510	665	540	680

9. Draw a Hierarchical diagram of the work packages A.00 and B.00. Finally, compute the total percent complete of given, using the WBS code values. Calculate the BCWP and the project percent complete for all codes and work packages to include A.00 and B.00. Finally, compute the total percent complete of the project (see Table P5.4).

TABLE P5.4 Data for Problem 9

		Work Hours			Quantities		
Code	Description	EST	ACT	Forecast	EST	ACT	Forecast
A.00	E/W Duct	440					
A.10	Partitions	230	150	225	25	14	25
A.20	Hangers	210	130	220	3	2.2	3.8
B.00	N/W Duct	645					
B.10	Partitions	370	75	390	50	12	48
B.20	Hangers	275	85	260	16	4.5	16

10. Given the diagrams in Figure P5.1 of progress on individual work packages of a project, answer the following questions:

- For Case 1, is the project ahead or behind schedule?
- For Case 2, is the project over or under cost?
- For Case 3, is the Cost Performance Index greater than 1?
- For Case 1, is the SPI greater than 1? Explain by calculation.
- For Case 4, is the project on schedule and budget or not? Explain.

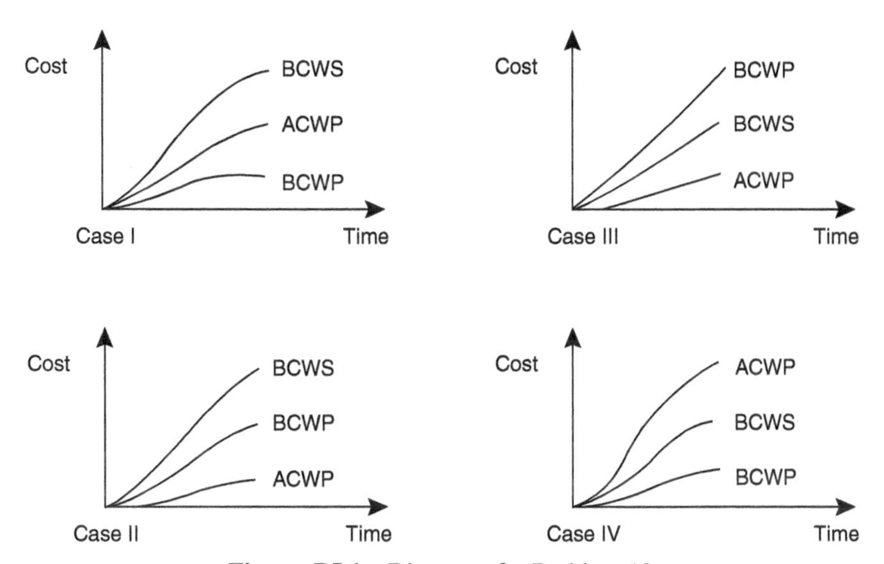

Figure P5.1 Diagrams for Problem 10.

CHAPTER 6

FORECASTING FINANCIAL NEEDS

IMPORTANCE OF CASH MANAGEMENT

A construction company may be rich on paper and yet go bankrupt if it does not have enough liquidity to pay for its financial obligations. More companies fail for lack of liquidity than for any other reason. It can be argued that cash is the most important resource that any contractor can manage.

A key issue in managing cash is the forecasting of cash needs over time. The construction industry is extremely project-oriented, and, therefore, much of the planning involves the forecasting of cash at the individual project level. The cash requirements for the overall company is the summation of the cash needed for all projects, plus the cash used by the central office and any strategic purchases that the company decides to undertake.

UNDERSTANDING CASH FLOW

The progression from the moment at which a project item or resource[1] is acquired by a contractor to the moment that the project owner pays for it can be traced as follows:

1. The item or resource is purchased or acquired. From an accountant's viewpoint, the contractor incurs the cost of the item or resource at this point.

[1] The term "Resource" covers payroll, payments associated with equipment usage, subcontract payment requests and other overhead related charges in contrast to materials or equipment procured to be incorporated into the finished project.

2. The item is placed in inventory or the resource is assigned to the project. From the owner's perspective, no value is achieved until the purchase is incorporated into the project or the resource performs work related to the project.
3. The item or resource is used in the project. The contractor earns value. This value can be billed to the client. In the vast majority of cases, however, this value sits idle until the next progress payment is submitted to the project owner.
4. Value is billed to the project owner. The owner takes some time to review and prepare the payment.
5. Payment is made to the supplier, subcontractors, worker, and so forth. This is when the contractor disburses cash to offset the cost that has been incurred.
6. Payment is made by the project owner. The contractor receives cash for the value incorporated in the project.

This sequence is shown in Figure 6.1. Notice that the cash available to the contractor is not affected by any but the last two steps. If all transactions were paid in cash immediately, the last two steps would become the first two. A cash-only business environment is, as a practical matter, difficult to achieve. In the preceding sequence, the contractor generates value, and the project owner sees the project becoming more valuable before any cash is exchanged.

The difference between the point in time at which the items or resources are paid for by the contractor and the point in time at which payment is received from the client is called *payment lag*. Payment lag determines the amount of time that the item or resource will have to be financed with the contractor's cash.

To illustrate this concept, suppose that an HVAC rooftop unit (RTU) costs $10,000 and is received on April 10th. Its cost is added to the cost to date of the project immediately. When it is installed on the 20th, its value (the price charged to the client) of $11,000, is counted as revenue right away. The

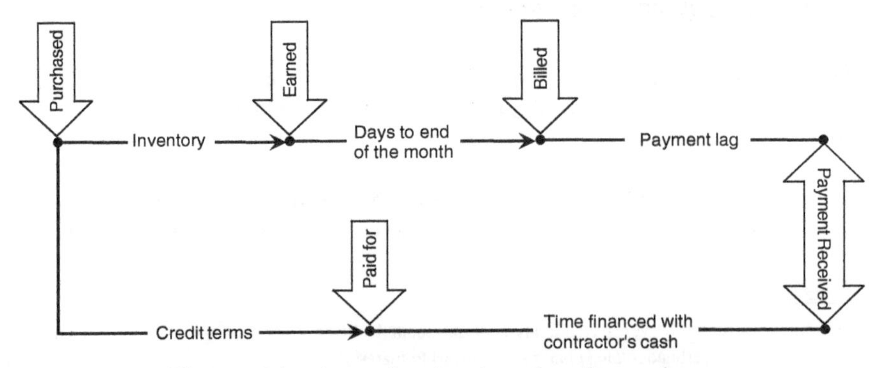

Figure 6.1 Accrual and cash cycles of a purchase.

contractor pays the supplier the full cost of $10,000 on May 15th and receives progress payment from the client on May 25th, of which $9,900 is for the RTU. The payment lag for $9,900 out of the $10,000 paid by the contractor is 10 days (May 15th to May 25th).

RETAINAGE

The contractor in the example above earned $11,000 of the total contract when the RTU was installed but received only $9,900 from the project owner for this line item. $1,100 (10% of $11,000) was retained by the client until a point in time specified in the contract, such as the date of the project's substantial completion. It is common practice for owners to keep or hold back a percentage of the progress payment amount requested. The amount held back in escrow is called a *retainage* or *retention*. Retainage is intended to serve as a guarantee that the contractor will not walk away from the contract without a substantial effect on income. This is a rather large amount (e.g., often 5 to 10% of the bid price) held by the owner until the project is satisfactorily completed as required by the contract. Retainage is an incentive and motivation for the contractor to cooperate in completing the project in a timely and professional manner.

Retainage has a considerable effect on the amount of cash that the contractor has available to build a project. In the case of the RTU, the $1,100 kept as retainage will have a much longer payment lag than the $9,900 included in the progress payment. It is viewed as a *deferred account receivable* to be collected when the entire project is satisfactorily completed (see examples in Appendix B, Transactions 2 and 4).

PROJECT COST, VALUE, AND CASH PROFILES

Curves for the total project value and project costs to date are shown in Figure 6.2. It is common that a project's execution begins at a relatively slow pace, then speeds up to a steady rate for most of the construction, and decreases its tempo in its last stages. These changes in the rate of execution result in cumulative curves having a shape approximating a letter S. Consequently, these graphs are called "lazy S" curves.

The cumulative value curve is normally higher than the cost incurred curve because (hopefully) at any given point the value generated by a project is higher than its cost to date. The disbursements flow tends to lag the cumulative cost curve because of the payment lag.

Figure 6.3 shows the disbursement flow and includes two additional curves. The receipts flow is a stair-step plot which reflects the cumulative amount of money received to date. The cash position (also called net cash flow) curve

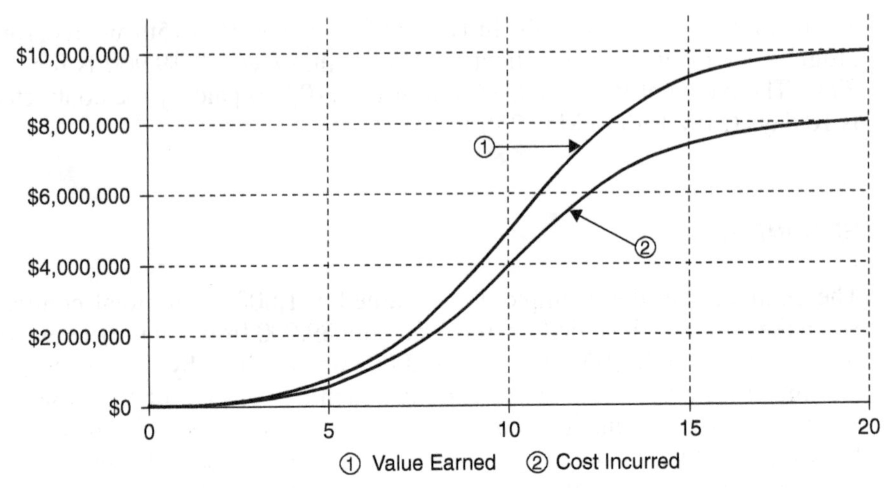

Figure 6.2 Value earned and cost incurred.

shows the difference between the receipts and disbursements flows. If the cash position is negative, it is called an *overdraft*.

The disbursements flow is shown as a lazy S curve for simplicity. An actual disbursement curve would have a more jagged appearance (due to discrete day to day cash outflows). The curve would, however, have the same general shape. As will be demonstrated later in this chapter, disbursements continue to occur even after the project execution is complete, since the contractor can delay payment for items taken on credit. The disbursement flow is also called the outflow curve.

As noted above, the receipts flow has a stair-step shape, as shown in Figure 6.3. The payment amount is depicted by the jump at the beginning of

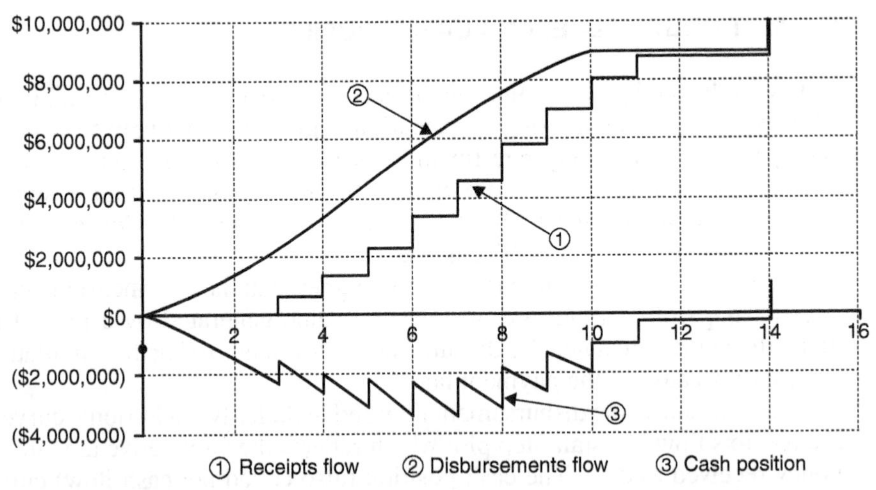

Figure 6.3 Receipts, disbursements and cash position/overdraft.

each month. Due to the payment lag in any contract, each received payment pays for work earned several weeks or even months before. The billing-payment cycle leads to this payment lag. Figure 6.3, for example, shows the first payment occurring at the beginning of the third month. This would be a reasonable payment lag for a contractor requesting payment of the work done in the first month. The request is submitted at the beginning of the second month. The remaining of the second month is spent preparing the payment request (which is reviewed by the owner's representative and then sent to the client, who will take days to prepare the payment check).

The project's net cash flow is the lower, saw tooth curve shown in Figure 6.3. It is the difference between the receipts and disbursements, and shows the contractor's cash position throughout the project execution. This curve can be visualized as the balance of a "checking account" for the project. Because of "payment lag" and retainage, the balance of this accounting account is negative for most of the project until the cumulative progress payments are sufficient to offset expenditures. This normally doesn't occur until the retainage, held by the owner, is released at the end of the job. This negative balance must be financed by the contractor. The contractor establishes a line of credit at the bank to cover this negative balance. The line of credit operates much like a personal credit card. Expenditures are made against the line of credit and interest is paid for the outstanding negative balance, as shown in the following simple example.

CASH FLOW CALCULATION—A SIMPLE EXAMPLE

Consider the simple four-activity example shown in Figure 6.4.

The base cost of the project is \$200,000. Assume that the contractor originally included a profit or markup in the bid of \$10,000 (i.e., 5%) so that the total bid price was \$210,000. The owner retains 10% of all validated progress payment claims until one half of the contract value (i.e., \$105,000 × 0.10) has been built, approved, and accepted as complete. The progress payments will be billed at the end of the month, and the owner will transfer the billed amount minus any retainage to the contractors account 30 days later. The amount of each progress payment can be calculated as:

$$\text{Pay} = 1.05 \times (\text{indirect expense} + \text{direct expense})$$
$$-0.10 \times 1.05\,(\text{indirect expense} + \text{direct expense})$$

The minus term for retainage drops out of the equation when 50% of the contract has been completed. Because of the delay in payment of billings by the owner and the retainage withheld, the income profile lags behind the expense S-curve, as shown in Figure 6.5.

The calculations required to define the overdraft profile are summarized in Table 6.1. The information in the table is plotted in Figure 6.6. The table indicates that the payment by the owner occurs at the end of each month, based on the billing at the end of the previous month. It is assumed that a

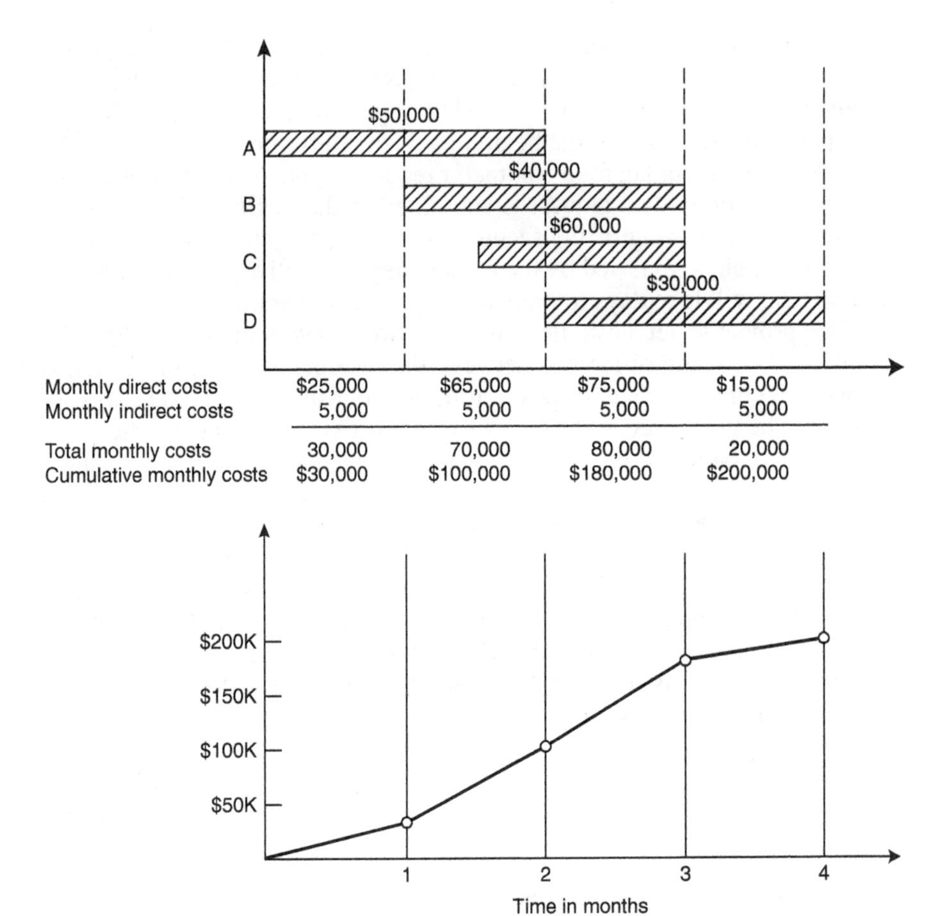

Figure 6.4 Simple time-scaled budget.

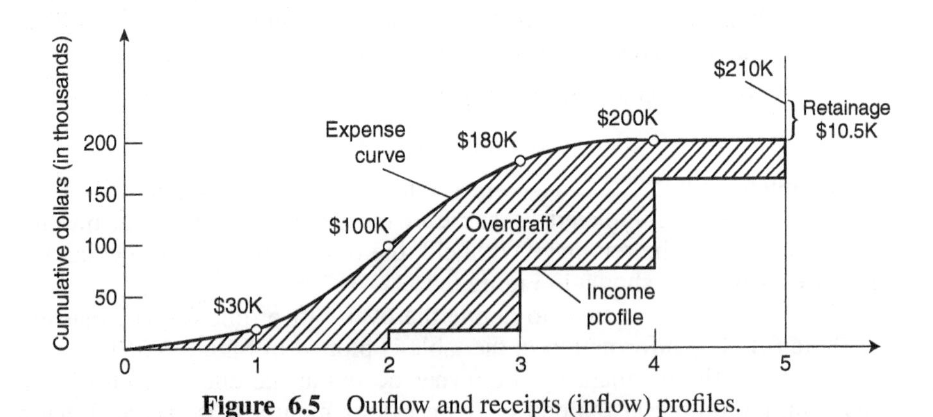

Figure 6.5 Outflow and receipts (inflow) profiles.

TABLE 6.1 Overdraft Calculations

	Month								
	1	2		*3*		*4*		*5*		
Direct cost	$25,000	$65,000		$75,000		$15,000				
Indirect cost	5,000	5,000		5,000		5,000				
Subtotal	30,000	70,000		80,000		20,000				
Markup	1,500	3,500		4,000		1,000				
Total Billed	31,500	73,500		84,000		21,000				
Retainage withheld	3,150	7,350		0		0				
Payment										
Received			$28,350		$66,150		$84,000		$31,500	
Total cost to date	30,000	100,000	↓	180,000	↓	200,000	↓	200,000	↓	
Total amount billed to date	31,500	105,000		189,000		210,000		210,000		
Total paid to date			28,350		94,500		178,500		210,000	
Overdraft end of month	30,000	100,300		152,953		108,333		25,416		(−)5830
Interest on overdraft balance[a]	300	1,003		1,530		1,083		254		0
Total amount financed	$30,300	$101,303		$154,483		$109,416		$25,670		

[a] A simple illustration only. Most lenders would calculate interest charges more precisely on the amount/time involved, employing daily interest factors.

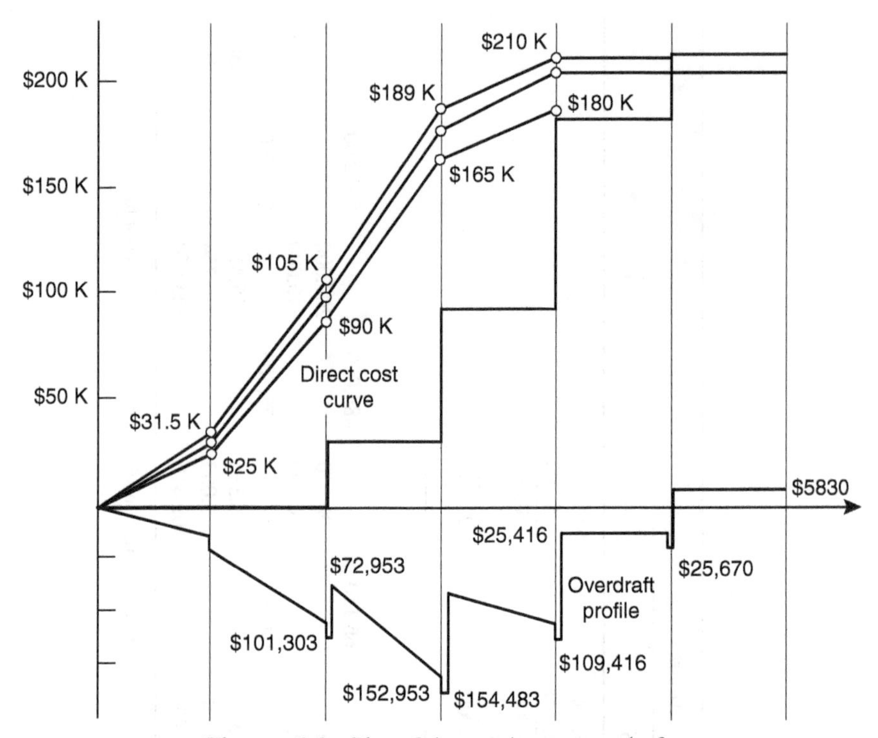

Figure 6.6 Plot of the maximum overdraft.

simple interest of 1% is applied to the overdraft to obtain the total amount financed ($101,303). To obtain the overdraft at the end of the second month, the progress payment of $28,350 is applied to reduce the overdraft at the beginning of the third month to $72,953. The overdraft at the end of the period is, then, $72,953 plus the costs for the period. Therefore, the overdraft is $72,953 plus $80,000, or $152,953. The overdraft profile is shown below the base line. This profile shows that the maximum requirement for borrowing by the contractor is $154,483. Therefore, for this project the contractor must have a line of credit of at least $155,000 plus a margin of safety, say $170,000 to cover expenses.

PEAK FINANCIAL REQUIREMENTS

Since the owner pays progress payments, the contractor can successfully complete a project without paying the total cost of the project from his own funds. As shown in the four-activity project in Figure 6.4, the contractor must pay interest at the bank to temporarily finance the project until full payment is received. For the simple example project, the maximum amount that must be borrowed is $154,483. It is important for the contractor to be aware of this *peak financial requirement* so that the available line of credit at the bank is high enough to cover this maximum. (The limit of the line of credit should be greater than say $170,000).

For the $10 million job shown in Figure 6.3, the maximum overdraft that the contractor of that job needs to cover is about $3.3 million (see the maximum point of the saw tooth curve). The project total cost is $8 million. Luckily, this peak amount is needed only in months 5 through 8. However, interest is being paid throughout the project, since there is some level of overdraft (see the sawtooth plot) until the end of the project. The good news is that the progress payments received from the project owner can be leveraged to cover project expenditures to reduce the overdraft amount. The bad news, however, is that the contractor must borrow from the bank to cover the overdraft throughout the project. The amount of interest paid to the bank for this financing must be estimated and included in the project bid amount. Otherwise, the contractor is paying the interest (i.e., financing the project) without being reimbursed by the owner. This interest charge is a cost of doing business and industry protocols assume that the owner must pay for this project cost. Similar examples of this so-called *inventory financing* can be found in many cyclic commercial undertakings. Automobile dealers, for instance, typically borrow money to finance the purchase of inventories of new car models. The borrowing is repaid as cars are sold over time.

GETTING HELP FROM THE OWNER

The best cash flow scenario for any project, from the contractor's viewpoint, would be to reduce the amount of contractor financing to zero. For example, if the project owner agrees to directly pay all payroll and provide advance funding for all materials and subcontractor charges, the contractor would be able to avoid financing the project overdraft. Some publicly funded projects allow the contractor to charge a *mobilization* item at the beginning of the project. For instance, if the owner agrees to pay the contractor of the simple example project of Figure 6.4 a mobilization payment, this has the effect of reducing the amount of the overdraft. As can be seen in Figure 6.7, the area between the revenue profile and the expense curves is reduced.

In some public projects, certain special material items or unique assemblies can be more efficiently purchased by the owner. This leads to the idea of government provided equipment or components. These items are specified in the contract documents as being provided by the owner. This relieves the contractor from the requirement of purchasing or procuring certain materials or components that will be incorporated into the final constructed facility. For instance, on large and complex NASA projects, tracking and electronic gear can be purchased by the government and provided to the contractor for assembly and inclusion into the project. The assumption is that the owner, the

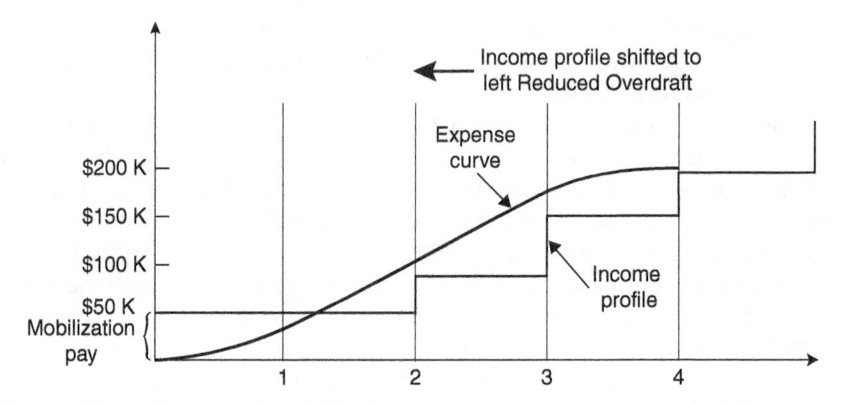

Figure 6.7 Influence of mobilization payment on payments and disbursements profiles.

government, can procure these items at a better price or more easily than the contractor constructing the project.

Thought Problem. Can you think of a privately owned project where the owner would provide materials to the contractor, rather than having the constructor shop for these items on the open market?

OPTIMIZING CASH FLOW

Realistically, a project's cash flow is optimized, from its contractor's perspective, when it ties up as little of the contractor's money as possible. This means that the amount of the project overdraft would be as close to zero as possible throughout the project execution (i.e., the sawtooth curve would virtually disappear).

Since the Cash Position curve is the distance between the disbursements and receipts flow curves, it follows that a project has an optimum cash flow when the disbursements curve is as close as possible to the payments curve. The issue is, then, to develop strategies that would minimize the distance between these two curves. Figure 6.8 helps understand the basic rationale underlying a cash flow optimization strategy.

One strategy consists of minimizing the distance between the payment and disbursements curves. Overdraft is reduced if the payment curve is shifted as much as possible to the left. This implies that money is received as soon as possible or as an advance payment (e.g., a mobilization payment.)

When the disbursements curve is moved as much as possible to the right, the distance between the two curves is reduced, also resulting in a reduction of the maximum cash overdraft. This is achieved when payments (by the

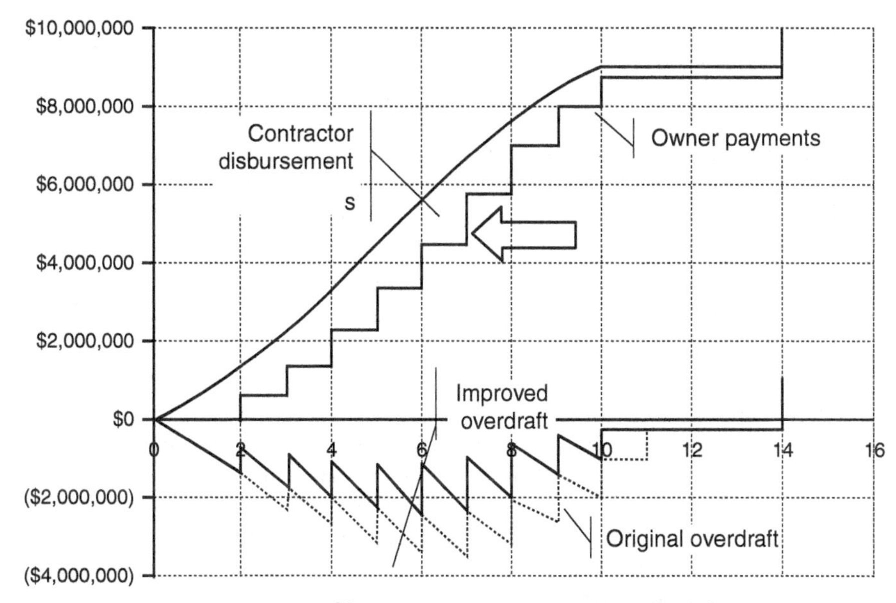

a. Move owner payments curve to the left

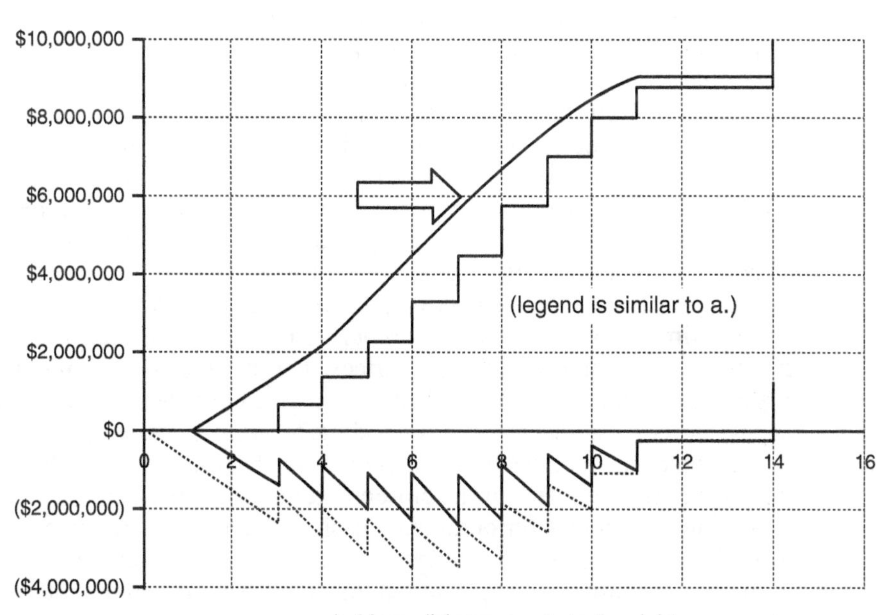

b. Move disbursements to the right

Figure 6.8 Cash flow optimization strategies.

contractor) are delayed as much as possible (e.g., delaying payment to vendors and subcontractors).

Some measures to optimize cash flow include the following.

One effective measure shifts the payments curve to the left by implementing *an efficient collection system*. Construction contractors have notoriously poor collection practices. If the contractor can reduce the delay between work placement and cash payment by the client, this immediately reduces the area between the disbursement curve and the receipts profile. Many contractors are not aggressive enough in policing outstanding accounts receivable. In contrast, credit card companies are very aggressive in policing card holders when their accounts are in arrears.

Payments to vendors can be done in a timely manner, to take advantage of discounts. Vendors provide incentives for timely payment in the form of trade credit. For example, a vendor invoice may offer a 2/10 net 30 discount. This contract terminology establishes that a discount of 2% can be taken if payment is received within 10 days. After 10 days, full payment is required and must be paid by the 30th day. A contractor may decide to pass on the 2% discount and wait until the 30th day to send payment. These discounts are discussed further in Chapter 8.

Similarly, contractors can arrange to *pay subcontractors when progress payment covering their work is received*. This pushes the payment lag down to the subcontractor level. Subcontractors are increasingly subject to these "pay when paid" terms, in which payment is delayed until the contractor receives payment from the owner. This can be difficult to implement since subcontractors may choose not to work for "pay when paid" contractors.

Another way to move the receipts flow curve to the left is to *unbalance the bid*. This is possible only with unit price contracts. On unit price contracts, a schedule of unit prices for the work is agreed to by the owner at bid time. This schedule becomes the basis for payment of work placed.[2]

Essentially, for items that occur early in the project, inflated unit prices are quoted at the time of bid submittal. *Closeout* items such as landscaping and paving will be quoted at lower than cost prices. This has the effect of moving reimbursement for the work forward in the project construction period. It unbalances the cost of the bid items, leading to what is called *front-end loading*. The increase in prices for the early items is balanced or offset by the lower than cost prices quoted for the closeout items on the project.

The preceding strategies for optimizing cash flow are not mutually exclusive, and they are often used in tandem. Moreover, there are drawbacks to most of them. For example, using external financing requires paying interest, which must be recovered in the bid price. Failure to correctly estimate the interest costs at bid time can lead to insufficient coverage of these costs. This directly affects the contractor's bottom line. Delaying payment to suppliers (within stipulated credit limits) results in forgoing discounts offered for early

---

[2] For more information, see section 4.6, Halpin, *Construction Management 3rd ed.*, 2006.

payment. An aggressive collection policy can have a negative effect on obtaining repeat business with the affected client. On the other hand, the contractor should avoid bidding future work with owners who do not pay in a timely manner. As noted, delaying payment to subcontractors may lead to the prime being "blackballed" by the better subcontractors. The prime contractor's reputation will suffer in such cases. Even reducing in-house costs requires an organizational effort that some contractors are not willing to undertake.

PROJECT CASH FLOW ESTIMATES

We have already developed a simple cash flow estimate, in which all costs were assumed to be paid immediately as incurred. While this assumption is helpful for understanding the conceptual framework of cash flow analysis, a more realistic procedure must incorporate the impact of contractor payment delays. Contractors may delay payment of part or all of their costs in order to be competitive. Therefore, a real-world cash flow estimate must consider these delays. Such an analysis requires the following information:

Cost incurred and contract value earned per time period. These two elements are critical and accuracy of the cash flow analysis depends on the method used to derive period (e.g. monthly) values for these two parameters.

Contractor credit conditions. The contractor's policy for payment of payroll, vendors, subcontractors and other creditors can greatly influence cash flow. It is common to use historical percentages to characterize this policy. For example, a contractor may figure that on average 40% of its costs are paid immediately after they are incurred, 50% are paid the following month, and the remaining 10% are paid two months later.

Table 6.2 shows an example of the process and computations required to study the impact of payment delays. The table simplifies some elements of the calculation and does not consider the interest costs associated with overdraft financing as the project proceeds.

The table indicates that the contract value (bid price) of the job under consideration is $3,000,000.[3] The projected cost of the job is $2,800,000. Retainage of 10% is held back throughout the project and is released to the contractor after the last progress payment is made. The table shows that the cash flow on this project occurs across a 7 month period (M1 to M7) and appropriate columns for each month are provided to show:

1. Monthly percent executed
2. Monthly earned value and payment received
3. Monthly costs and disbursements
4. Monthly net cash flow and cash position

[3] Numerical data given in Table 6.2 are shown in thousands of dollars.

TABLE 6.2 Comprehensive Cash Flow Analysis Example

1. General project information							
Contract value:	$3,000	(All dollar amounts are in thousands)					
Contract cost:	$2,800						
Retainage	10%	Returned the month following the last regular progress payment					
Credit terms (paying suppliers and subs)	60%	Paid immediately		40% Paid next month			
Payment lag (owner payments)	Two months (value earned in month 1 is available in month 3)						
2. Construction schedule and monthly percent executed							
Month	M1	M2	M3	M4	M5	M6	M7
Percent executed	15%	30%	35%	20%			
3. Monthly earned value and payments received							
Monthly earned value	450	900	1,050	600	—	—	—
Monthly payment received	—	—	405	810	945	540	300
4. Monthly costs and disbursements							
Monthly costs	(420)	(840)	(980)	(560)	—	—	—
Monthly Immediate expenditures	(252)	(504)	(588)	(336)	—	—	—
Monthly expenditures @ 1-month credit	—	(168)	(336)	(392)	(224)	—	—
Total monthly expenditures	(252)	(672)	(924)	(728)	(224)	—	—
5. Monthly net cash flow and cash position							
Net cash Flow	(252)	(672)	(519)	82	721	540	300
Cash Position (Overdraft)	(252)	(924)	(1,443)	(1,361)	(640)	(100)	200
6. Most negative net cash flow and maximum overdraft							
Most negative net cash flow	(672)						
Most negative cash position (Max overdraft)	(1,443)						

The actual work on the project occurs during a four month period (M1 to M4) and percent estimates for each of these months is shown in section 2 of the table.

The earned value during each month is calculated as $3,000,000 × (Monthly percent executed). In the first month, the earned value is $3,000,000 × 15 percent = $450,000. These monthly earned value amounts, as well as payments received are shown in section 3 of the table.

Client payments received lag the month in which work is performed by two months. Therefore, payment for month 1 is received in month 3. The actual payment received is reduced by the amount retained. That is:

$$\text{Payment}\,(i) = \text{Value Earned}\,(i) - ((0.10) \times (\text{Value Earned}\,(i))$$

The first payment received in M3 is, therefore:

$$\text{Payment}\,(1) = \$450{,}000 - (0.10 \times (\$450{,}000)) = \$405{,}000$$

Monthly payment continues to occur through month 6 (M6) and the retainage is released in month 7. The amount of the retainage is $300,000 × 0.10 = $300,000.

Section 4 of the table deals with the costs and disbursement made period by period from month 1 to month 4. These are calculated by multiplying the total cost, $2,800,000 by the monthly percent executed as shown for each month in section 1. For example, the costs and disbursements associated with work performed in month 1 are $2,800,000 × (0.15) = $420,000. The contractor credit policy is assumed to be - Pay 60% of the cost incurred immediately and pay 40 % (the remainder) in the following month. This means that $420,000 × (0.60) is paid immediately (i.e. in month 1) and $420,000 × (0.40) is paid in month 2.

This procedure is followed for costs and disbursements for months 2 through 4:

Month	Cost Incurred	Paid Immediately	Paid in Following Month
2	$840,000	$504,000	$336,000
3	$980,000	$588,000	$392,000
4	$560,000	$336,000	$224,000

The final disbursement of $224,000 occurs in month 5. Adding all disbursements results in a total of $2,800,000.

With this information in hand, it is possible to calculate the net cash flow for each month as shown in section 5 of the table. The net cash flow or "check book balance" for the first two months is the same as the amount of expenditures in these two months since there has been only outflow during these months and no payments received. Therefore, the net cash flows for

M1 and M2 are \$252,000 and \$672,000. The cash position is the cumulative value of net cash flows to date or \$252,000 + \$672,000 = \$924,000.

As we proceed to month 3, there is a cash inflow (payment) of \$402,000. The expenses during this period are \$924,000. These expenditures are offset by the \$405,000 yielding a net outflow for the month of \$519,000. This increases the negative value of the cash position (i.e. the overdraft) to \$1,443,000 (i.e. \$924,000 + \$519,000).

As a self study activity, the reader should verify the entries in this table and develop a plot of the overdraft profile similar to that shown in Figure 6.6.

Section 6 of the table notes that the maximum net cash flow occurs in the second month and that the *peak financial requirement* (maximum overdraft) occurs in month 3.

As might be expected, the negative cash position continues until month 7 at which time the retainage is released. This results in a positive cash position of \$200,000 which is also the profit (i.e. value earned minus cost) expected for the project. As was noted earlier in this chapter, the contractor must typically finance the project throughout its duration only achieving a positive return on the job upon its completion.

It should be emphasized that this analysis does not consider interest paid on the overdraft. These interest charges will further reduce the project profit so that the final take home amount for this job will be less than \$200,000. Even, if the overdrafts were to be financed out of company retained earnings, the overdraft financing is a de facto charge for the contractor. If internal company funds are involved, they could have been invested elsewhere in some interest bearing instrument. Under this scheme, the company funds committed to cover overdrafts are not available for investment (say, in a money market account). There is an opportunity cost involved in carrying the project overdraft. To cover this cost, most contractors include a charge for overdraft financing in their bid for the project.

USING SOFTWARE FOR CASH FLOW COMPUTATIONS

Computer scheduling software can automate much of the cash flow computations, but nearly none of the existing scheduling packages provides the complete set of capabilities required for a project-level cash flow estimate. At a relatively basic level, electronic spreadsheets such as Microsoft Excel are widely used for calculating the repetitive formulas discussed here. Most scheduling packages such as *Primavera Project Planner, Suretrak* (both by Primavera Systems, Inc.), and *MS Project* (by Microsoft Corporation) include the capability to cost-load project activities and can create a cost/month report (which is the same report used for the BCWS computations in Earned Value Management). Few programs provide for the simultaneous loading of costs and contract values, and even fewer allow the specification of credit terms for paying the various resources included in each activity. Software packages combining estimating and scheduling provide the most complete capabilities for estimating a project cash flow.

COMPANY-LEVEL CASH FLOW PLANNING

Construction companies have complex cash requirements since they have a relatively small number of ongoing projects, and as discussed in the previous sections, the cash requirements for each project vary widely, depending on factors such as overhead. Moreover, a company must plan for salary increases, the acquisition of fixed assets and similar strategic expenditures. Finally, any well-managed company must have cash reserves to provide for revenue downturns and unexpected events. Table 6.3 shows an example of a company-level cash needs forecast (values are shown in 100s).

STRATEGIC CASH FLOW MANAGEMENT: "CASH FARMING"

A hallmark of well-managed construction companies is their close attention to cash management. The top management of a large Midwest contractor, for example, has monthly meetings to plan their upcoming operations and purchases based on their forecast of cash availability. When a company is so

TABLE 6.3 Company-Level Monthly Cash Flow Estimate

Fudd Associates, Inc Cash Requirements Estimate			
Month **April 20XX**			
1. Active Projects			
Project	**Estimated payments**	**Estimated expenditures**	**Net cash flow**
109–02	21,900	24,336	(2,436)
112–11	79,500	55,733	23,767
212–01	45,333	500	44,833
215–01	25,850	—	25,850
222–00	35,045	42,750	(7,705)
335– 11	90,560	95,666	(5,106)
	Net operating cash flow from projects		**79,203**
2. General and Administrative Expenses			
Salaries			(45,355)
General office expenses			(10,775)
		Total G&A expenses	**(56,130)**
			—
		Total cash requirements	**23,073**

keenly aware of their cash movements, it can practice what has been called *cash farming*. Cash that is not immediately needed can be invested in money market funds or short term certificates of deposit (CDs) to generate income. The profitability of this approach is not trivial. An analysis of Australian contractors in the 1990s[4] found that their available cash was approximately 16.7% of their revenue. This means that a contractor generating one billion dollars of annual revenue would have $167 million to invest. At a 5% return, that is $8.3 million, made on top of their regular operating income.

PROJECT AND GENERAL OVERHEAD

As discussed in Chapter 5, three categories of cost are of interest to a contractor: direct costs, production support costs, and general and administration costs. Production support costs are directly related to on-site project support and are often referred to as project indirect costs. The home office charges are referred to as general or home office overhead. All of these costs must be recovered before income to the firm is generated. The home office overhead, or general and administrative (G&A) expense, can be treated as a period cost and charged separately from the project (direct costing). On the other hand, these costs may be prorated to the job and charged to the job cost overhead accounts and the work-in-progress expense ledger accounts (absorption costing).

Job-related indirect costs such as those listed in the labor cost report of Figure 6.9 (e.g., haul trash) are typically incurred as part of the on-site related cost associated with realizing the project. As such, they are charged to appropriate accounts within the job cost system. The level and amount of these costs should be projected during the estimating phase and included in the bid as individual estimate line items. Although it is recommended that job indirect costs be precisely defined during estimate development, many contractors prefer to handle these charges by adding a flat rate amount to cover them. Under this approach, the contractor calculates the direct costs (as defined in Chapter 5) and multiplies these charges by a percentage factor (a *markup*) to cover both project indirect cost items and home office fixed overhead. To illustrate, assume that the direct costs for a given project are determined to be $200,000. If the contractor applies a markup of 20% to cover field indirect costs and home office overhead, the required flat charge would be $40,000. If 10% is added for profit, his total bid amount would be $264,000.

The estimate summary shown in Figure 5.1 establishes line items for indirect charges and calculates them on an item-by-item basis (rather than applying a flat rate). Typical items of job-related indirect cost that should be estimated for recovery in the bid are those listed in Figure 5.4 as project overhead accounts (700–999). This is the recommended procedure when it is felt that

[4] Kenley, R (1999). Cash farming in building and construction: a stochastic analysis. *Construction Management and Economics*, 17:3, 393–401.

CENTURY CENTER BLDG #5 ATLANTA, GA				LABOR COST REPORT HALCON CONSTRUCTORS, INC. ATLANTA DIVISION					WEEK 57 PAGE 1	WEEK ENDING 10/11 PROJECT NUMBER 13-5265	
	Cost Code Information			Quantity		Unit	Price		Cost	Projected Cost	
Cost Code	Description	Units	% Comp	Estimated	Actual	Estimated	Actual	Estimated	Actual	To Date Over/Under	To Complete Over/Under
111	**This Week**		2		1		248.00		248	48	
	Haul Trash	Wk	68	50	34	200.00	92.76	10,000	3,154	3,646-	1,716-
112	**This Wcek**		1		1		543.00		543	326	
	Daily Clean	Wk	68	69	47	217.39	311.59	15,000	14,645	4,428	2,072
115									1,747	1,747	No Budg.
130	**This Week**		1		1		13.00		13	204-	
	Safety	Wk	81	69	56	217.39	206.30	15,000	11,553	621-	144-
131	Protect Trees	Ls						500	31		
132	Shoring	Ls	100	18	18	1,388.88	1,345.38	25,000	24,217	783-	Comp.
307	**This Week**		1		5		19.60		98	4	
	Hand Exc	Cy	97	725	705	18.79	19.56	13,625	13,796	547	15
310	Dewater	Ls	100					2,000	2,060	60	Comp.
312	Bkfl Hand	Cy	83	6,000	5,000	1.50	1.29	9,000	6,455	1,047-	209-
316	Fine Gr	Sf	99	15,000	14,830	.16	.13	2,500	1,999	501-	Comp.

Figure 6.9 Labor cost report (some typical line items).

sufficient information is available to the contractor at the time of bid to allow relatively precise definition of these job-related indirect costs. The R. S. Means method of developing overhead and profit represents a percentage rate approach that incorporates a charge into the estimate to cover overhead on a line item-by-item basis. This is essentially a variation of the flat rate application.

FIXED OVERHEAD

Whereas the project indirect charges are unique to the job and should be estimated on a job-by-job basis, home office overhead is a more or less fixed expense that maintains a constant level not directly tied to individual projects. In this case, the application of a percentage rate to prorate or allocate home office expense to each project is accepted practice, since it is not reasonable to try to estimate the precise allocation of home office charges to a given project. Rather, a percentage prorate or allocation factor is used to incorporate support of home office charges into the bid.

The calculation of this home office overhead allocation factor is based on:

1. The general and administrative (G&A) (home office) expenses incurred in the past year.
2. The estimated sales (contract) volume for the coming year, and
3. The estimated gross margin (i.e., markup) for the coming year.

This procedure is illustrated in the following example (Adrian, 1978).

Step 1: Estimate of Annual Overhead (G&A Expense)
Last year's G&A $270,000
10% inflation 27,000
Firm growth 23,000
Estimated G&A $320,000
Step 2: Estimate of $ of Cost Basis for Allocation
Estimated volume $4,000,000
Gross margin 20% = $800,000
Labor and material $3,200,000
Step 3: Calculate Overhead Percent

$$\frac{\text{Overhead Est.}}{\text{Labor and Materials}} = \frac{\$320{,}000}{\$3{,}200{,}000} = 10\%$$

Step 4: Cost to Apply to a Specific Project
Estimated labor and material costs $500,000
Overhead to apply (@ 10%) 50,000
Total $550,000

In the example, the anticipated volume for the coming year is $4,000,000. The G&A expense for home office operation in the previous year was $270,000. This value is adjusted for inflation effects and expected expansion of home office operations. The assumption is that the overhead allocation factor will be applied to the direct labor and materials costs. These direct costs are calculated by factoring out the 20% gross margin. Gross margin, in this case, refers to the amount of overhead and profit anticipated.

Direct costs amount to $3,200,000. The $320,000 in G&A costs to be recovered indicate a 10% prorate to be applied against the $3,200,000 of direct costs. This means that an overhead amount of $50,000 would be added to a contract bid based on $500,000 of direct cost to provide for G&A cost recovery. The profit would be added to the $550,000 base recovery amount.

CONSIDERATIONS IN ESTABLISHING FIXED OVERHEAD

In considering costs from a business point of view, it is common to categorize them either as variable costs or fixed costs. Variable costs are costs directly associated with the production process. In construction, they are the direct costs for labor, machines, and materials as well as the field indirect costs (i.e., production support costs). These costs are considered variable since they vary as a function of the volume of work underway. Fixed costs are incurred at a more or less constant rate independent of the volume of work-in-progress. In order to be in business, a certain minimum of staff in the home office, space for home office operations, telephones, supplies, and the like must be maintained, and costs for these items are incurred. These central administrative costs are generally constant over a given range of sales/construction volume. If volume expands drastically, home office support may have to be expanded also. For purposes of analysis, however, these costs are considered fixed or constant over the year. Fixed costs are essentially the general and administrative costs referred to earlier.

The level of G&A (fixed) costs can be estimated by referring to the actual costs incurred during the previous year's operation. The method of projecting fixed overhead as a percentage of the estimated total direct costs projected for the coming year is widely used. Since the fixed overhead incurred in the previous year is typically available as a percentage of the previous year's total sales volume, a simple conversion must be made to reflect it as a percentage of the total direct cost. The formula for this conversion is:

$$P_C = \frac{P_S}{(100 - P_S)}$$

where

P_C = percentage applied to the project's total direct cost for the coming year, and

$P_S =$ percentage of total volume in the reference year incurred as fixed or G&A expense.

If, for instance, $800,000 is incurred as home office G&A expense in a reference year in which the total volume billed was $4,000,000, the P value would be 20% ($800,000/$4,000,000 x 100). The calculated percentage to be added to direct costs estimates for the coming year to cover G&A fixed overhead would be:

$$P_C = \frac{20}{100 - 20} = 25\%$$

If the direct cost estimate (e.g., labor, materials, equipment and field indirect costs) for a job is $1,000,000, $250,000 would be added to cover fixed overhead. Profit would be added to the total of field direct and indirect costs plus fixed overhead. The field (variable) costs plus the fixed overhead (G&A) charge plus profit yield the bid price. In this example, if profit is included at 10%, the total bid would be $1,375.000. It is obvious that coverage of the field overhead is dependent on generating enough billings to offset both fixed and variable costs.

Certain companies prefer to include a charge for fixed overhead that is more responsive to the source of overhead support. The assumption here is that home office support for management of certain resources is greater or smaller, and this effect should be included in charging for overhead. For instance, the cost of preparing payroll and support for labor in the field may be considerably higher than the support needed in administering materials procurement and subcontracts. Therefore, a 25% rate for fixed overhead is applied to labor and equipment direct cost, while a 15% rate on materials and subcontract costs is used. If differing fixed overhead rates are used on various subcomponents of the field (variable) costs in the bid, the fixed overhead charge will reflect the mix of resources used. This is shown in Table 6.3 in which a fixed rate of 20% on the total direct costs for three jobs is compared to the use of a 25% rate on labor and equipment and a 15% rate on materials and subcontracts.

It can be seen in Table 6.4 that the fixed overhead amounts using the 25/15% approach are smaller on jobs 101 and 102 than the flat 20% rate. This reflects the fact that the amount of labor and equipment direct cost on these projects is smaller than the materials and subcontract costs. The assumption is that support requirements on labor and equipment will also be proportionately smaller. On job 102, for instance, it appears that most of the job is subcontracted with only $200,000 of labor and equipment in house. Therefore, the support costs for labor and equipment will be minimal, and the bulk of the support cost will relate to management of materials procurement and subcontract administration. This leads to a significant difference in fixed overhead charge when the 20% flat rate is used, as opposed to the 25/15% modified rates (i.e., $440,000 versus $350,000).

TABLE 6.4 Comparison of Fixed Overhead Rate Structures

			20% on Direct Costs	25% on Labor & Equipment; 15% on Materials and Subs
	Labor and Equipment	$800,000	160,000	200,000
Job 101	Materials and subcontracts	1,200,000	240,000	180,000
			400,000	380,000
	Labor and Equipment	200,000	40,000	50,000
Job 102	Materials and subcontracts	2,000,000	400,000	300,000
			440,000	350,000
	Labor and Equipment	700,000	140,000	175,000
Job 103	Materials and subcontracts	700,000	140,000	105,000
			280,000	280,000

On job 103, the fixed overhead charge is the same with either of the rate structures, since the amount of labor and equipment cost is the same as the amount of the materials and subcontract cost.

It should be obvious that in tight bidding situations use of the stylized rate system, which attempts to better link overhead costs to the types of support required, might give the bidder an edge in reducing the bid. Of course, in the example given (i.e., the 25/15% rate versus 20%), the 20% flat rate would yield a lower overall charge for fixed overhead on labor- and equipment-intensive jobs. The main point is that the charge for fixed overhead should be reflective of the support required. Because the multiple rate structure tends to reflect this better, some firms now arrive at fixed overhead charges by using this approach rather than the flat rate applied to total direct cost.

BREAKEVEN ANALYSIS

Both variable and fixed costs (as described in the previous section) must be recovered by billing to the client. A certain percentage of each billed dollar must cover the variable costs incurred, while another portion of the billed dollar must cover the fixed overhead costs (general and administrative expense). This is shown conceptually in Figure 6.10. If billings exceed a certain amount (a certain level of volume is achieved), profit will be generated by the billings. However, up to a certain level of billings, all of the dollars billed are essentially covering the combined fixed and variable costs. The point at which billed volume is equal to the fixed and variable cost incurred to generate it is called the *breakeven point*.

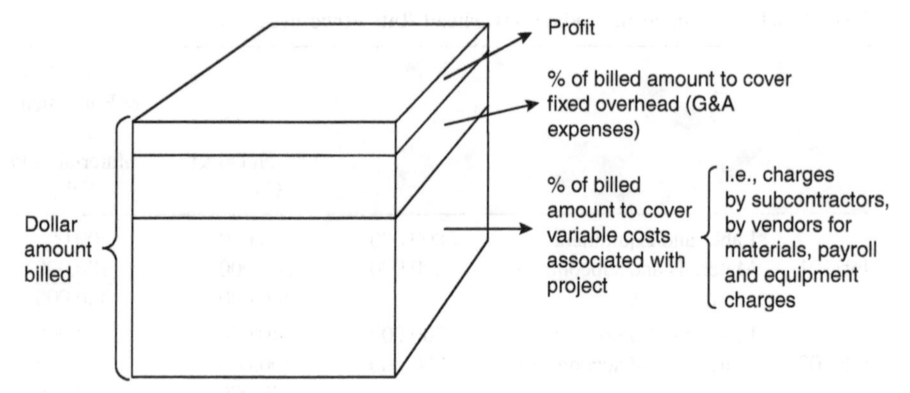

Figure 6.10 Distribution of Billed Dollars to Cover Costs Incurred.

To better understand this concept, consider the portion of each dollar billed that is to be applied to fixed overhead. As described in the previous sections, the amount of fixed overhead (G&A) expense is normally projected from the previous year's operations. Assume that the amount of fixed overhead corrected for inflation and expanded operations for the present year is \$400,000. Assume further that the percentage of each billed dollar required to cover operations costs is 80%. This means that 80 cents of each billed dollar is expended on variable costs. Twenty cents is left to cover fixed overhead. How many dollars must be billed to cover the \$400,000 of projected fixed overhead? Based on 20 cents per dollar for fixed overhead, the minimum amount of billings required to cover the projected G&A expense would be \$400,000 divided by 0.20, or \$2 million. That is, until \$2 million has been billed to clients, the fixed overhead has not been recovered. If more than \$2 million is billed over the year, profit will be generated at the rate of 20 cents on the dollar for every dollar in excess of \$2 million. On the other hand, if the volume of billings falls below \$2 million, the fixed overhead will not be fully covered, and a loss will occur. A graphical representation of this situation is given in Figure 6.11. It is this point, at which billings just cover fixed and variable costs, that is called the *breakeven point*. The chart shown is referred to as a breakeven chart. The slope of the variable cost line is \$0.80/billed dollars. The variable cost line starts on the Y-axis at the level of projected fixed cost (i.e., projected G&A costs for the year). The broken line shows the billings amount profile. As the amount billed increases, the variable costs also increase. At \$2 million billed, the "billings amount" line intersects the total cost line (fixed + variable), indicating that billings equal fixed and variable costs. The chart is based on a few simple equations. At the breakeven point, billings volume (*BV*) must equal fixed costs (*FC*) plus variable costs (*VC*).

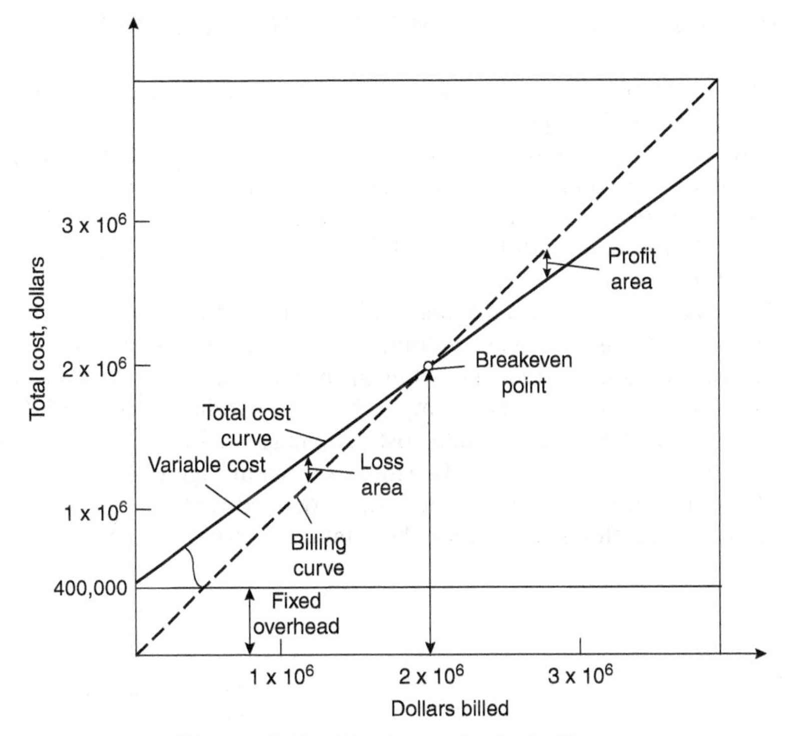

Figure 6.11 Breakeven Analysis Chart.

The BV can be found as follows:

$$BV = FC + VC$$

If variable costs (VC) are X percent of the billed dollar (BV), then:

$$VC = (X/100)(BV)$$

$$\text{Replacing } VC \text{ in (1), } BV = FC + (X/100)(BV)$$

$$\text{or } BV - (X/100)(BV) = FC$$

Solving for BV, we obtain:

$$BV = (FC)/(1 - (X/100))$$

In the example given, the breakeven point is calculated as:

$$BV = (\$400{,}000)/(1 - (80/100)) = \$400{,}000/0.20 = \$2{,}000{,}000$$

BASIC RELATIONSHIPS GOVERNING THE BREAKEVEN POINT

Several relationships are clear from a consideration of the breakeven chart in Figure 6.11 and the supporting equations.

Lowering the fixed overhead amount will move the breakeven point to the left (reduce the volume require) if the variable cost percentage is constant. If the fixed overhead amount in the example is reduced to \$300,000, the new breakeven point will be \$300,000/0.20 = \$1,500,000. This is shown in Figure 6.12.

If the percentage of each billed dollar required to cover variable costs is reduced, the breakeven point volume is reduced. This is obvious, since the total cost curve slope is reduced, while the billings curve slope is constant (i.e., always 45 degrees). If in the example the fixed overhead cost is \$400,000 as originally stated, but the variable cost percentage is 75%, the new breakeven point volume will be \$1,600,000. This is shown in Figure 6.13.

Below the breakeven point, the loss associated with a given construction volume is the difference between the total cost curve and the billings curve.

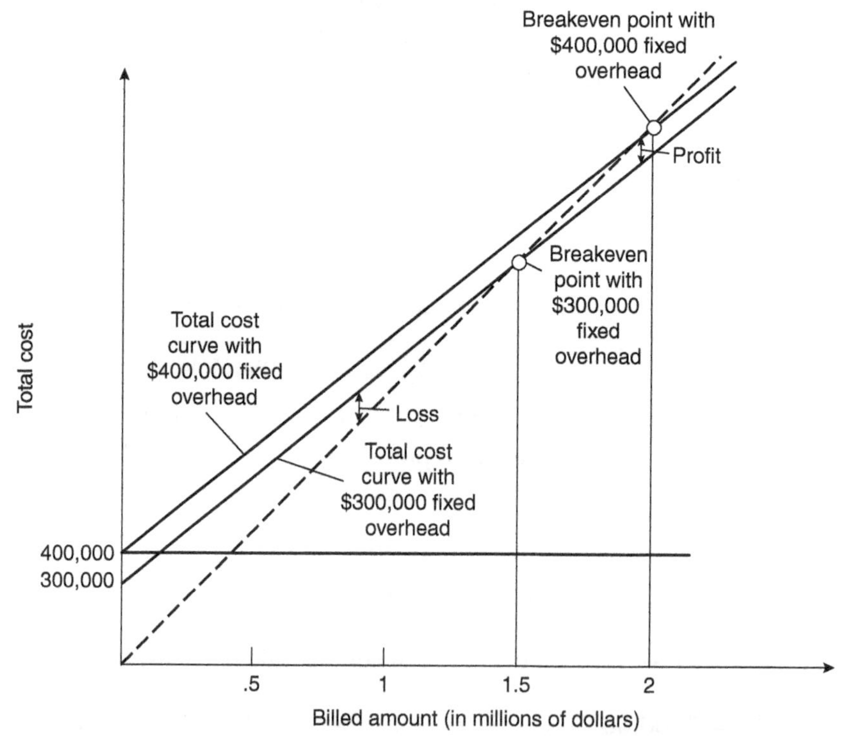

Figure 6.12 Variation in breakeven point based on reducing fixed overhead.

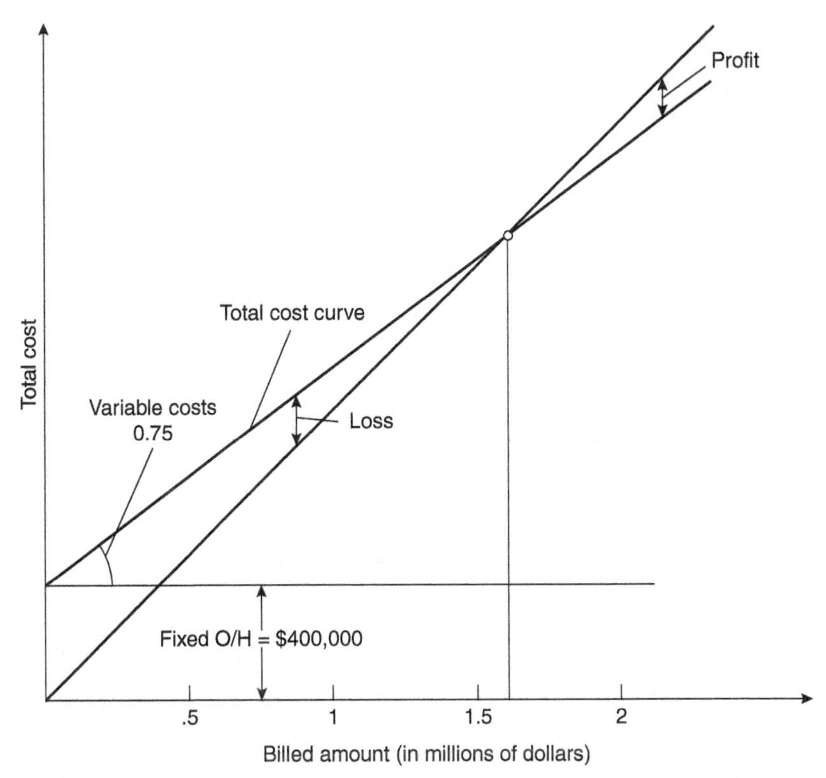

Figure 6.13 Variation in breakeven point based on reducing variable cost.

Above the breakeven point, this difference represents the profit associated with a given volume. These relationships are illustrated in Figure 6.11.

REVIEW QUESTIONS AND EXERCISES

1. Given the following cost expenditures (see Table P6.1) for a small warehouse project (to include direct and indirect changes):
   A. Calculate the peak financial requirement.
   B. Sketch a diagram of the overdraft profile.
   C. What would be the impact on the maximum overdraft of giving the contractor a $25,000 mobilization payment? (Show calculation.)

   Assumptions:
   - The markup for this project is 12%.
   - Retainage is 10% throughout project.
   - Finance charge is 1.5% per month.

- Payments are billed at the end of the month and are received one month later.
- All expenses are paid immediately.

TABLE P6.1 Monthly Cost

Month	1	2	3	4
Monthly cost*	69,000	21,800	17,800	40,900

* Direct cost + indirect cost

2. The contract between Ajax Construction Co. and Mr. Jones specifies that the contractor will bill Mr. Jones at the end of each month for the amount of work finished that month. Mr. Jones will then pay Ajax a specified percentage of the bill the same day. The accumulated retainage is to be paid one month after project completion. The latest cumulative billing was $5 million, of which Ajax has actually received $4.5 million. The project is to be finished two months from now. Ajax estimates the bill for the remaining two months will be $100,000 and $50,000. Mr. Jones, being short of cash at present, proposed the following alternative: Rather than follow the contract and make the three payments required, he will make one final payment (for the two months' work plus the retainage) five months from now. Mr. Jones will also pay a 4% monthly interest rate because of the delay in payment.

   Find what would be the total final payment according to the actual contract and the new final payment according to the new proposal. Should the contractor accept the new proposal? Why?
3. Can you think of additional strategies to bring closer the payments and disbursements curves (as discussed for Figure 6.8)?
4. Direct costs of a project are shown on the following bar chart (see Table P6.2). Assume $500 per month indirect cost and 10% retainage with payment time lag of one month.

   Calculate the project's net cash flow for a 10-percent markup. Estimate the proper markup for 1%, 1.5%, and 2% rate of return per month, using methods described in Chapter 7.
5. Given the following cost expenditures (see Table P6.3) for a small bridge job to include direct and indirect charges (but not bank interest):

   A. Calculate the peak financial requirement.

   B. Sketch a diagram of the overdraft profile.

   Assumptions:

   - The markup for this project is 8%.
   - Finance charge is 1% per month.

TABLE P6.2 Project Bar Chart

Activity	Month				
	1	2	3	4	5
1. Survey	$2,000				
2. Mobilization		$1,200			
3. Trench		$3,000	$6,000		
4. Lay pipe			$3,000		
5. Concrete				$500	$500
6. Backfill					$500
7. Move out					$500
8. Prepare valves				$1,000	
9. Install valves				$1,000	
10. Test					$1,500

- Retainage is 6% throughout project.
- Payments are billed at the end of the month, and are received one month later.

TABLE P6.3 Monthly Costs for Bridge Job

Month	1	2	3
Monthly cost*	29,000	48,900	16,400

* Direct cost + indirect cost

6. A contractor is preparing to bid for a project. He has his cost estimate and the work schedule. Table P6.4 gives his expected expenses and their time of occurrence. Other expenses such as insurance, bonds, and payroll taxes are included. For simplicity of analysis, he assumed that all expenses are recognized at the end of the month in which they occur.

The contractor is planning to add 10% to his estimated expenses to cover profits and office expenses. The total will be his bid price. He is also planning to submit for his progress payment at the end of each month. Upon approval the owner will subtract 5% for retainage and pay the contractor one month later. The accumulated retainage will be paid to the contractor with the last payment (i.e., at the end of month 13). What is the peak financial requirement and when does it occur?

TABLE P6.4 Table of Expenses

Month	*Mobilization / Demobilization*	*Subcontractors*	*Materials*	*Payroll*	*Equipment*	*Field Overhead*
0	$40,000	$0	$0	$0	$0	$0
1	0	10,000	10,000	10,000	20,000	1,000
2	0	30,000	20,000	15,000	10,000	5,000
3	0	30,000	30,000	20,000	20,000	6,000
4	0	40,000	30,000	20,000	30,000	6,000
5	0	50,000	40,000	40,000	20,000	6,000
6	0	50,000	40,000	40,000	15,000	6,000
7	0	40,000	30,000	40,000	10,000	6,000
8	0	40,000	10,000	20,000	10,000	6,000
9	0	70,000	10,000	10,000	10,000	6,000
10	0	30,000	5,000	5,000	10,000	6,000
11	0	30,000	5,000	5,000	5,000	6,000
12	20,000	50,000	0	5,000	5,000	5,000
Total	$60,000	$470,000	$230,000	$230,000	$165,000	$65,000

Total cost = 60,000 + 470,000 + 230,000 + 230,000 + 165,000 + 65,000 = $1,220,000
Profit plus overhead = (10%) (total cost) = $122,000
Bid price = total cost + profit + overhead = $1,342,000

7. Based on the definition of fixed cost, list costs in the main office that are not really fixed, and field office costs that are fixed.
8. Categorize the following costs as (a) direct, (b) project indirect, or (c) fixed overhead.
   - Labor
   - Materials
   - Main office rental
   - Tools and minor equipment
   - Field office
   - Performance bond
   - Sales tax
   - Main office utilities
   - Salaries of managers, clerical personnel, and estimators
9. The following data are available on Del Fabbro International, Inc. What bid price should be submitted to insure proper coverage of fixed overhead?
   - The fixed (home office) overhead for the past year was $365,200.
   - Total volume was $5,400,000.

- Del Fabbro uses a profit markup of 10%.
- The estimating department has indicated that the direct and project (field) indirect costs will be $800,000.
- Assume a 5% inflation factor and a 12% growth factor.

10. Using the data of the previous problem, calculate the breakeven point for Del Fabbro International in the present year.

CHAPTER 7

TIME VALUE OF MONEY AND EVALUATING INVESTMENTS

INTRODUCTION

The topics that we have covered so far do not consider the fact that the use of money costs money. If you use your own money you forgo any interest you might achieve by investing it. If you need to borrow it, you must pay for the use of the money. The time required to make or spend money has not been addressed so far in our discussion of financial management. In fact, the cost of money and the time during which money is tied up in any business decision process are crucial factors. To illustrate this point, consider the following cases.

1. Assume that you are buying a new car. The purchase price is $26,050, and you plan to pay this amount over a 60-month time period. One option involves a $1,000 cash discount and a loan at the rate of 6.94%. A second option is to finance the car at with a 5.9% interest loan, but no discount? Which option is better? *(Answer: the two are the same—you'd pay around $500 per month taking either one).*
2. What will be the balance of your mutual funds account when you retire at 65 if you are 25 now and begin depositing $100 per month right away? Assume a 12% annual interest. *(Answer: Over a million dollars).*
3. What would be the balance of the same account, if you begin saving when you are 35? *(Answer: $350,000).*
4. What is the down payment for a house worth $250,000, if you can get a 30-year fixed interest rate of 7.50% and want to cap the payments to $1,500 per month? *(Answer: Exactly $35,473.56).*

Similar decisions are frequently faced by any business enterprise. Should my company continue leasing this building or should it build a new one? Which piece of equipment from among these choices should I choose? Is this 3-year project with a profit of $1,000,000 better than this other that can be finished in a year with a $250,000 profit?

Engineering Economy considers the time and cost of borrowing as they relate to these and many other questions involving money as a construction resource. In this chapter, we will discuss the time value of money. The management of expensive, long-term construction projects definitely requires a thorough understanding of the principles and techniques discussed here. Money management on complex long term projects is too important to be dealt with using "horseback questimates." Engineering Economy is increasingly useful even for small, personal decisions like those described above. Calculators, software, and thousands of websites provide support in dealing with the topics covered here. Construction managers must have a very good working understanding of the concepts discussed in this chapter.

TIME VALUE OF MONEY

If someone asked you to decide between receiving $1,000 right now or receiving it 1 year from now, wouldn't you choose to get the money right away? You could invest the money, pay back a debt, or simply keep it until there is a good reason to spend it.

If you were lending $1,000 for 1 year to a perfect stranger, wouldn't you expect to receive more than $1,000 back? After all, you could have invested this money in a Certificate of Deposit or in your own business. The borrower would have had the potential to invest this money, while you'd have been deprived from this possibility. In short, it is reasonable that anyone lending money asks for "rent," and anyone borrowing money must expect to pay "rent" for its use. This rationale is at the center of the concepts presented in this chapter.

INTEREST

Interest is the fee that a lender charges for the use of their money. From the borrower's viewpoint, it is the "rent" paid for the use of someone else's money. Normally, it is directly proportional to the amount of money loaned or borrowed. That is, a $2,000 loan almost always earns twice the interest amount of a $1,000 loan. Interest is also proportional to the time that money has been loaned or borrowed. A three-year loan earns more interest than a single year loan for the same amount.

The money that one dollar earns for each unit of time that it is loaned is called its interest rate. This rate is expressed as a percentage: If $20.00 earns $2.40 over a year, then the interest rate in this case is $2.40 / $20.00 = 0.12, or 12% per year. It is so common to deal with interest rates that often they are referred simply as "interest," with the "rate" being implied by the context of the situation. To avoid confusion, the money earned by charging interest is frequently called *total* interest. It should be easy to notice which of these terms is being used in a specific context. If the interest refers to a percentage, then the complete term should be *interest rate*.

The interest rate established for any business transaction is of paramount importance. Let's suppose that you take a $200,000 mortgage loan to purchase a house. If you agree to pay back the loan over 30 years at a 15% interest[1], your monthly payment toward repaying the loan will be $2,528.89. If the interest is 7%, your payment would be $1,330.60, approximately half the previous amount. While the total interest paid over the 30 years would be $710,399.70 at 15%, it would be $279,017.80 at 7%. In both cases, you end up paying more for interest than the original borrowed amount. But, why would anyone agree to pay 15% instead of 7%? The degree of investment risk, the inflation rate, and the stability of the market where the money will be used are important factors determining the interest rate of a particular loan. For instance, lenders charge higher interest rates to borrowers with a higher risk of defaulting on the loan. Other factors can also influence rates, such as the level of competition among lenders and the cost of the money to them (since often lenders, in turn, borrow their money).

SIMPLE AND COMPOUND INTEREST

There are two fundamentally different ways to compute interest. Let us suppose that you need $10,000, which you will pay back at the end of 3 years, plus all the interest accumulated over the 3 years. If you have to pay a 10% interest rate, how much money will you have to pay back? The original borrowed amount, or loan *principal*, is $10,000. The total interest, as previously mentioned, depends on the principal, the time that the money was borrowed, and the interest rate. Moreover, it depends on the type of interest: whether it is simple or compound, as discussed here.

Simple Interest. You ask Uncle Fudd to lend you the $10,000. He agrees to allow you to pay simple interest, as a concession to one of his favorite family members. Paying simple interest means that you pay interest on the principal but *not* on the interest accumulated from previous periods. Therefore, your debt at the end of each year is computed as shown in Table 7.1.

[1] In this context, it is assumed that 15% means 15% per year.

TABLE 7.1 Simple Interest Calculation

Year n	Base for applying i	Interest for period	Total owed, F
0 (now)			10,000
1	10,000	$10{,}000 \times 0.10 = 1{,}000$	11,000
2	10,000	$10{,}000 \times 0.10 = 1{,}000$	12,000
3	10,000	$10{,}000 \times 0.10 = 1{,}000$	13,000

For simple interest, the future value F_n of a principal P borrowed over n periods at an interest rate i can be computed as $F_n = P + P * i * n$, or more directly,

$$F_n = P \times (1 + i \times n)$$

In this case, we can compute $F_3 = 10{,}000 \times (1 + 0.10 \times 3) = 10{,}300$ without developing the table above.

Compound Interest. You decide to use a bank to borrow the \$10,000 instead of asking your uncle. Banks never use simple interest to compute your debt. Instead, the interest owed after each time period is computed by adding the total interest accumulated from previous periods to the loan principal, and then multiplying this total by the interest rate. The computations in Table 7.2 help to illustrate how this process works.

Using the same notation as for simple interest above, you can see that:

$$F_n = P \times (1 + i)^n$$

Using this formula, we can find F_3 directly:

$$\mathrm{F}_3 = 10{,}000 \times (1 + 0.10)^3 = 13{,}310$$

The formula above also works in the opposite direction. One can find the principal that would result in a future value, F, after n periods at an interest rate of i using the formula:

$$P = F_n/(1 + i)^n$$

TABLE 7.2 Compound Interest Calculation

Year n	Base for applying i	Interest for period	Total owed, F
0 (now)			10,000
1	10,000	$10{,}000 \times 0.10 = 1{,}000$	11,000
2	11,000	$11{,}000 \times 0.10 = 1{,}100$	12,100
3	12,100	$12{,}100 \times 0.10 = 1{,}210$	13,310

The P found here is usually called the *Present Value*, P, of the future amount, F.

Compound interest is more logical and equitable than simple interest. Why would the interest accumulated to date be exempt from consideration, as in simple interest? In fact, as you can appreciate in the mortgage loan example of the previous section, many times the earned interest is higher than a loan's principal. Simple interest is never used for business transactions.

NOMINAL AND EFFECTIVE RATE

A lender could advertise in big letters a 1% interest rate and then add in very small letters that it refers to the monthly rate, not the annual rate as many people would assume. Technically, the lender would be telling the truth: an interest rate can refer to any time unit, from seconds to decades, but many people would be deceived by such advertising strategy. To protect consumers, Congress passed in 1968 the Truth in Lending Act, which, among other issues, requires that lenders clearly indicate interest rates as an annual percentage of each received dollar. This *Annual Percentage Rate (APR)* also includes upfront commissions and fees converted to equivalent annual percentages.

Requiring that lenders disclose their APR was an important step for consumer protection. However, an APR can still be misleading. The APR simply adds the interest rates charged in 1 year, instead of compounding these intermediate periods. If your credit card charges 1% per month, then its APR is (1%/month) × (12 months/year) = 12%. This is the *nominal interest rate* of your card, but it is not its *effective interest rate*. If you have \$10,000 in credit card charges, the APR suggests that you will owe $\$10{,}000 \times (1.12) = \$11{,}200$ a year from now (assuming that you make no payments during the year). However, since the card company computes financial charges every month, you would owe $\$10{,}000 \times (1 + 0.01)^{12} = \$11{,}268.25$ at the end of the year, using the compound formula for F that we examined in the previous section. The \$68.25 difference is substantial, and perfectly legal for the credit card company to charge. Almost all credit cards charge interest by the day, which means that the total annual interest that you'd owe in the above example almost certainly would be $\$10{,}000 \times (1 + {}^{0.12}/_{365})^{365} = \$11{,}274.75$.

It is possible to find the effective annual rate that you pay on a loan (usually called its Annual Percentage Yield, or APY) if you know its APR and the number of times per year that the interest is compounded. The following formula finds a loan's APY when its APR and the number of times per year that the APR is compounded (the C in the formula below) are known.

$$APY = (1 + {}^{APR}/_{C})^{C} - 1$$

If the interest is compounded monthly, then $C = 12$, the number of months in a year.

In the previous example, APR = 12% compounded monthly. Its APY is:

$$APY = (1 + {}^{0.12}\!/_{12})^{12} - 1 = 0.126825 \text{ or } 12.6825\%$$

An APY is not too different from its APR compounded C times, but a seemingly small difference can be significant. If you take $100,000 to repay after 15 years at $i = 12.0000\%$ and at12.6825%, the future amounts owed are:

$$\text{For } i = 12.0000\% : F = \$100{,}000 \times (1 + 0.120000)^{15} = \$547{,}356.57$$
$$\text{For } i = 12.6825\% : F = \$100{,}000 \times (1 + 0.126825)^{15} = \$599{,}580.20$$

The difference of $52,223.63 is due exclusively to the difference between APR and APY.

EQUIVALENCE AND MARR

If you have an offer to receive $100 now or $110 a year from now, which option would you choose? Assume that you have enough resources to think about this offer in terms of investing the money, as opposed to spending it on some "spur of the moment" purchase. Even with enough money to subsist, many people will choose the $100 now, for the reasons previously discussed in the section on the time value of money. But, what happens if instead of $110, you are offered $500 a year from now? What if the future amount offered is $400? Or $300? Or $150? Deciding which offer to take gets more tricky. It is quite possible that you'd choose the $500 a year from now, but some of the lower amounts would probably be attractive enough to prefer them over the $100 offered now in this hypothetical example.

Let us take this mental exercise to the point that you are offered an amount that is just high enough to make you indifferent to having it now or in the future. The two amounts are said to be *equivalent* to you. The "you" in the previous sentence is important. Someone else could have a different idea of what is equivalent to $100 now. You would certainly require a higher amount two years from now than what you would get a year from now to declare equivalence in preference.

Equivalence is usually considered in terms of interest rates instead of total interest. If your equivalent amount for $100 is $150 received a year from now, then the total interest would be $150 − $100 = $50. The interest rate would be $50 / $100 = 50%. This particular interest rate is called your *Minimum Attractive Rate of Return*, or *MARR*. In an actual business environment, MARR is not found as arbitrarily as shown here. If you are in heavy construction and in your experience a reasonable Return on Equity is 20%, then you could not aspire to a 50% MARR. If you are paying 10% interest on your borrowed money, then your MARR cannot be less than 10%. If your MARR is less than 10%, you would lose money when borrowing at 10%—you would pay more borrowing costs than you would make on the return from your MARR. If your Uncle Fudd makes 25% on his money and is willing to let you invest in his company, then you could establish this rate as your MARR. If Uncle Fudd can consistently offer you 25%, why should you accept less?

Computing MARR is a sensitive and sophisticated issue for any company, and we will not cover this topic in detail here. However, keep in mind that when we use an interest rate i in any problem here, usually this rate is the MARR for at least one of the stakeholders in the problem.

DISCOUNT RATE

What we have called so far *interest rate* is sometimes called *discount rate*. When we discussed the concept of interest, we approached it from the perspective of someone beginning with an amount P (at time present) and ending in the future with an amount F. When we know the value in the future of a given sum and want to know what its value in today's currency would be, one speaks of discounting the future amount, F, to the present at some "i" value. We want to find P given F. P is equivalent to the known value F when F is discounted using the formula $P = F_n / (1 + i)^n$. This formula is the inverse of $F_n = P \times (1 + i)^n$.

The equivalence of P and F can be extended to include several Fs. Let us suppose that we are offered \$55.00 a year from now and \$60.50 2 years from now. If our MARR is 10%, we can find that the first future amount is equivalent to a present amount of $\$55.00/(1.10)^1 = \$50$, and the second amount is equivalent to $\$60.50/(1.10)^2 = \$50$ now. We will be willing to accept $\$50 + \$50 = \$100$ now, instead of the two amounts in the future.

IMPORTANCE OF EQUIVALENCE

Equivalence is a very handy way moving values such as P and F from point to point in time. For instance, we can use this principle to determine payment or payment required to purchase merchandise on credit. If a seller offers a 10% credit, we can purchase a \$100 music player and arrange to pay for it in two payments, one of \$55.00 a year from now and another of \$60.50 2 years from now, since these two payments are equivalent to \$100 today, as you saw previously. The seller will be willing to let us take home the merchandise with our written promise to pay back these two amounts above the specified points in time. (More likely, we would prefer to pay two equal payments of \$57.62—such equal amounts are called *annuities*, as we will discuss later)

The concept of equivalence is central to understanding the rest of this chapter, so let us work through a final mental exercise about this concept. The music you listen in your earphones could sound exactly as loud *to you* as the systems in your car or living room, although the earphones are, in decibels, much quieter than the car or room systems. In our money equivalence, the dollar amount now is the decibel volume delivered right in your ears; the distance between you and the speakers of the other systems is the time between receiving different dollar amounts. And equivalence is equivalence. You can be as happy quietly listening to a good pair of earphones as to a good set of speakers. You can be equally satisfied receiving one dollar now as opposed to $1.10 a year from now or $1.21 2 years from now, provided that your MARR is 10%.

INFLATION

The cost of goods and services tend to increase over time. This phenomenon is called *inflation* and is subject to much consideration by macroeconomists. The purchasing power of a dollar is progressively reduced by inflation, and, therefore, a year from now we may need $105.00 to purchase the merchandise that we can buy today for $100.00. In 2 years, we would have to pay $110.25 for the same article (more than $110.00, because the inflation for the second year is 105% of $105, not of $100). In general, an inflation rate of "f" increases the cost of goods over n periods in accordance with the formula $F = P \times (1 + f)^n$. You can see that this formula is similar to our familiar F and P relationship when applying a compound interest rate i over n periods.

Inflation undercuts the practical effect of charging interest. An investor charging a 5% interest on $100 for a year should be able to purchase an additional $5 worth of goods at the end of the year. If inflation runs at an annual 5%, this investor would not be able to purchase any additional goods, since a $100 today would cost $105 in a year. Inflation would have offset all the real gains from the interest collected.

The effective interest rate "i'," that is, what an investor is actually making after considering an inflation rate of f is found as follows:

$$i' = [(1 + i)/(1 + f)] - 1$$

If $i = f$, as in the previous example, the effective interest rate is zero. Applying the formula above, a nominal interest of 9% in an economy with 3% inflation would yield an effective interest rate of 5.825%. This result is close to the difference between the nominal interest rate and the inflation rate: 9% – 3% = 6%. They are not the same—and worse, this incorrect computation underestimates the effect of inflation. An investor would believe that money will yield 6.000% when in fact it is yielding less, 5.825%.

Inflation is not considered in many engineering economy analyses, because all costs, including those charged by the investor, can be passed through to the purchaser or client. That is, as prices and costs raise, the selling price is adjusted to pass the inflation increase on to the purchaser. As an example, consider the economic analysis of a new toll highway. The analysis must address a period of 15 or 20 years, and estimating the inflation for such an extended period would be futile. Tolls, however, can be expected to rise with inflation, and their price in 20 years is immaterial as long as the analysis clarifies that figures are in today's "constant dollars." The toll will be adjusted (increased) to pass inflation for operation and other costs through to the user. Some financial arrangements, however, are fixed and cannot be adjusted over time for inflation. A typical example is the case of a 30-year *fixed interest* home mortgage loan.

SUNK COSTS

We have discussed situations in which all factors involved in the business decision occur in the present or the future. Engineering economy is not concerned with past costs or lost benefits. Any cost in the past is called a *sunk cost*, and as the name implies, it should be irrelevant to the decision at hand.

Let us illustrate the concept of sunk costs with the following scenario. You purchased a $2,000 brand new laptop computer 3 months ago. However, a new model with the same features is available now for $1,000. Cousin Elmer offers you $1,200 for the old computer. Should you sell the old computer and purchase the new model?

The cost of the original laptop is irrelevant to the decision of keeping or selling it. Many people have difficulty in selling a possession at a loss compared to the price they paid for it. In this case, keeping the computer would avoid the sense of having made a bad decision by purchasing it in the first place. But, that is faulty logic. The key point to consider should be whether you are better off by your decision. What is relevant is that selling the computer would give you an extra $200, and keeping it would prevent you from realizing this profit.

There could be other factors not mentioned previously that could sway your decision toward keeping the computer. Maybe the laptop was a gift of your dear Uncle Dewey, and you don't want to offend him by selling it. Or, not having the hassle of setting up the new computer is worth more than the extra $200. These are true considerations, not easily quantifiable. But in summary, using engineering economy principles, you should sell your current laptop and buy the new model given the scenario of this example.

CASH FLOW DIAGRAMS

So far, we have used only words to describe our examples. In most real situations, this is a cumbersome approach at best, especially when compared to using the power and simplicity of graphics to convey information. *Cash flow diagrams* are a graphical means used to describe engineering economy

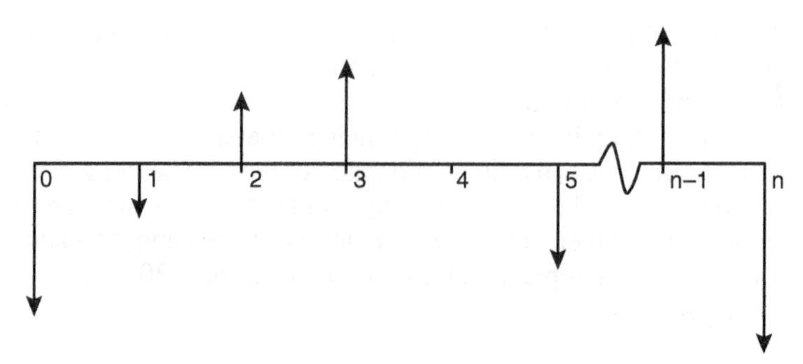

Figure 7.1 Structure of a cash flow diagram.

applications. Developing the appropriate cash flow diagram is often the critical step in solving an engineering economy analysis. After the proper diagram is developed, the numerical solution of the problem is straightforward.

A cash flow diagram has the structure shown in Figure 7.1. Specifically, it consists of the following:

1. A timeline consisting of a horizontal line, which begins at present time, usually represented as time zero. The end of each future period is drawn as a tick mark on the line. The timeline normally ends at period n, that is, the last time period in the analysis. Since n can be relatively large, the line is frequently broken as shown in Figure 7.1.
2. Arrows placed on the tick marks show each sum of money. An inflow of money is shown as an arrow pointing up at the point the money or value of an asset was earned; similarly, the arrow for a money outflow is drawn pointing down. One person's inflow is another person's outflow (e.g., lender/borrower; seller/buyer), and therefore it is important to keep track of which side of a transaction applies to the situation at hand. If a person buys a car and pays for it in two payments of $15,000, then the value of the car appears as an arrow pointing up (it is a new asset for the person), and the two payments are shown as arrows pointing down. From the seller's perspective, the arrows would point exactly opposite to the buyer's.
3. Comments and additional data are written near the main timeline and arrows. The interest used is usually included in this way.

It is tempting to skip the diagram and jump into the numerical solution just by reading the problem. This is a relatively safe way to approach a simple problem, including many of the examples included in this chapter. However, real-life situations tend to be complicated, with multiple amounts of money flowing in and out of the scenario. A cash flow diagram is indispensable in these real-life analyses, since it is quite easy to forget one of the amounts that must be included. A cash flow diagram is the easiest way of making sure that the whole cash flow sequence is in view and each individual amount is included in the computations.

ANNUITIES

An *annuity* is a series of equal payments (or receipts) paid out or received in a sequence over a period of time. Our examples so far have consisted of single amounts of money in the present or the future. We have discussed situations such as a \$100 article purchased at a 20% interest, which can be paid as \$120 a year from now or \$144 in two years. But one-time payments are the exception, not the norm in a real business environment. Many common business transactions consist of purchasing an article or service by paying a number of equal amounts of money ("*installments*") over a number of time periods. For example, a person rarely purchases a \$30,000 car on credit by promising to pay 36 months from now a single payment of \$40,446, based on an APR of 10% compounded monthly. Much more frequently, the arrangement will be that the person pays \$968 per month for 36 months, assuming the same APR and compounding frequency.

You know by now that the preceding \$40,446 can be found with the formula $F_n = P \times (1 + i)^n$ $(F = 30{,}000 \times (1.008333)^{36} = 40{,}446)$[2] But, how do we know that the 36 payments of \$968 are equivalent to \$30,000 in the present with the lending terms presented above? We could check that this is the case by finding the Present Value, P, of each payment, which we will call A_n, and then adding all these individual A_ns. Since the APR is 10% compounded monthly, the effective monthly interest is $i = 10\%/12 = 0.83333\%$ or 0.008333.

$$P_1 = \$968/(1 + 0.008333)^1 = \$960.00$$
$$P_2 = \$968/(1 + 0.008333)^2 = \$952.06$$
$$P_3 = \$968/(1 + 0.008333)^3 = \$944.19$$
$$P_4 = \$968/(1 + 0.008333)^4 = \$936.40$$
$$[\ldots]$$
$$P_{36} = \$968/(1 + 0.008333)^{36} = \$718.00$$

The total present value P is:

$$P_{total} = P_1 + P_2 + P_3 + P_4 + [\ldots] + P_{36}$$
$$P_{total} = \$30{,}000$$

This present value P is *exactly* the amount of merchandise that can be purchased by n installments A at an interest i. In this case, the above computations check that 36 installments of 968 and an APR of 10% compounded monthly can purchase *exactly* \$30,000 of merchandise.[2]

[2] $F = 30{,}000(1 + 0.1/12)^{36} = 30{,}000 \times (1.008333)^{36} = 30{,}000\ (1.34820) = 40446$

It is important to remember when using "i" in a formula, it has decimal value—for example, 10% is 0.1 and 0.1/12 (to covert APR to a monthly interest rate) is 0.008333.

Finding the Present Value of a series of equal payments this way is totally impractical. Instead of a number of individual F to P conversions, we can find the Present Value of a series of equal payments (for a given interest rate and a specified number of payment periods) using a simple, single formula. To develop this formula, we need to consider once more the example above. You can see that what we did can be expressed as the following mathematical series:

$$P = A/(1+i)^1 + A/(1+i)^2 + A/(1+i)^3 + [\ldots] + A/(1+i)^n$$

You may remember from algebra that a series like this can be converted into a single equivalent formula. In this case, the resulting formula is:

$$P = A \times [(1+i)^n - 1]/[i \times (1+i)^n]$$

For the preceding example,

$$P_{total} = 968 \times [(1+0.008333)^{36} - 1]/[0.008333 \times (1+0.008333)^{36}]$$

$$P_{total} = 968 \times 30.99142 = \$30{,}000$$

The preceding formula is a bit more complicated than the basic P from F conversion, but it is much more straightforward, especially when a high number of Fs need to be converted.

***Finding* A *given* P.** The inverse of the above formula can be used to find A given P, i, and n. The most common business situation is to know the price of a merchandise and then to compute the monthly (or more generically, the "periodic") amount to be paid for the merchandise. For $i = 0.008333$ and $n = 36$, the preceding *A to P* formula will always result in the value 30.99142. The *P to A* computation consists of using the inverse of this factor. Therefore, to find the monthly payment of a $30,000 car paid with 36 equal payments over 36 months at an APR of 10% compounded monthly, the following computation applies:

$$\text{A} = \$30{,}000/30.99142 = \$968$$

More generally, the formula to convert P to A can be written as follows, which is simply the inverse of the formula for A to P:

$$A = P \times [i \times (1+i)^n]/[(1+i)^n - 1]$$

CONDITIONS FOR ANNUITY CALCULATIONS

Let us recapitulate the conditions that a series of payments or receipts must fulfill to be considered an annuity.

1. All payments must be equal. Figure 7.2a is an annuity, but Figure 7.2b is not one, because not all payments are equal.
2. Payments must take place uninterruptedly throughout a number of periods and the periods must be of the same length or duration—for example, week, month, and so forth. It is not acceptable to have one period of 2 weeks followed by a period of 2 days, and so on. Figure 7.2c is not an annuity, since one payment is missing. This causes the payment periods to vary in duration.
3. Payments normally begin at the end of the first period. Figure 7.2d begins after the first period, and therefore, cannot be directly considered an annuity.
4. All payments must carry the same interest.

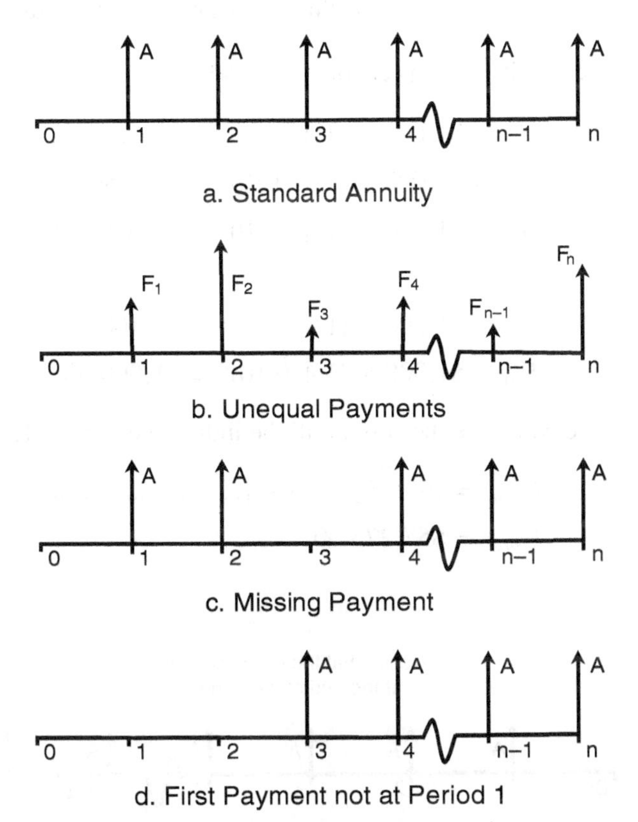

Figure 7.2 Standard and defective annuity schemes.

Annuities do not necessarily refer to annual payments. Although the semantics are contradictory (think of *annual*), any series of payments fulfilling the above conditions is an annuity regardless of its time unit. An alternative expression for an annuity sequence is *uniform payment series*, which although more technically correct, is not quite as easy to use in everyday language.

CALCULATING THE FUTURE VALUE OF A SERIES OF PAYMENTS

A similar mathematical process can be used to find the future value of an annuity. Consider the following situation. Assume that a person saves $1,000 per year over 20 years at 10% annual interest. What would be the balance of this savings account at the end of the 20 years?

In this case, we are interested in the amount of the account at the end of the n periods instead of at present. The first $1000, deposited at the end of year 1, would have accumulated interest over 19 years. The last one, at the end of the 20th year, would have no interest gain (in this example, depositing $1,000 at the end of the 20 years to have it paid back right away does not make much sense, but in many other situations this is reasonable). This situation is shown in Figure 7.3.

The balance of the savings account would be:

$$F_1 = \$1{,}000 \times (1 + 0.10)^{19} = \$6{,}115.91$$
$$F_2 = \$1{,}000 \times (1 + 0.10)^{18} = \$5{,}559.92$$
$$F_3 = \$1{,}000 \times (1 + 0.10)^{17} = \$5{,}054.47$$
$$[\ldots]$$
$$F_{19} = \$1{,}000 \times (1 + 0.10)^{1} = \$1{,}100.00$$
$$F_{20} = \$1{,}000 \times (1 + 0.10)^{0} = \$1{,}000.00$$

The balance would be the sum of all the individual deposit future values:

$$F_{total} = F_1 + F_2 + F_3 + [\ldots] + F_{19} + F_{20}$$
$$F_{total} = \$57{,}275.00$$

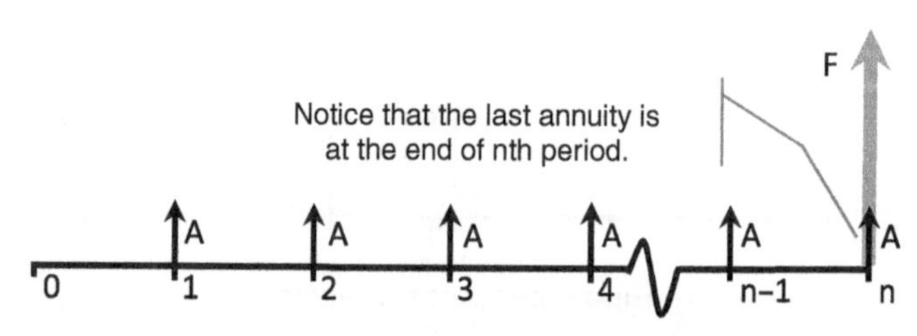

Figure 7.3 Future value of an annuity.

Following a line of reasoning parallel to the one discussed for the present value of an annuity, we find that a single formula can take care of the 21 individual computations above. The formula for the future value of an annuity (e.g., to convert A to F) is:

$$F = A \times \{[(1 + i)^n - 1]/i\}$$

We would find the same result for F_{total} by directly applying the above formula:

$$F = \$1{,}000 \times \{[(1 + 0.10)^{20} - 1]/0.10\} = \$1{,}000 \times 57.2749995$$

$$F = \$57{,}275.00$$

As with present value, we can use the inverse of the above formula to compute an annuity given a goal future value, that is to convert F to A. To find the annual payment that would result in a balance of \$50,000 after 20 years at 10%,

$$A = \$50{,}000/57.2749995 = \$872.98$$

Annuity calculations address many practical situations such as:

1. Given an assumed MARR, how much should you save each month to accumulate \$400,000 over 20 years? This is a sinking fund type of calculation. Assuming an MARR of 10% APR, the answer is \$526.75.
2. What is the amount of the monthly mortgage payment you should expect to pay for a 15-year mortgage in the amount of \$100,000 at a fixed borrowing rate of 6.0%? The answer is \$843.85. In a typical mortgage agreement, taxes and insurance would be added to this interest amount. If taxes are 2,500 per year and insurance is \$1,000 annually, your mortgage payment would be around \$1135.52 per month.

Real estate and financial professionals are involved with calculations similar to these examples on a daily basis.

SUMMARY OF EQUIVALENCE FORMULAS

Table 7.3 gives a summary of the formulae and factors we have discussed linking P to F, F to P, A to F, F to A, A to P and P to A. The factors relating the principal variables (P, F, and A) have the standard "full" names shown. Moreover, they are frequently referred to by abbreviations shown in parentheses below their full names. The abbreviations are expressed as "fractions,"

TABLE 7.3 Interest Factors

Type of cash flow	Single Payment		Uniform Series (Annuities)			
Name	Compound Amount Factor	Present Worth Factor	Compound Amount Factor	Sinking Fund Factor	Present Worth Factor	Capital Recovery Factor
Description	Find F given P (F/P)	Find P given F (P/F)	Find F given A (F/A)	Find A given F (A/F)	Find P given A (P/A)	Find A given P (A/P)
Formula	$(1+i)^n$	$\frac{1}{(1+i)^n}$	$\frac{[(1+i)^n-1]}{i}$	$\frac{i}{[(1+i)^n-1]}$	$\frac{[(1+i)^n-1]}{i^*(1+i)^n}$	$\frac{i^*(1+i)^n}{[(1+i)^n-1]}$

and are borrowed from nomenclature used in statistics. An abbreviation such as *F*/*P* is read "Find F given *P*." *P*/*A* is read "Find P given *A*."

These six conversion factors along, with the concept of equivalence, are the building blocks that make complex engineering economy calculations possible. They are sufficient for most financial situations involving time value of money. Given an interest rate (e.g., a MARR), we can find the amount of a future payment *F* based on a *P* and interest accruing at the rate, *i* over some number, *n*, of time periods (in shorthand, we can convert *P* to *F*). Conversely, given *F*, *i*, and *n*, we can calculate *P*. We have discussed that annuities (*A*) are so common in business that formulas have been developed to allow the conversion of these sets of payments to an equivalent present value (that is, converting *A* to *P*) or to a point in the future (converting *A* to *F*). We also found that the formulas for these conversions work in either direction. That is, the formula converting *P* to *A* is the inverse of the formula converting *A* to *P*.

The formulas we have discussed consist of an amount of money times an expression involving *i* and *n*. For example, $F_n = P \times (1 + i)^n$. The expression $(1 + i)^n$ will always result in the same number for a given *i* and *n* (e.g., for $i = 10\%$ and $n = 2$, the result is always $(1 + 0.10)^2 = 1.21$). If we make a table listing the result of this expression for many combinations of *i* and *n*, we can simply look up its value for a given *i* and *n*. Since the formula relating *F* and *P* is relatively simple, the advantage of using tables is not so obvious. However, in cases where more complicated situations must be analyzed, such as the one for taking *A* to *P* or combinations of calculations are required in the same problem (e.g., *F* to *P* plus *A* to *P*, and the result converted to *F*), the advantage of using tables is more apparent. Tables for the various interest factors shown in Table 7.3 are given in Appendix C.

WORTH ANALYSIS TECHNIQUES: AN OVERVIEW

Financial options can be compared by calculating the Present Worth analysis of each option and determining which is most attractive. Consider the following situation, which requires comparison of options. Cousin Thaddeus offers to pay you $3,000 every year for the next 10 years if you give him $20,000 now. Let us suppose that your Minimum Attractive Rate of Return is 10%, and that a local bank is offering this rate. How can you determine whether Cousin Thaddeus' offer is better than putting the money in the bank?

There are three techniques which can be used to compare investments involving time and money. This offer would be attractive:

1. If the present value of a $3,000 annuity over 10 years at your MARR is greater than the $20,000 that Thaddeus will borrow from you, or
2. If the $3,000 per year is greater than the annuity resulting from investing $20,000 over 10 years at your MARR, or
3. If the interest that yields 10 payments of $3,000 from an initial investment of $20,000 is greater than your MARR.

The three options above should result in the same decision to accept or reject your cousin's offer, since, after all, they are variants of Present Worth analysis.

1. *Present Worth Analysis (PW)* consists of adding up the present value (or present worth, PW) of all the amounts involved in the investment. If this sum, which is the net present worth of the entire investment, is less than $20,000, then you would better off if you placed the $20,000 in a bank for 10 years at your MARR.
   In this case:

$$\begin{aligned} PW &= -20{,}000 + 3{,}000\ (\text{P/A},\ 10\%, 10) \\ &= -20{,}000 + 3{,}000 \times 6.1446 = -1{,}566.30 \end{aligned}$$

   Since the present worth of the entire investment is negative (i.e. less than $20,000), we should reject Thaddeus's proposition.
2. We can determine the *Equivalent Annual Worth Analysis (EAW)* of Thaddeus' proposal . All the money is converted into equivalent annuities and these annuities are added up. If the resulting net annuity is a negative number, you would be better off if you placed your money in a bank for 10 years at your MARR and withdrew $3,000 every year.
   For our example,

$$\begin{aligned} \text{EAW} &= -20{,}000\ (\text{A/P},\ 10\%,\ 10) + 3{,}000 \\ &= -20{,}000 \times 0.1627 + 3{,}000 = -254.91 \end{aligned}$$

Consistent with the previous finding, and since the annual equivalent is negative (i.e. less than \$3,000), we conclude that the proposed arrangement is not attractive. We can deposit the \$20,000 at the bank and receive \$254.91 more than Thaddeus is paying.

3. The method of determining of the effective rate of interest given n, and two of the values P, F, or A (i.e., P and F, or P and A, or F and A) is called *Internal Rate of Return*, frequently abbreviated as IRR.

Based on Uncle Thaddeus' proposal, you are offered a 10-year annuity of \$3,000 per year based on a payment to him of \$20,000. Using this technique, you can use an interpolation approach to determine which rate of interest you will be receiving. If this interest rate is less than your MARR, then, by definition, the investment opportunity yields less than you require. In such case, you should consider placing your money with an investment firm or bank offering an interest rate that meets or exceeds your MARR.

To find the investment IRR, we use either of the equations above, with two differences: We begin with the assumption that *PW* or *EAW* is zero, and second, the interest rate we are calculating is the yet unknown *IRR*. In other words, we could use *any* of the following equations:

$$PW = 0 = -20{,}000 + 3{,}000\ (P/A,\ IRR,\ 10)$$

$$EAW = 0 = -20{,}000\ (A/P,\ IRR,\ 10) + 3{,}000$$

$$\text{If we use } PW:\ 20{,}000 = 3{,}000(P/A,\ IRR,\ 10)$$

$$(P/A,\ IRR,\ 10) = 20{,}000/3{,}000 = 6.6667$$

Knowing that (*P*/*A*, *IRR*, 10 yr) is 6.6667, we can proceed by looking up in a compound interest table (see Appendix C) the two values, between which we find 6.6667. We can then interpolate to find our unknown value of "*i*.": Consulting the tables, we attempt to bracket the (*P*/*A*, *IRR*, 10 yr) value of 6.6667. The value appears to be near 10%, so we could select the Table values for 11, or 12% (assuming a positive return), or Tables for 10% or lower. Since we know there was a negative return based on FW and EAW, we would expect the IRR to be below 10%. Looking at 10% and lower, we see that 6.667 falls between the Table value for (*P*/*A*, 9%, 10) and (*P*/*A*, 8%, 10). In other words, our IRR will be between 8 and 9%.

$$(P/A,\ 10\%,\ 10) = 6.1446$$

$$(P/A,\ 9\%,\ 10) = 6.4177$$

$$\text{Our value} = 6.667$$

$$(P/A,\ 8\%,\ 10) = 6.7101$$

Interpolating between 9% and 8%, the IRR in this case is 8.15%. That is,

$$IRR = 8\% + ((6.7101 - 6.667)/(6.7101 - 6.4177)\% = 8.15\%$$

Since this interest rate is lower than our MARR of 10%, we conclude that the offer made by Thaddeus is not attractive for us.

The following sections discuss the use of PW, EAW, and IRR in greater detail.

PRESENT WORTH ANALYSIS

There are many situations in which it is important to know whether we are paying too much upfront for something, or how much should we pay for something. In such cases, PW is a great way to proceed, as we will see in the following examples. This technique does, however, have its limitations. First, we need to estimate our MARR, which is a relative quantity, and can change depending on many circumstances. Second, this is not the simplest approach when we need to choose between alternatives with different life span: A 10-year project with present worth of $100,000 is obviously less attractive than a 1-year project whose present worth is $90,000. We will discuss the case of comparing investments of different duration later. Finally, the technique may be difficult to explain to a layperson.

Example: Car Down Payment

How much is the down payment for a $30,000 car, if you want to limit your payments to $450 per month for 60 months? You will pay a 12.00% APR, compounded monthly (therefore, $i = 1\%$)

The car dealer doesn't care whether the car is paid for using a $30,000 check right away or a down payment and a check from the bank financing your loan. The bank will pay the dealer exactly the present worth of your 60 installments of $450 each compounded at 12%. If you can figure this present worth, then you can find how much you must pay up front to complete the $30,000. In other words:

$$\text{(Down payment)} + \text{(PW of installments)} = \text{Value of your purchase}$$

$$\text{(Down payment)} + 450 \times (P/A,\ 1\%,\ 60) = 30{,}000$$

$$\text{Down payment} = 30{,}000 - (450 \times 44.9550) = 9{,}770.23$$

This example is very useful. It applies to any similar situation involving a down payment and installment payments.

Example: Small Excavator

Fudd and Nephews, Inc. purchased a small excavator for $100,000 2 years ago. The excavator was supposed to generate a net benefit of $30,000 per year, but it has generated only $5,000 per year. It is expected to generate the same amount of net benefit of $5,000 for the next 4 years, with a negligible market value after these 4 years. Another contractor offers to purchase the machine for $15,000 now. Should Fudd sell the machine?

In this case, we have to decide the better of two undesirable options. Regardless of whether or not Fudd sells the machine, it will end up losing money. Since the machine has not lived up to expectations, the first impulse for many people would be to sell the machine and move on. But, the original plans are water under the bridge. The chief consideration is whether or not the $15,000 offered for the excavator is more than the Present Worth of the annuity of $5,000 over 4 years. This Present Worth, of course, is not $5,000 × 4 = $20,000, since the money is received over a period of time. Using the formula for the Present Worth of an annuity that we previously discussed, we find that:

$$\text{PW at a MARR of } 13.0\% = \$14,872$$

$$\text{PW at a MARR of } 12.6\% = \$15,000$$

$$\text{PW at a MARR of } 12.0\% = \$15,187$$

If Fudd's MARR is less than 12.6%, then the best deal is to keep the excavator, since the Present Worth of the money stream that it will generate over its remaining economic life is higher than the $15,000 offered for it now.

In this case, as before, there could be other factors that are not readily visible that may influence the decision. If Uncle Elmer's ulcer acts up each time he receives a repair bill from another breakdown of this excavator, it may be better to get rid of it. Notice, however, that the whole rationale considers only what will happen from now on.

INVESTMENTS WITH DIFFERENT LIFE SPANS

If an investment opportunity A results in a benefit of $1,000 per year for each of the next 15 years, its Present Worth at a MARR of 15% ($5,847) will be higher than another investment B generating a benefit of $1,500 for the next 5 years ($5,710). However, B can be intuitively seen to be more profitable than the first. Fifteen hundred dollars per year is better than $1,000/year. One could reinvest the money at the end of the 5 years and keep making $1,500/year.

The Present Worth method in the above case leads to the wrong choice because of the difference in the time span of the two alternatives. The evaluation must assume that we are evaluating each scenario over comparable periods of time. To compare these two alternatives, the second one, B, must be assumed to operate across the same time span for A of 15 years To accomplish this apples-to-apples comparison, it is necessary to assume that B will be repeated two more times, as shown in Figure 7.4. The Present Worth of

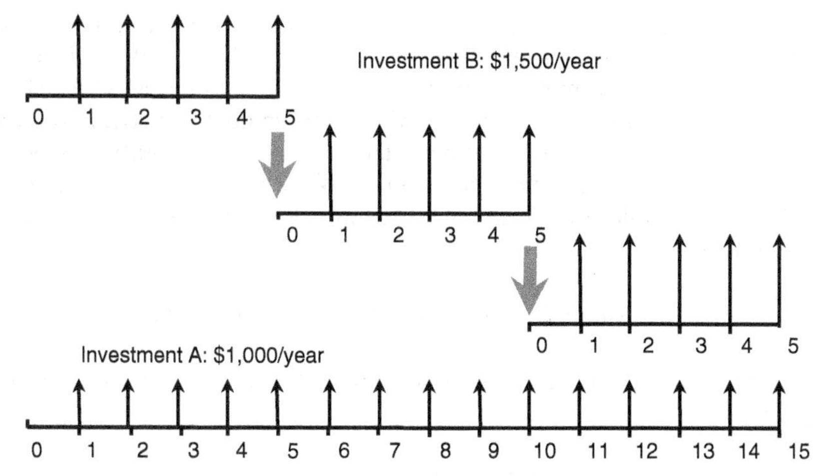

Figure 7.4 Common time frame for two investments with different life spans.

three consecutive investments B is $8,771, more than the Present Value of A. The intuitive preference for B is now reflected in the numerical comparison.

While it may be possible to invest in three consecutive Bs, sometimes it is impossible to repeat the business opportunity. If a third alternative, *C*, becomes available at the end of the fifth year, and it yields a better return than *B*, it would be illogical to reinvest in *B* instead of *C*.

This approach can be extended to any situation comparing alternatives with different life spans. To compare alternative *X* spanning 3 years with alternative *Y* having a duration of 4, we must find the least common multiple for the three, which is 12 years in this case. We then have a new scenario, in which alternative *X* is repeated 4 times, and alternative *Y* is repeated three times.

A variant of the strategy for dealing with unequal investment time spans is to shorten the scope of the longer investment to the same length of the shorter one. This requires an estimation of the future value that the investment would bring if it were sold at the end of the shorter time span. This approach is often used when comparing investments with reasonably predictable future values, such as heavy construction equipment.

EQUIVALENT ANNUAL WORTH (EAW)

The *Equivalent Annual Worth (EAW)* of an investment is its value expressed as a uniform dollar amount over a time period. More concisely, it is the equivalent annuity of an investment. When the cost of an investment is the main issue under consideration, then the term Equivalent Annual Cost (EAC) instead of EAW is frequently used. Both terms refer to the same concept.

If the PW of an investment's cost at, say 8%, is $15M and the PW of its revenue is $16M, then the investment is desirable. It will result in more income than the minimum required by its investors at their MARR: $16M – $15M = $1M. If you "spread out" $1M Present Value dollars over 10 years at an interest rate of 8% (8.14%, to be exact), the resulting annuity is $150,000. Any PW value can be expressed as an EAW by just converting the present worth of the whole investment into an equivalent annuity. The formula for this conversion is:

$$EAW = P \times ([i \times (i + 1)^n] / [(i + 1)^n - 1])$$

Using factors from a compound interest factor table (CRF column in App. C).

$$EAW = P \times (A/P, \ i, \ n)$$

$(A/P, i, n)$ is called the *capital recovery* factor.

A future value F can also be converted into equivalent annuities, using a similar procedure:

$$EAW = F \times ([i]/[(i + 1)^n - 1])$$

Or, using factors from a compound interest factor table,

$$EAW = F \times (A/F, \ i, \ n)$$

$(A/F, i, n)$ is called the *sinking fund* factor.

A given annuity does not need to be converted at all. It is already in the desired format.

The EAW of an investment involving initial costs, annuities and future values is the sum of the EAW of each component.

Example: Best Job Offer

Greg is undecided between two job offers. Fudd Associates, Inc. offers a $10,000 sign-up bonus and $50,000/year, which will be revised after 3 years. ABC Contractors, Inc. offers an annual salary of $54,000, which will also be revised after 3 years. Greg has made 12% on the money invested in the stock market and believes that he can keep getting the same return if he invests the entire $10,000 bonus. Both companies offer comparable benefits, and Greg likes both of them. Based on the salary offer, which offer is better?

Solution: The time frame for the comparison is three years, and the interest is 12%. The question is whether the extra $4,000/year is more attractive than the initial $10,000 bonus. The EAW for the $10,000 is:

$$EAW = 10{,}000 \times (A/P, 12\%, 3) = \$10{,}000 \times 0.41635 = \$4{,}163.50$$

Based exclusively on salary considerations, Fudd Associates, Inc.'s offer is slightly better ($163.50/yr) than ABC Contractors, Inc.'s offer. Since this is such a small difference, other considerations will play a role (e.g., office location, job assignments, etc.), given that the two alternatives are practically the same.

INTERNAL RATE OF RETURN

The Internal Rate of Return (IRR) of an investment is the interest rate at which the investment has a Present Worth of zero. The PW of any investment depends on the interest used for the computations. More specifically, the higher the interest, the lower the investment's PW. You can recall that an investment with a positive PW is desirable and one with a negative PW is not. An investment with a PW of zero is exactly breaking even at the interest rate used in the PW calculations. (When an investment's PW is zero, its EAW is also zero. The IRR is usually defined as the interest resulting in PW = 0, but it is also perfectly correct to define it as the interest at which the investment's EAW = 0)

The concept of IRR is best understood in the context of an example. Suppose that an investment consists of the following:

Initial cost: $100,000

Net benefits: $27,057/year

Investment span: 5 years

The PW of this investment can be expressed as:

$$PW = -100{,}000 + 27{,}057 \times (P/A, i, 5)$$

If we look this *P/A* for several interest rates, we can develop the following table:

$$(P/A, 12\%, 5) = 3.60478 \quad \text{PW } 12\%$$

$$0 = -\$100{,}000 + \$27{,}057 \times (3.79079) = -\$2,465$$

$$(P/A, 11\%, \ 5) = 3.69590 \quad \text{PW } 11\%$$

$$0 = -\$100{,}000 + \$27{,}057 \times (3.69590) = \$0$$

$$(P/A, 10\%, 5) = 3.79079 \quad \text{PW}10\%$$

$$0 = -\$100{,}000 + \$27{,}057 \times (3.79079) = \$2{,}567$$

The preceding results indicate that if we want to get a return of 12% on our money, this investment is not desirable, while if we want a 10% return, the project is generating money above and beyond the desired return. Eleven percent is the IRR for this project because it is the interest rate that leaves exactly no shortfall or surplus money.

The weak point of the IRR is that it is much more difficult to calculate than the PW or EAW of an investment. The critical effort for computing these two latter methods is in understanding the problem and creating a correct cash flow diagram. Once the diagram is made, computations are quite mechanical, consisting of bringing to time zero (in the case of the PW method) or spreading over the investment time frame (for the EAW) each arrow in the diagram. The IRR, in contrast, has the extra requirement that the PW (and therefore the EAW) of the investment must be zero.

The IRR of very simple investment scenarios can be computed by manipulating the Time Value of Money formulas that we have discussed. For example:

Cousin Wolfgang wants to borrow \$1,000 and pay you back \$1,500 five years from now. What is the IRR of this investment?

Since $F = P \times (1 + i)^n$, we can solve for i:

$$\$1{,}500 = \$1{,}000 \times (1 + i)^5$$

$$i = (\$1{,}500/\$1{,}000)^{1/5} - 1 = 8.45\%$$

Unfortunately, most cases involve the use of factors for $(P/A, i, n)$ and $(P/F, i, n)$. The mathematical formulas for each of these factors involve i multiple times. It is practically impossible to find a neat solution for i as in the preceding example for even one of these complex formulas, let alone a combination of them. The only practical way to find the IRR of most investments is by trial and error, as in the following example.

Assume that a contractor has purchased a piece of heavy equipment for \$2,000,000. The equipment has a service life of 5 years and a salvage (resale) value at the end of 5 years of \$400,000. The operating cost is \$70/hour. The owner has decided to rent the machine and it is assumed that the machine will be rented at \$350/hour for 2000 hours each year. This will yield revenues in the amount of \$350/hour x 2000 hours/year = \$700,000 per year. Operating cost is \$70/ hour x 2000 hours/year or \$140,000/year. To determine the IRR, PW is set to zero.

$$\text{PW} = 0 = (\text{Initial investment}) - (\text{PW (Annual Revenue - Annual Cost)}) - (\text{PW (Salvage Value)})$$

$$\text{Therefore } 0 = \$2{,}000{,}000 - \$560{,}000\ (\text{P/A, IRR, 5}) - \$400{,}000\ (\text{P/F, IRR, 5})$$

Let use evaluate this expression for I = 15%, 16%, and 17%. (This range is selected as a rough guess as to the IRR, we could have selected the range I = 10%, I = 15%, and I = 20%)

To simplify the calculations, we will work in thousands of dollars so than \$2,000,000 is taken as \$2,000.

$$I = 15\% \quad 0 = \$2{,}000 - \$560\ (3.35216) - \$400\ (0.49718)$$
$$= \$2{,}000 - \$1876 - \$198.9$$
$$= -\$75{,}8 \text{ This is not zero, so the equation is not satisfied.}$$
$$I = 16\% \quad 0 = \$2{,}000 - \$560\ (3.27429) - \$400\ (0.47611)$$
$$= \$2{,}000 - \$1833.60 - \$190.44$$
$$= -\$24.0 \text{ Again, the value is not zero.}$$
$$I = 17\% \quad 0 = \$2{,}000 - \$560\ (3.19925) - \$400\ (0.45611)$$
$$= \$2{,}000 - \$1735.64 - \$182.44$$
$$= +\ \$81.9$$

Since the expression for 17% yields a positive value, the IRR is between 16% and 17%. The IRR that yields a value of zero is between 16% and 17%.

$$\text{Interpolating we find IRR} = 16\% + (17.0 - 16.0) \times (24/(24 + 81.9))\%$$
$$= 16\% + 0.226\% \text{ or } 16.226\%$$

A much simpler solution is offered by electronic spreadsheets such as Microsoft Excel. Among the considerable number of financial functions this program provides is the IRR function. It requires the payment-expenditure sequence be listed in order, from 0 to n. In this case, this means that six entries are required

The function is written as follows:

$$= IRR\{(-2000, 560, 560, 560, 560, 960)\}$$

The result is immediately computed as 16.22%.

LIMITATIONS OF THE IRR METHOD

The IRR method has some limitations. If we need to choose among two investments A and B, it could happen that A had the higher IRR, indicating that it is the most desirable. However, B could have the larger PW of the two when computed using MARR as the interest rate. This is frequently the case when there are large differences in the size of the investments. For example, investment A may involve \$100,000 and have an IRR of 17%, while investment B can have an IRR of 15% but involve one million dollars in approximately the same time frame. If the company's MARR is 12% and the PW of both projects is computed using this interest, it is likely that B will have the largest PW. The results of the two methods are contradictory in such case, and most analysts would consider B as the best choice. There are other more obscure shortcomings, such as the capital reinvestments assumptions of the method, which are not appropriate for discussion at this introductory level. In general, IRR is not suited for selecting mutually exclusive options unless their size and time frame are similar.

The IRR, in summary, is very intuitive and proceeds from having the expected amounts in an investment to finding which interest rate is implicit in the investment. In contrast, the PW and EAW methods check these amounts against a goal interest rate (normally the company's MARR), and their results are more difficult to interpret by a nonspecialist. The IRR, as discussed above, has its share of drawbacks, and must be part of the analysis arsenal instead of outright replacing the other two methods.

AN EXAMPLE INVOLVING COST RECOVERY

In the purchase of equipment, contractors are confronted with how to recover the cost of purchase and operation over a given life span. The following example takes an engineering economy approach to solving this using a MARR of 10%.

Assume that we buy a small hauler for $80,000. The following data regarding anticipated costs for the operation of the equipment piece are available.

1. Initial Cost $80,0000.
2. Operational Costs (Operator, Fuel, Oil, etc.) $20,000/year.
3. Tire Replacement at the end of years 2 and 4 $11,000.
4. Major Overhaul at the end of year 3 $15,000.
5. Sale at the end of year 5 (Salvage Value) $8,000.

The cash flow diagram for this situation is shown in Figure 7.5.

The first thing we must do is to refer all costs to Present Value. That is, we calculate the PW for all costs and the resale value. The following Present Values are calculated:

1. The cost of the unit, $80,000, is at Present Value.
2. The operational costs each year of $20,000 are considered as an annuity. Therefore A is $20,000, and P is calculated. Find P, given A.

$$P = A(P/A, 10\%, 5) = \$20{,}000\ (3.7905) = \$75{,}820.$$

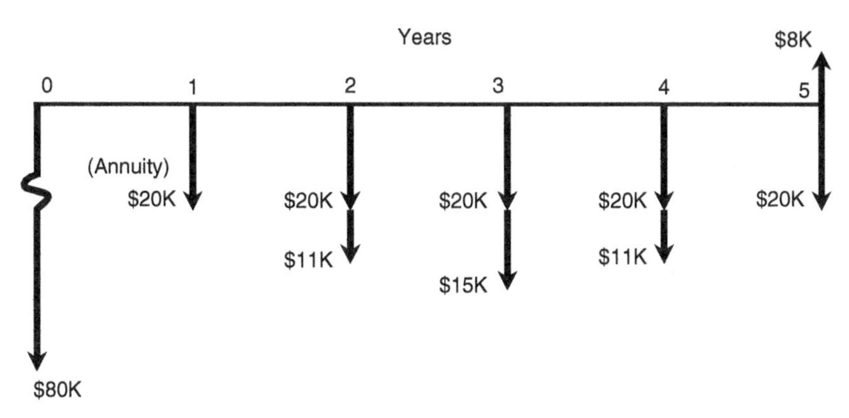

Figure 7.5 Cash flow diagram for equipment problem.

3. The tire replacement coasts of $11,000 are assumed to occur at the end of year 2 and 4. Since the payments are not uniform, we use Find *P*, given *F* for each expenditure. (Why can't we use Find *P*, Given A?)

$$P = F((P/F, 10\%, 2) + (P/F, 10\%, 4))$$
$$= \$20{,}000\ (0.8264 + 0.6830)) = \$16{,}603$$

4. The major overhaul occurs at the end of year 3. Therefore $P = F\ (P/F,$ $10\%, 3) = 0.7512\ (\$15{,}000) = \$11{,}200$
5. Since the resale or salvage value represents revenue, it is considered a minus value in this calculation (i.e., costs are plus and revenues are negative).

Then, $P = -8{,}000\ (P/F, 10\%, 5) = -\$8{,}000\ (0.6209) = -\$4{,}967$

Summation of the values calculated in (1) through (5) yields

$$P = \$80{,}000 + \$75{,}820 + \$16{,}693 + \$11{,}270 - \$4{,}967 = \$178{,}726$$

To find the amount that must be recovered each year from clients, we spread the Present Value calculated above across the 5-year period as a annuity or uniform series using a capital recovery factor (*A*/*P*). To find *A* Given *P* (see Figure 7.6), then:

$$A = P\ (A/P, 10\%, 5) = \$178{,}726(.2638) = \$47{,}100.$$

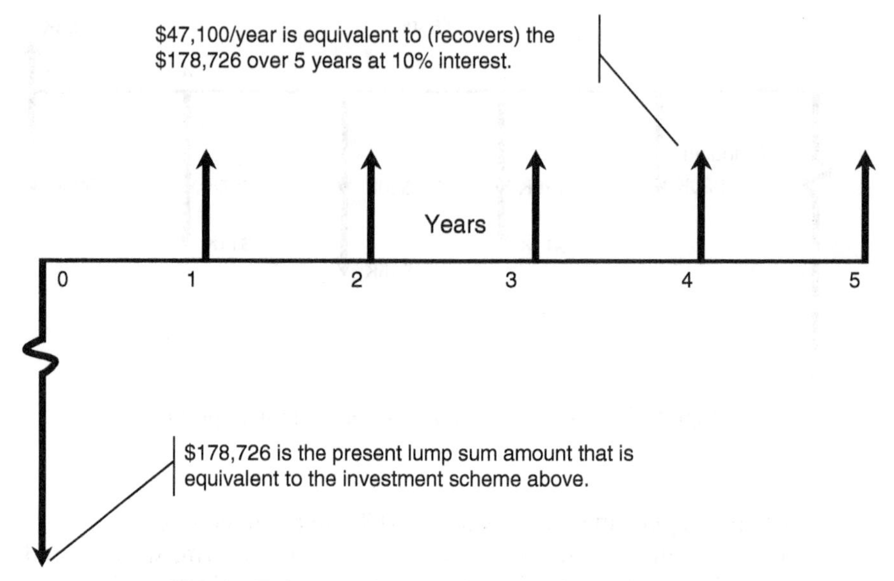

Figure 7.6 Distribution of costs/revenue using A/P.

COMPARISON USING EAW

Let us consider a situation comparing two alternatives using an MARR of 7 %. A new reservoir has been constructed providing water to a midsized town called Littleville (see Figure 7.7).

Two options are available for transporting water from the reservoir to Littleville town. Option A involves building a 10-mile long gravity pipeline at an initial cost of $2,800,000. Option B envisions using a pumping station to lift water over some barrier high ground. This option reduced the length of the pipeline to a length of 2 miles. A 40-year service life is used to compare these two alternatives.

	Cost for Pipeline A	Cost For Pipeline B
Initial Investment Pipeline	$2,800,000	$1,500,000
Cost of Pumping Station	none	$500,000
Operation and Maintenance (O & M) Costs	$30,000/year	$60,000/year
Power Costs during 1st 10 yrs	0.00	$40,000/year
Power Costs After 10 yrs	0.00	$120,000/year

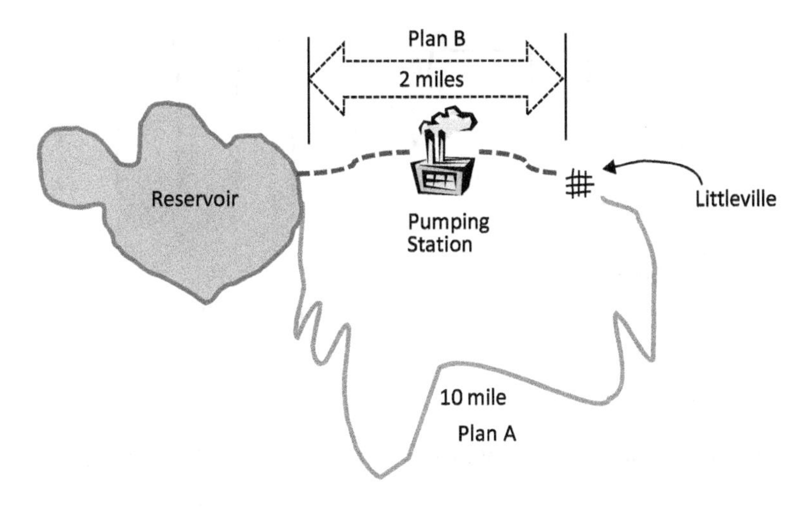

Plan A - Gravity Line Plan; Plan B – Line Using Pumping Station

Figure 7.7 Alternatives for Littleville water supply.

To compare these two options, we refer all costs to Present Value. For Plan A, the cash flow diagram is shown in Figure 7.8

$$P(1) = \$2,800,000$$

$$P(2) = \$30,000\ (P/A, 7\%, 40) \text{ Using the Tables in Appendix C, Therefore,}$$

$$P(2) = \$30,000\ (13.332) = \$399,960.$$

$$P(tot) = \$2,800,000 + \$399,960 = \$3,199,960$$

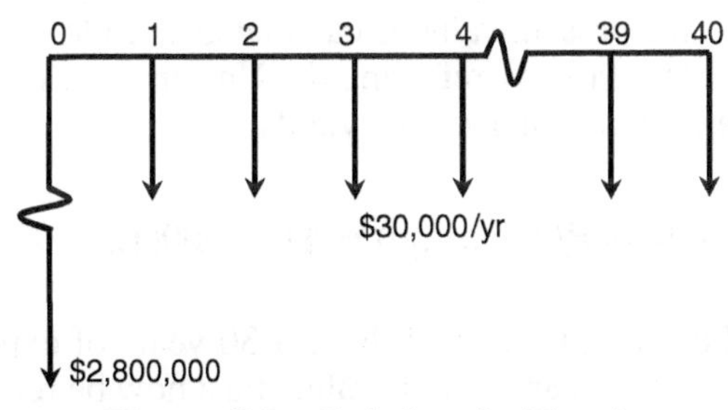

Figure 7.8 Cash flow for Plan A.

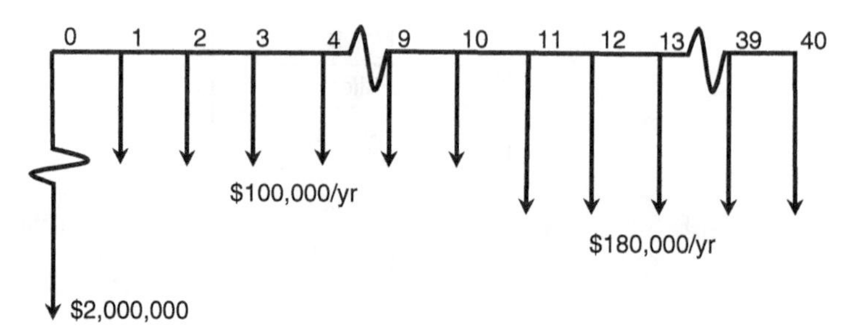

Figure 7.9 Cash Flow for Plan B.

$$\text{The Equivalent Annual Worth is EAW} = \$3,199,960\ (A/P, 7\%, 40)$$
$$= 3,199,960(.07501)$$
$$= \$240,029.00$$

For Plan B, the cost of operating the pumping station must be included. The diagram above describes this option as a cash flow diagram (see Figure 7.9).

The present value of the cost of building the pipeline with pumping state is:

$$\$1,500,000 + \$500,000 \text{ or } \$2,000,000. \text{ Therefore } P(1) = \$2,000,000$$

The cost of O & M plus power for the first 10 years is $60,000 + $40,000 or $100,000/year. The present value of this as a uniform series over 10 years is:

$$P(2) = \$100,000\ (P/A, 7\%, 10) = \$100,000\ (7.024) = \$702,400$$

Cost of O & M and power for the last 30 years is $60,000 + $120,000 or $180,000/year. This present value is calculated as follows:

First we calculate P(3) by referring the Uniform Series of payments from years 11 through 40 to year 10. This yields:

$$P(3) = \$180,000(P/A, 7\%, 30) = \$180,000(12.409) = \$2,233,620$$

This gives the Present Value of the last 30 years of expense consolidated as a single payment at year 10. This value must now be referred from year 10 to year 0 (i.e., Present Worth at time 0).

Therefore we have a Given F, Find P situation.

$$\text{Therefore } P(4) = \$2{,}233{,}620\,(P/F, 7\%, 10)$$
$$= \$2{,}233{,}620\ (.5084) = \$1{,}135{,}572.$$
$$P(tot) = P(1) + P(2) + P(4)$$
$$= \$2{,}000{,}000 + \$702{,}400 + \$1{,}135{,}572$$
$$= 3{,}837{,}972$$

Converting this to EAW over the 40 year life span,

$$EAW = \$3{,}837{,}972(A/P, 7\%, 40) = \$3{,}837{,}972(0.07501)$$
$$= \$287{,}886.$$

Since \$240,029 is < \$287,886, Option A is less expensive and should be selected.

AN IRR EXAMPLE—OWNER FINANCING USING BONDS

Large corporations and public institutions commonly use the procedure of issuing bonds to raise money for construction projects. A bond is a kind of formal IOU issued by the borrower promising to pay back a sum of money at a future point in time. Sometimes this proviso is supported by pledging some form of property by way of security in case of default by the borrower. A series of bonds or debentures, issued on the basis of a prospectus, are the general type of security issued by corporations, cities, or other institutions, but not by individual owner-borrowers.

In this discussion, owner financing means financing arrangements made by those corporations or institutions that are the owners of the project property. In this illustrative material that follows, "Joe" stands as a surrogate for "any borrower" ("Joan" would have served as well). During the period in which he has use of the money, the borrower promises to pay an amount of interest at regular intervals. For instance, Joe borrows \$1000 for 10 years and then pays back the amount borrowed. The rent is payable at the end of end year. The sequence of payments for this situation would be as shown in Figure 7.10.

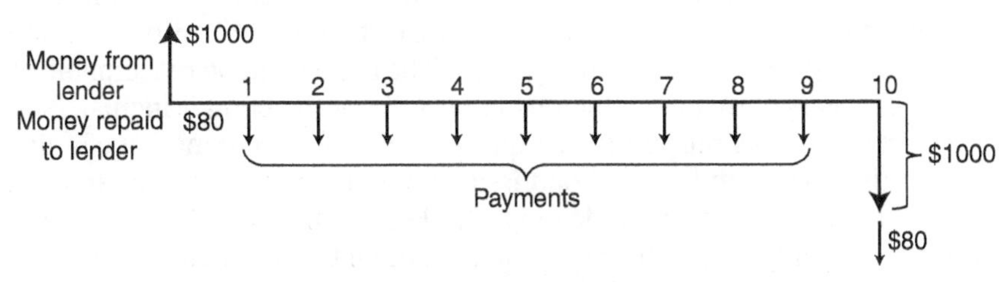

Figure 7.10 Sequence of payments for a bond.

When a series of bonds is issued, there may be a commitment to pay the interest due in quarterly installments rather than in one amount at the end of the year. A bond, as a long-term promissory note, may take any one of a variety of forms depending on the circumstances; mortgage bonds involve the pledging of real property, such as land and buildings; debentures do not involve the pledging of specific property. Apart from the security offered, there is the question of interest rates and the arrangements to be made for the repayment of the principal sum. Sometimes a sinking fund may be set up to provide for the separate investment, at interest, of capital installments that will provide for the orderly retirement of the bond issue. Investors find this type of arrangement an attractive condition in a bond issue.

In preparing for a bond or debenture issue, financial statements must be drawn up, and sometimes a special audit may be required. A prospectus for the issue may need to be drawn up, and this will involve settling the terms of issue and of repayment, the interest rates payable, and the series of promises or conditions related to the issue, such as its relative status in terms of priority of repayment, limitations on borrowing, the relative value of the security, and the nomination of a trustee to watch the interests of bond or debenture holders. These details are usually settled with the aid of specialists such as a CPA firm or mortgage broker.

Public bodies may need the approval of some local regulatory authority, and corporations may have to file and have approved a prospectus for the proposed bond issue. Charters or other constitutional documents must, of course, confer on the public body or corporation the power to borrow money in this way; this power is exercised by the council or by the board of directors or governors. For public offerings that are particularly attractive, banks bid for the opportunity to handle the placement of the bonds. The banks recover their expense and profit by offering to provide a sum of money slightly less than the amount to be repaid. As noted previously, this is called discounting the loan. The fact that more will be repaid by the borrower than is lent by the lender leads to a change in the actual interest rate. The rate is established through competitive bidding by the banks wishing to provide the amount of the bond issue. The bank that offers the lowest effective rate is normally selected; this represents the basic cost incurred for the use of the money.

Consider the following situation in which a city that has just received a baseball franchise decides to build a multipurpose sports stadium. The design has been completed, and the architect's estimate of cost is $40.5 million. The stadium building authority has been authorized to issue $42 million in bonds to fund the construction and ancillary costs. The bonds will be redeemable at the end of 50 years with annual interest paid at 5% of the bond principal. Neither the term nor its rate purport to be representative of current market conditions. At this time the term for any bond issue would tend to be shorter and its rate higher. In some commercial dealings "index number" escalation clauses are also occasionally seen. The banks bid the amounts for which they are willing to secure payment support. Suppose the highest bid received is $41 million.

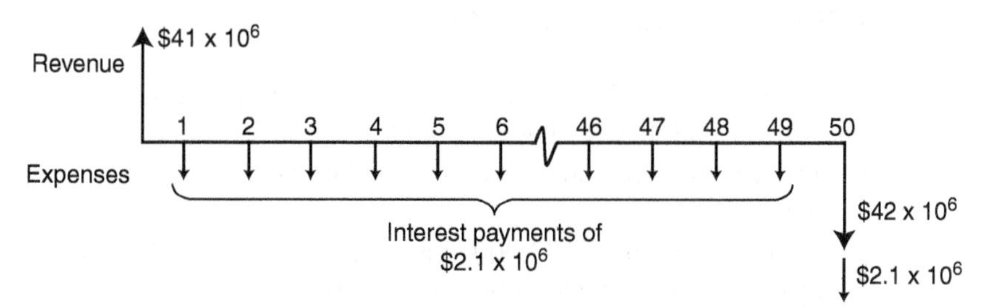

Figure 7.11 Cash flow diagram for sports stadium example.

In order to determine the effective rate of interest, a IRR analysis may be used. The profile of income and expense is shown in Figure 7.11. The effective rate of interest is that rate for which the present worth of the expenses is equal to the present worth (PW) of the revenue (in this case, $\$41 \times 10$). That is,

$$\text{PW (revenue)} = \text{PW (expenses)}$$

Utilizing the preceding information, this expression for the bond issue problem becomes $\$41 \times 10^6 = \$2{,}100{,}000\ (P/A, i, 50) + \$42{,}00{,}000\ (P/A, i, 50)$.

The annual interest is \$2.1 million, and this is a uniform series of payments for 50 years. The \$42 million must be repaid as a single payment at the end of 50 years.

In making this approach to a solution, a trial-and-error method (similar to that used in the Internal Rate of Return section) must be employed to solve the equation. That is, values of i must be assumed and the equation solved to see if the relationship [e.g., PW(revenue) – PW(expenses) = 0] is satisfied. In this case, two initial candidates for consideration are $i = 0.05$ and $i = 0.06$. Consulting appropriate tables for the present worth factors, the right side of the equation becomes:

$$i = 0.05\ PW = \$2.1 \times 10^6\ (18.256) + \$42 \times 10^6 (0.0872)$$

$$= \$42 \times 10^6 \text{ difference} = +1.0 \times 10^6$$

$$i = 0.06\ PW = \$2.1 \times 10^6 (15.762) + \$42 \times 10^6 (0.0543)$$

$$= \$35.38 \times 10^6 \text{ difference} = -\$5.62 \times 10^6$$

Since the equation balance goes from plus to minus, the value satisfying the relationship is between 5 and 6%. Using linear interpolation, the effective interest rate is found to be:

$$i = 0.05 + (0.06 - 0.05) \times [1.0 \times 10^6/(1.0 + 5.62) \times 10^6)]$$

$$= 0.515 \text{ or } 5.15\% \text{ as an approximation.}$$

REVIEW QUESTIONS AND EXERCISES

1. If your credit card's APR is 24% compounded daily, what is the effective annual interest rate that you are paying?
2. Explain why i sunk costs are not considered in the analysis of investment alternatives.
3. You want to accumulate $500,000 in a savings account in 20 years. If the bank pays 6% compounded annually, how much should you deposit in the account?
4. What is the future value 8 years from now of $2,000 invested today at a periodic interest rate of 12% compounded annually?
5. What would be the result of the previous problem, if simple interest were used?
6. You must decide between job offers from CPM Construction and Fudd Associates. Both companies concentrate on sustainable commercial construction, have similar fringe benefits, and offer comparable salary increases. CPM Construction offers you a $20,000 sign-up bonus, while Fudd Associates will give you $30,000 after 3 years in the job. Which offer would you take? Why? Consider engineering economy principles in your explanation.
7. You want to purchase a car with a sticker price of $25,000. The car dealer offers you a $2,000 discount and a 48-month, 8.5% APR compounded

TABLE P7.1 Worksheet for Problem 9

APR	Monthly Payment	Difference
0%		
4%		
8%		
12%		
16%		

TABLE P7.2 Worksheet for Problem 10

A	B	C	D	E
APR	Monthly Payment (From previous problem)	Total payment (1.15 × Column B)	Annual Payment (12 × Column C)	Minimum Annual Income (Column D / 12)
0%				
4%				
8%				
12%				
16%				

monthly, or no discount with a 4.0% APR on a 48-month loan. Which offer is better?

8. Repeat the preceding problem, considering an inflation of 4.0%. How does inflation affect the results of the previous problem?
9. Develop Table P7.1 comparing the monthly payments for a 30-year $200,000 mortgage loan with its APR compounded monthly.
10. The total monthly payment for a home mortgage loan is affected by the property tax and insurance. This amount is around 15% of the principal plus interest payment that you calculated in the previous problem. Moreover, as a rule of thumb, a family should not devote more than 28% of its gross income for home loan payments. With this information, find the amount that a family must earn to afford a $200,000 mortgage loan using the APRs of the previous problem. Table P7.2 should help you in your computations.
11. Develop Table P7.3 comparing the monthly payments for a $200,000 mortgage loan with an 8.00% APR compounded monthly.
12. Find the minimum amount that a family should earn to afford the mortgage loan in the preceding loan scenarios. Use the same assumptions of the previous similar problem, and Table P7.4.
13. CPM Construction plans to buy a truck for $150,000 and sell it for $15,000 at the end of five years. The annual operating cost of the vehicle is $60,000. What is the equivalent annual cost of this truck, if the company uses a MARR of 12%?
14. What is the present value of the truck described in Question 13?
15. CPM Construction plans to charge $100,000/year for the truck described in Question 13 (since the annual operating cost is estimated

TABLE P7.3 Worksheet for Problem 11

Years	Monthly Payment	Difference
100		
30		
15		
10		
5		

TABLE P7.4 Worksheet for Problem 12

A	B	C	D	E
Years	Monthly Payment (From previous problem)	Total payment (1.15 × Column B)	Annual Payment (12 × Column C)	Minimum Annual Income (Column D / 12)
100				
30				
15				
10				
5				

at $60,000, this means that the annual income will be $40,000). What is the IRR of this investment? Given CPM's MARR, is this an attractive investment?

16. Suppose that for the multipurpose sports stadium example considered in the *Owner Financing Using Bonds* section, the bond issue was

40 years and the annual interest rate is 9% of the bond principal. Using the architect's estimate of cost of $41 million, determine the effective interest rate.

17. Suppose in the preceding problem that the bonds are financed by a bank that discounts the bond issue to $40 million. What is the new effective interest rate?

CHAPTER 8

CONSTRUCTION LOANS AND CREDIT

INTRODUCTION

The ability to borrow or access funds plays a critical role in the construction industry. Construction loans and credit can be thought of as the air that inflates the balloon we call the economy. Without it, the economy shrinks, business dries up and projects shut down. This chapter addresses the subject of credit and borrowing and how they provide the monetary flexibility to keep the industry moving forward.

As has been emphasized in Chapter 1 and throughout this text, construction is a project-based industry. This strongly influences how construction is financed. Construction financing is mainly concerned with (1) project financing and (2) company financing. Project financing is effectively a short-term activity tied to "line of credit" issues and protocols. Short-term financing, as the name indicates, has to do with loans or credit, which must be repaid in the "near future." This can be defined as anywhere from a week to a year, depending upon stipulations established by the lender.

Company financing is handled mainly using commercial bank loans and retained earnings from within the firm or organization. This is in strong contrast to other major industries in which development of capital for company operations and expansion is based upon the issuance of stock and the generation of new funds from capital markets. Except in the in the very largest construction companies, generation of new capital by issuing stock is not the norm. Most construction firms are small to medium-sized closely held companies that attempt to avoid dilution of ownership by the issuance of additional stock.

Therefore, development of new external capital for expanded operations comes through borrowing from commercial banks in the form of long-term financing. Long-term financing refers to borrowing and loans that are to be repaid over a multiyear period. It is used for the acquisition of long-term assets such as equipment or real property.

Rising construction costs have increased the pressure on the industry to carefully monitor costs and predict cash timing to better balance the flow of receipts and disbursements. To remain competitive, contractors are being forced to manage liquidity by using new credit and borrowing strategies. In this chapter, development of external financing in the form of both short-term project-oriented financing and long-term financing will be discussed. The topic of project development will be presented in order to become familiar with some of the basic terminology and concepts associated with both short- and long-term financing. This will provide a basis for an in-depth discussion of credit and borrowing from the contractor's perspective.

THE CONSTRUCTION FINANCING PROCESS

Money is often referred to as one of the 4 M's of construction, the other three being manpower, machines, and material. It is a cascading resource that is encountered at various levels of project development. The owner or developer must have money available to initiate construction. The contractor (as you saw in Chapter 6) must have cash reserves available to maintain continuity of operations since progress payments lag actual construction expenditures. The major agents involved in the flow of cash in the construction process are shown in simple schematic format in Figure 8.1.

The *owner's financing* of any significant undertaking typically requires two types of credit: short-term financing during the development and construction period and long-term financing during the operational life of the project. The short-term financing usually occurs in the form of a *construction loan*. Long-term financing involves a mortgage loan or similar funding scheme (e.g., issuance of bonds, etc.) over an extended period of time (10, 15, 30, or more years).

Short-term loans may provide funds for items such as facility construction, land purchases, or land development. Typically, these short-term loans extend over the construction period of the project. A short-term loan is provided

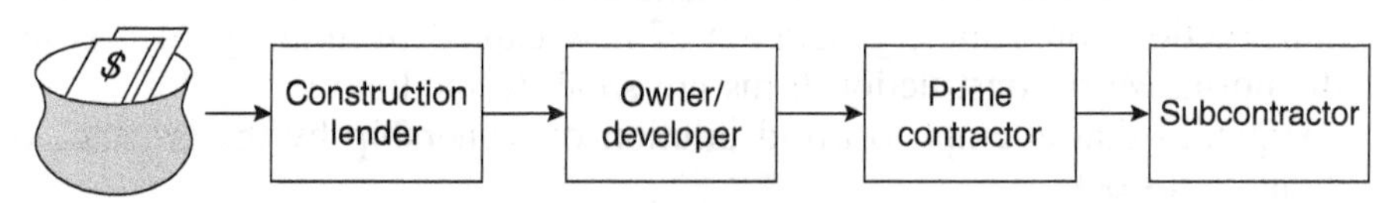

Figure 8.1 Project money flow.

by a lending institution, based on the assurance that it will be repaid with interest, by some other loan. This subsequent mortgage loan constitutes the long-term financing. Therefore, the first objective of any entrepreneur is to seek a commitment for long-term, or permanent, financing from a mortgage lender. Regardless of the type of project, this commitment will permit the construction loan, and any other funding required to be obtained with relative ease or, at least, more easily.

Unless the development company is in a position to raise the funds required directly by the issuance of their own securities (e.g. stock, bonds, etc.), it will seek to obtain a commitment from one of several alternate sources, including real estate investment trusts (REITs), investment or merchant banks, commercial banks, savings and loan associations, insurance companies, governmental agencies (VA, FHA), or, in special cases, from one of the international development banks. Public institutions, and to an increasing extent private entities, are raising funds by the sale of bonds.

The choice of lender depends on the type and size of the project. The choice of the form of security employed depends on a number of factors such as relative cost, the time period for which the funds will be available, the degree of flexibility involved (e.g., the freedom to pay out or refinance) and whether there are any restrictions involved which result in significant sacrifice of control to the lender. The funding of some larger projects may be handled by a consortium of international bankers (e.g., the English Channel Tunnel).

Notwithstanding recent developments in financial markets, lending institutions tend to be cautious: They are not interested in financing failure or in owning partially completed projects. Therefore, they will undertake a great deal of research and evaluation prior to providing a commitment for funding. At a minimum, an entrepreneur will be expected to provide the information given in Table 8.1.

In response to the recent financial crisis, many lending institutions are also requiring up to 75% of developmental projects to have lease agreements or commitments from major occupants in place (i.e., to support the pro forma

TABLE 8.1 Business Plan Documentation

Documents included in a typical Business Plan
1. Company financial statements
2. Company principals financial statements
3. Clear title to land involved in business proposal (if any)
4. Zoning approvals (if appropriate)
5. Preliminary architectural drawings (as appropriate)
6. Preliminary cost estimates
7. Preliminary implementation estimate
8. Market Research
9. Pro forma of expected income and expenses

Market rent for subject property (unfurnished) 55 two-bedroom A, B, or C units—1167 sq ft @ 41.0 cents/sq ft = \$478.47/mo or \$480 × 55	\$26,400.00
20 three-bedroom A, B, units—1555 sq ft @ 37.3 cents/sq ft = \$580.00/ mo \$580 × 20	11,600.00
Total estimated monthly income	\$38,000.00
Other income: Coin laundry, vending machine	150.00
	\$38,150.00
× 12 = annual total	457,800.00
Less vacancy factor of 5% (based on historical data)	−22,890.00
Adjusted gross annual income	434,910.00
Less estimated expenses @ 29.45%	−128,080.00
Net income before debt service	\$306,830.00

$$\text{Capitalized value @ } 9.5\% = \$3{,}229{,}789.00 = \left[\frac{\$306{,}830}{(0.095)}\right]$$

Requested loan value	= \$2,422,000.00
Loan/value ratio	= 75% (high) governed by low
Long-term debt service @ 9.75% constant	= \$236,145.00
Debt service coverage ratio	= 1.3
Loan per unit	= \$32,293.33
Loan per square foot	= \$25.42

Figure 8.2 Pro forma for 75 apartment units.

of expected income and expenses) before entering into a long-term financial agreement. This, of course, reduces the speculative nature of the income flow associated with the project. The concept of "Build it, and they will come" is a thing of the past.

A SAMPLE DEVELOPMENTAL PROJECT

Let's consider a typical project. An example of a pro forma for a venture involving the construction and leasing of a 75-unit apartment complex is shown in Figure 8.2. This document indicates that the annual income from the proposed apartment project will be \$306,830. The requested loan is \$2,422,000, and the annual debt service (i.e., interest) on this amount is \$236,145, realizing an income after debt service of approximately \$70,000. The ratio between income and debt service is 1.3. Lenders normally wish this ratio to be below 1.3.

The basis for the loan amount is given in Figure 8.3. Items 1 to 34 are construction-related items and are developed from standard references (e.g., R. S. Means, Co., *Building and Construction Cost Data* [published annually]) based on measures such as square footage. The lender normally has a unit-price guide for use in verifying these figures. Items 35 to 46 in Figure 8.3 cover non-construction costs that are incurred by the entrepreneur. It should

Construction-Ralated Costs	1. Excavation and grading	$ 67,500
	2. Storm sewers	48,000
	3. Sanitary sewers	84,030
	4. Water lines	28,000
	5. Electric lines	14,000
	6. Foundations	31,000
	7. Slabs	96,000
	8. Lumber and sheathing	185,000
	9. Rough carpentry	185,000
	10. Finish carpentry	81,362
	11. Roofing and labor	20,035
	12. Drywall and plaster	70,000
	13. Insulation	28,888
	14. Millwork	140,556
	15. Hardware	8,813
	16. Plumbing	165,000
	17. Heating and air conditioning	95,025
	18. Electrical	90,350
	19. Linoleum and tile	17,752
	20. Carpeting	101,881
	21. Kitchen cabinets	62,075
	22. Painting and decorating	107,000
	23. Masonry, block	20,680
	24. Masonry, brick	100,200
	25. Ranges and hoods	29,638
	26. Disposals	3,139
	27. Exhaust fans	1,022
	28. Refrigerator	35,040
	29. Paving	20,915
	30. Walks and curbs	20,792
	31. Landscaping	30,000
	32. Fence and walls	36,792
	33. Fireplace	51,100
	34. Cleanup	29,200
Non-Construction Costs	35. Lender's fee	32,000
	36. Surveyor's fee	1,000
	37. Architect's fee	12,500
	38. Land cost	80,000
	39. Attorney's fee	7,500
	40. Title insurance premium	5,762
	41. Other closing costs	150
	42. Hazard insurance premium	4,780
	43. Construction loan interest	120,000
	44. Appraisal	750
	45. Building permit	1,500
	46. Tax	50,000
	Total	$2,422,000

Figure 8.3 Construction cost breakdown for 75 apartment units.

be noticed that the interest for the construction loan is included in the costs carried forward in the long-term financing.

The method used to calculate the actual dollar amount of the loan is of great interest to the developer. The interest paid for the use of the borrowed money is an expense, and it is generally considered prudent business policy to minimize expenses. One way to minimize the interest expense would be to borrow as little a possible. The developer is, however, interested in putting as little of his own capital as possible at risk. Therefore, there is an incentive to use monies provided by others. This means borrowing as much as possible. That is, the developer tries to expand his own small initial capital input to start the project into a large amount of usable capital provided by others. This is the principle of *leveraging*. The entrepreneur takes a small amount and expands or leverages it into a large amount. The more he invests, the more he stands to lose.

Prudent lending organizations are normally careful, however, to ensure that the developer (i.e. the borrower in this case) has a sufficient amount of capital in the project so that if it fails, the developer's losses will be significant. This is the motivation for the borrower to push the project forward to a successful and profitable result. It is not prudent to allow the prime mover for a project to assume that if the project fails, he or she can simple walk away with minimum damage. There must to painful consequences for the borrower in the event of a failure.

Lack of attention to this basic concept was one of many elements that led to failure and worse during the so-called "subprime" financial crisis of 2008. Loans were made with no down payment and very little or none of the borrower's capital at risk. In addition, many borrowers were not properly vetted and were not able to pay the terms of the loan. The borrowers defaulted (i.e., failed to make repayment) and lenders were confronted with the ownership of assets that could not cover the amount of the borrowing.

THE AMOUNT OF THE LOAN

The amount of the mortgage loan should be a happy medium between too much and too little. If the loan is too small, there will not be enough to cover the costs of the project. On the other hand, if the loan is too large, the developer will find that the individual loan payments will exceed the available revenue, and he or she may be unable to meet their obligations.

The amount the lender is willing to lend is derived from two concepts: the economic value of the project and the capitalization rate (cap rate). The economic value of the project is a measure of the project's ability to earn money. One method of predicting the economic value is called the *income approach* to value and is the method shown in Figure 8.2. Simply stated, it is the result of an estimated income statement of the project in operation. Like any income statement, it shows the various types of revenue and their sum. These are matched against the predicted sums of the different expenses.

Although the predicted net income is a function of many estimated numbers, commonly, a fairly reasonable degree of accuracy is achieved. The expected net income divided by the cap rate produces the economic value of the project. The cap rate used in this example is 9.5%. The capitalized economic value of the project is obtained by dividing the net income ($306,830) by the cap rate factor (0.095). This yields an economic value of $3,229,789.

HOW IS THE CAP RATE DETERMINED?

Of course, the critical factor in the determination of the economic value is the cap rate. This must be determined for the situation at hand. First, the lender generally provides a loan that is about 75% of the estimated economic value of the project. This is done because 25% of the value, or thereabouts, must be invested by the developer to serve as an incentive for making the project a success. That is, a prudent lender furnishes 75% and the developer furnishes 25%. The lender must then decide what the interest rate will be and considers what will be the developer's rate of return. The sum of these numbers, times their respective portions, gives the cap rate.

Assume that the lender decides that an appropriate interest rate, given market conditions, will be 8.5% and the developer's planned rate of return will be 12%. Then the cap rate is taken as the weighted average of these two values. It is calculated as 8.5% times 75% plus 12% times 25%, which yields 9.375% or 0.09375 as the cap rate factor. Obviously, the value of the cap rate can be adjusted by the values that the lender places on his interest and the developer's rate of return. These numbers are a function of the existing economic conditions and thus fluctuate with the state of the economy. The lender, therefore, cannot exert as much influence on their values as might at first be expected. The lender is not in the business of pricing himself out of the lending market. He will attempt to establish a rate that is conservative (from the point of security of the loan) but attractive. The expected income divided by the cap rate gives us the economic value. The loan value may then be on the order of 75% of the calculated economic value.

Not every lender, however, will follow this formulaic type of approach. Some lenders may have a policy of lending a fixed proportion of what they view as the assessed valuation, which may not be based on the economic value. The assessment will be made upon the actual market value of the property as opposed to the potential economic value. This is the approach used for home and single-family residential mortgage loans.

The long-term financing for a project is critical. For this reason, the company or individual developing the project may exercise the right to hire a professional mortgage broker whose business is to find a source of funds and service mortgage loan activities. The broker's reputation is based on the ability to obtain the correct size mortgage loan at the best rate that is also fair to his client. The broker acts as an advisor to the client, keeping him or her apprised of all details of the proposed financing in advance of actually entering into the commitment. For this service, the mortgage loan broker receives a fee of about 2% of the loan, although the rate and amount will vary with the size of the loan.

MORTGAGE LOAN COMMITMENT

Once the lending institution has reviewed the venture and the loan committee has approved the loan, a preliminary commitment is issued. Most institutions reserve their final commitment approval until they have reviewed and approved the final construction plans and specifications.

The commitment issued is later embodied in a formal contract between the lender and borrower, with the borrower pledging to construct the project following the approved plans, and the lender agreeing that upon construction completion, and the achievement of target occupancy, he will provide the funds agreed upon at the stated interest rate for the stated period of time. As noted earlier, the actual amount of funds provided generally is less than the entire amount needed for the venture. This difference, called owner's equity, must be furnished from the entrepreneur's own funds or from some other source. The formal commitment will define the floor and ceiling amounts of the long-term loan.

During the construction period, no money flows from the long-term lender to the borrower. Funds necessary for construction must be provided by the entrepreneur or obtained from a short-term construction lender. Typically, the lender of the long-term will pay off the short-term loan in full, at the time of construction completion, thereby canceling the construction loan and leaving the borrower with a long-term debt to the mortgage lender.

CONSTRUCTION LOAN

Once the long-term financing commitment has been obtained, the negotiation of a construction loan is possible. Very often commercial banks make construction loans because they have some guarantee in knowing the loan will be repaid from the long-term financing. However, even in these situations, there are definite risks involved for the short-term lender. These risks relate to the possibility that the entrepreneur or contractor may, during construction, find themselves in financial difficulties. If this occurs, it may not be possible for the entrepreneur/contractor to complete the project, in which case the construction lender may have to take over the job and initiate action for its completion. This risk is offset by a discount (1–2%), which is deducted from the loan before any money is disbursed. For example, if the amount of construction money desired is $1,000,000, the borrower signs a note that he will pay back $1,020.000. The borrower, in effect, pays immediately an interest of $20,000. This is referred to as a discount and may be viewed as an additional interest rate for the construction loan. The current trend to minimize these risks is to require the borrower to designate his intended contractor and design architect. The lender may also require that all contactors involved in the construction be bonded as well. Some commercial banks evaluate and seek to approve the owner's intended contractor, his prime subcontractors, and the owner's architect, as a prerequisite to approving the construction loan. This evaluation extends to an evaluation of their financial positions, technical capabilities, and current workloads.

To minimize the risks involved, the banks will also base their construction loans on the floor of the mortgage loan, and only 75 to 80% of this floor will be lent. Of course, the developer may need additional funds to cover construction costs. One way to ensure this is to finance the gap between the floor and ceiling of the long-term mortgage loan. The entrepreneur goes to a lender specializing in this type of financing and obtains a standby commitment to cover the difference or gap between what the long-term lender provides and the ceiling of the long-term mortgage. Then, if the entrepreneur fails to achieve the breakeven rent roll or similar requirement, he still is ensured of the ceiling amount. In this situation, the construction lender will provide 75 to 80% of the ceiling rather than the floor. If the floor of the loan is $2,700,000 and the ceiling is $3,000,000, the financing of the gap can lead to an additional $240,000 for construction (i.e., 80% of $300,000). Financing of the gap is usually expensive, requiring a prepaid amount of as much as 5% to the gap lender. In the preceding example, this would be $15,000 paid for money that may not be required if the rent roll is achieved. Nevertheless, the additional $240,000 of construction funding may be critical to completion of the project and, therefore, the $15,000 is well spent in ensuring that the construction loan will include this gap funding.

Once the construction loan has been approved, the lender sets up a draw schedule for the builder or contractor. This draw schedule allows the release of funds in a defined pattern, depending on the site and length of the project. Smaller projects, such as single-unit residential housing, will be set up for partial payments based on completion of various stages of construction (i.e., foundation. framing, roofing, and interior), corresponding to the work of the various subcontractors who must be paid. For larger projects, the draw schedule is based on monthly payments. The contractor will invoice the owner each month for the work he has put in place that month. This request for funds is usually sent to the owner's representative or architect, who certifies the quantities and value of work in place. Once approved by the architect and representative, the bank will issue payment for the invoice, less an owner's retainage.

The owner's retainage is a provision written into the contract as an incentive for the contractor to continue his efforts, as well as a reserve fund to cover defective work that must be made good by the contractor before the retainage is released. Typically, this retainage is 10%, although various decreasing formulas are also used. When the project is completed, approved, cleared, and taken over by the owner, retainage is released to the builder.

In addition to the funds mentioned, the developer should be aware that some front money is usually required. These funds are needed to make a good-faith deposit on the loan to cover architectural, legal, and surveying fees and for the typical closing costs.

COMMERCIAL LENDERS

As noted in the Introduction to this Chapter commercial banks are the primary source of loans for the construction industry. Banks normally have a loan oversight group, which reviews loan applications. Despite recent developments in the credit and loan markets, bankers are typically conservative and study the financial position of potential borrowers very carefully. As noted above, both personal and company financial statements are key documents that are reviewed to establish the credit worthiness of potential borrowers. The supporting documentation required is similar to that required shown in Table 8.1.

Lending organizations analyze the so-called "Three Cs" of lending: credit, capacity, and collateral when considering loans to various individuals, companies and other organizations (i.e., the local church planning to borrow to expand facilities, etc.). Credit refers to the borrower's history of loan repayment. Capacity considers the borrower's ability to repay the loan given the borrower's financial situation both in terms of income and expenses. Collateral refers to the value and quality of the assets that the borrower offers to secure a loan in the event it is not repaid. These are relatively quantitative measures, which provide a good basis for evaluating the potential of a given borrower to repay the loan. During the recent financial crisis, it became apparent that these measures were subject to manipulation and a lack of clarity (i.e., transparency), which very nearly led to the shutdown of the national and international credit markets.

As noted in Chapter 6, retailers typically borrow money to purchase new inventory. As the inventory is sold, the revenue generated is used to retire the loan. Farmers borrow money to purchase the seed and equipment required to plant a crop or raise livestock. When revenues are generated through the sale of crops or livestock, the debt is retired and any profit is retained. If there is a fear that borrowers will not be able to repay the loan, the credit markets can be severely impacted. This occurred during the 2008 crisis to the degree that even banks were reluctant to lend to each other because of the fear that the borrowing bank might be on shaky ground.

In addition to the three Cs noted above, two additional and more judgment-based (i.e., less quantitative) measures are often used by lending institutions to evaluate potential borrowers. These relate to a qualitative evaluation of a potential borrower's credit worthiness. These two Cs are character and capital. Character refers to the financial reputation of the borrower and to more subtle qualities such as trustworthiness and issues related to reliability and reputation. Capital considers the borrower's total volume of and access to assets.

The actual process of applying for a loan, line of credit, or a credit account, although based on the preceding principles, is much more specific. Tax returns, credit ratings, job information, and detailed personal data are required in a personal loan application. A loan application for a real estate development project requires a business plan explaining the income projections that the developer anticipates will be used to repay the loan. There are many sources of information for positive and negative factors that must be taken into account by the bank when evaluating such loan applications.

LINES OF CREDIT

A line of credit is an arrangement for loosely secured credit up to a certain amount and operates in a manner similar to a personal credit card. This is a so-called "revolving" type of credit. This means that monies are withdrawn and repaid on a continuous basis over the period of the loan. The short-term financing used by contractors to carry the costs of a project over the period of a project is an obvious example of a line of credit loan. Interest is paid on the outstanding balance, with a smaller fee paid on the unused portion of the credit as discussed below. Payments can be variable, subject to conditions discussed in this section. Again as with a credit card, the borrower charges "purchases," which can include a wide variety of liabilities such as payroll, material purchases, and business related expenses.

On the other hand, the Maturity Matching Principle is a protocol that precludes the purchase of major assets such as a piece of heavy equipment or a major home office computer system using a line of credit. This protocol dictates that line of credit and similar short-term financing should be used to realize short-term and clearly defined business activities (e.g., inventory financing, crop planting, contractor project financing, etc.). As discussed in the Introduction to this chapter, the intent is that the loan will be repaid over a short term, (e.g., months as opposed to years). In other words, buying a house or a private airplane using a credit card is not acceptable. Major purchases of equipment and real property and similar long-term assets require the use of long-term financing.

The credit card analogy is useful for explanation purposes, but it must be noted that there are significant differences between a personal credit card and short-term financing using a line of credit. Major differences have to do with the following:

1. Interest paid on outstanding balances
2. Commitment fees paid
3. Maintenance of outstanding balances
4. Clean-up requirements
5. Collateral required

INTEREST PAID ON OUTSTANDING BALANCE

The interest accrued on a line of credit is computed by multiplying the balance by the agreed-upon interest rate. An APR of, say, 12% will result in charging a daily interest rate of 12%/365. This daily interest rate is then multiplied by the balance that the account had each day of the month. The monthly statement will be equal to the outstanding balance at the end of the month, plus all interest accrued for the days in the month.

Example: A line of credit has a balance of $500,000 during the first 15 days of July, and $400,000 during the remaining 16 days. If the agreed upon APR is 12% compounded daily, what is the total interest for the month?

Answer: the first day of the month accrues an interest of $500,000 × (0.12/365) = $164.38. The second day accrues interest on $500,000 + $164.38, or $500,164.38 × (0.12/365) = $164.44. The third day will have to consider ($500,000 + $164.38 + $164.44) = $500,328.82. The computation for the 31 days are summarized in Table 8.2. The total interest for the month is $4,626.81 for a payable balance of $404,626.81.

This total is very close to the one resulting from multiplying the account's average for the month by APR/12. That is:

$$\text{Average for the month} = (\$500{,}000 \times 15 + \$400{,}000 \times 16)/31$$
$$= \$448{,}387.10$$
$$\text{Interest for the month} = \$448{,}387.10 \times (12\%/12) = \$4483.87$$

Virtually all banks compute interest using the average daily balance, so this monthly amount would be technically incorrect.

TABLE 8.2 Interest Computations for Outstanding Balance

Day	Principal	Added Interest	Cumulative Interest	New Principal + Interest
1	500,000.00	164.38	164.38	500,164.38
2	500,000.00	164.44	328.82	500,328.82
3	500,000.00	164.49	493.31	500,493.31
4	500,000.00	164.55	657.86	500,657.86
5	500,000.00	164.60	822.46	500,822.46
...				
15	500,000.00	165.14	2,471.44	502,471.44
16	400,000.00	165.20	2,636.63	402,636.63
...				
30	400,000.00	132.94	4,493.83	404,493.83
31	400,000.00	132.98	4,626.81	404,626.81

COMMITMENT FEES

This requirement, as well as others discussed in the following sections, underline the differences between lines of credit from credit cards. Lenders impose a commitment fee on the unused portion of the line of credit. This fee usually has an APR of between $^1/_2$% and 1%, although it varies with the market and the client's credit credentials (large clients with excellent credit may even have this fee waived).

Example: A line of credit has a limit of $1,000,000 and had an average of $400,000 for July. How much is the commitment fee for the month, if it has an APR of 1% compounded daily?

Answer: The formula to find the amount, which is derived from topics discussed in Chapter 8, is:

$$\text{Fee} = P \times (1 + i)^n - P, \text{ where}$$

$$P = \$1{,}000{,}000 - \$400{,}000 = \$600{,}000;$$

$$i = 1\%/365, \text{ and } n = 31 \text{ days}$$

$$\text{Fee} = [600{,}000 \times (1 + 0.01/365)^{31}] - \$600{,}000 = \$509.80$$

This fee raises the effective interest paid on a line of credit. If a line of credit maintains a balance of 50% of its limit and has a commitment fee of 1%, the added annual effect on the effective interest paid would be computed as: $i' = 50\% \times (1+ 0.01/365)^{365} - 50\% = 0.5025\%$

COMPENSATING BALANCES

A line of credit may require that the borrower keeps an account in the same lending institution with a minimum balance of a given percentage of the credit limit. This percentage varies depending on the credit worthiness of the borrower and the lender policy. A 10% minimum is a common figure. The account can be interest bearing or non-interest-bearing, although lenders increasingly allow only non-interest-bearing accounts.

A compensating balance is a form of collateralization. It guarantees that the borrower will have at least some funds under the lending institution's control to pay any outstanding balance. From the borrower's perspective, it is an undesirable requirement. If the compensating balance is 10%, it means that the actual money contributed by the line of credit is, in fact, 90% of its stated limit.

The first 10% of the outstanding balance at any given time would be simply offsetting the unmovable money sitting in the account as a compensating balance. A compensating balance increases the effective interest paid by the borrower.

Consider the case of a contractor needing $400,000 to cover the peak of the fluctuations in his cash flow. Suppose that these fluctuations average $200,000. Assume that he wants to have an extra $100,000 as a cushion for emergencies. If a 10% compensating balance in a non-interest-bearing account is stipulated by the lender, it means that instead of the $400,000 + $100,000 = $500,000 that the contractor actually intends to use, he will have to ask for $555,000, so that the frozen 10% (approximately $55,000) does not hinder its cash availability. The average balance for the line of credit will be $255,000 instead of $200,000. This is $255K/$200K = 1.275 or 27.5% more than the intended average. If the line of credit has an APR of 12%, the contractor will end up paying, on the average, an APR of 12% × 1.275 = 15.3%.

CLEAN-UP REQUIREMENT

Most lines of credit are required to be paid in full and remain with a balance of zero for a number of consecutive days each year. This clean-up requirement is an elegant way of forcing the borrower to keep in mind that a line of credit is a short-term loan arrangement, not a long-term one. There are several good reasons to separate these two categories of loans, from the lender's perspective. An important reason is that lines of credit and long-term loans require different collaterals, and therefore, the lender is exposed to different risk levels. When a line of credit does not contain a clean-up requirement, it is known in the lending trade as an *evergreen* loan.

COLLATERALS

A line of credit is considered to be a loosely secured financial arrangement. The assets that are offered as a guarantee for its repayment are called *collateral*. For lines of credit, collateral requirements are not a stringent as in the case of long-term loans. In many cases, a borrower can show that his or her financial strength is sufficient to allow clean up of the line of credit without materially impacting the borrower's financial position. In some cases, the lender may establish additional requirements involving supplementary collateral.

A common requirement requires that a second party endorse the terms of the loan, in effect agreeing to repay the loan in the event that the borrower defaults on payment. This person or entity is referred to as a cosigner. The cosigner agrees to repay the interest and borrowed amount of the loan in the event that the primary borrower defaults.

An alternative to cosigning is to pledge a specific asset as collateral. If the asset is real estate, a mortgage can be filed benefiting the lender. A mortgage is a specific form of lien against a property, which will be voided when the loan is paid, or executed in foreclosure if the borrower fails to follow the loan agreement. Despite the common misuse of the word, a mortgage is not a loan. It is a claim allowing foreclosure on the subject property in the event of default by the borrower.

ACCOUNTS RECEIVABLE FINANCING

In very unusual cases, contractors experiencing an urgent cash flow problem can use accounts receivable as collateral. A lender may accept accounts receivable as collateral in special situations. This arrangement is called *accounts receivable pledging*. Since this type of collateral presents a higher level of risk (Can collection of the accounts receivable, in fact, be made?), the lender will charge a high interest for such loans. The maximum amount of the loan depends on the amount and age of the accounts receivable.

A lender may agree to provide funds up to 75% of the value of accounts receivable due between 0 and 30 days. Older outstanding balances usually indicate that the account may be difficult to collect. The lender would view these accounts as being of higher risk and might lend up to 60% of the balance of accounts aged between 31 and 60 days.

In extreme financial emergencies, a firm may even sell its accounts receivable at a discount. This method of raising funds is called factoring. As discussed previously, the discount depends on the age of the accounts. Funds generated from factoring are not a loan, but a final transfer of funds to the buying agency. Clearly, this is a risky business for both seller and purchaser, and is seldom utilized by construction firms.

One advantage accruing to the seller is that responsibility for collection of the accounts is transferred to the purchasing entity—the factor. Just the cost of collecting money from slow or delinquent clients may justify the use of factoring. It is a quick way to obtain cash, and does not require the administrative burden of seeking a bank loan.

TRADE CREDITS

Trade credits are, in effect, free money provided to a company by its suppliers. It has a significant effect on a construction contractor's cash flow. Suppliers and subcontractors extend credit to contractors to encourage quick and timely payment. From the vendor's perspective, incentives for timely payment lead to a reduction in the carrying costs associated with inventory financing.

Suppliers have clearly defined credit terms, including the number of days allowed for payment before interest or fees apply and discounts for early payment. The typical credit account will stipulate 30, 60, or even 90 calendar days for payment, although in some cases credit can be for as few as 10 or even 7 days. It is important to verify whether credit terms are specified in calendar days, which is the most common case, or in business days. Lack of timely payment can lead to a negative reading of the financial "character" of the purchaser on behalf of creditors.

The leverage provided by trade credit on the availability of cash can dramatically improve the viability of any company. These considerations underscore the importance of having a pristine payment record with all merchants extending credit to a company. An impeccable record serves as the best business card to get further credit with other lenders. It also is the best argument to get an extension of payment terms or to increase the limit of credit when needed. Simply stated, timely payment leads to an excellent "credit rating."

Credit arrangements with subcontractors tend to be less stringent than those afforded by commercial suppliers. All subcontractors want to be considered for future jobs by the contractor, and seldom stipulate credit terms such as those discussed above. Subcontractors are also inclined to accept subcontracts that include "pay when paid" clauses. With such clauses, the subcontractor extends credit to the prime contractor for an indefinite period until the project owner supplies the funds to pay the subcontractor.

Vendors and suppliers deal with many customers and are much more demanding in encouraging and requiring timely payment. Invoices from vendors typically include compact information about payment terms, such as "2/10, net 30." The "net 30" means that the merchant expects that the invoice will be paid in full by the 30th calendar day after the date printed in the invoice. The "2/10" indicates that if payment is made within 10 calendar days, a 2% discount will be applied.

For an invoice totaling $1,000, this yields a $20 discount. This amount may seem insignificant, and a fair proportion of contractors simply do not take advantage of the offer. This is a great mistake. The annualized interest for early payment can be computed as:

$$\begin{aligned}\text{Effective annualized discount rate} &= (365/(30 - \text{grace period})) \\ &\quad \times [\text{Discount\%}/(100\% - \text{Discount\%})]\end{aligned}$$

For "2/10, net 30," using the above formula, the discount rate is:

$$\begin{aligned}\text{Effective annualized discount rate} &= (365/(30 - 10)) \times (2\%/(100\% - 2\%)) \\ &= 37.24\%\end{aligned}$$

This represents a 37.24% recovery on the average outstanding balance of invoices from vendors offering this discount.

Terminology relating to trade discounts is as follows:

ROG/AOG:	The discount period begins upon receipt of goods (ROG)
2/10 NET 30 ROG:	This expression appearing on the invoice means 2% can be deducted from the invoiced amount if the contractor pays within 10 days of the arrival or receipt of goods. Full payment is due within 30 days of AOG/ROG.
2/10 PROX NET 30:	A 2% cash discount is available if invoice is paid not later than the 10th of the month following ROG. Payment is due in full by the end of the following month.
2/10 E.O.M.:	The discount (2%) is available to the 11th of the month following ROG. Payment in full is due thereafter.

It is a longstanding "rule of thumb" in the construction industry that payments to vendors offering discounts should always be made in a timely fashion to take advantage of such trade credits.

LONG TERM FINANCING

Banks and other financial institutions can provide term loans. A term loan must be repaid by a given date, called its maturity date, and is repaid under a prearranged schedule. Although technically a term loan can be classified as a short-term loan if it is repaid within a year, such loans are primarily used for long-term financing. The term can be up to 15 years in most cases and even more in noncommercial settings. The 30-year loan that many people take to finance their homes is a term loan, albeit more regulated than normal commercial loans.

Term loans require an application process which, as noted previously, is usually quite elaborate. The prearranged timetable established for repayment of a term loan is called an amortization schedule. Most lenders require equal monthly payments throughout the life of the loan, although there are exceptions.[1]

The equal payments to amortize a term loan are computed as an annuity (paid every month, despite the name suggesting yearly payment). For illustration purposes, and using formulas discussed in Chapter 7, it can be shown that a $200,000 loan to be repaid in 30 years at a 7.00% APR compounded monthly results in 360 payments of $1,330.60. The same loan paid over 15 years would have payments of 1,797.66. The difference in payments is relatively small (35.1%), considering that the payment time is halved for the 15 year option. The reason for this seemingly illogical proportion can be found

[1] Production Credit Associations allow the payment of principal in equal amounts, with varying interest depending on the remaining principal owed at any given time.

in the interest paid, which grows at a much faster rate than the repayment period.

A term loan normally has fees that must be paid up front or deducted from its principal. Some of these fees are quite justifiable, but others may strike the borrower as subject to negotiation. There are loan origination fees, application, credit investigation fees, and closing fees, among others. Higher fees are expressed in *points*. Each point is 1% of the stated principal and, therefore, increases the nominal amount of the loan. In the case of the $200,000 loan at 30 years and APR of 7%, a 2-point fee translates into an extra $4,000, which usually is paid by increasing the nominal amount borrowed. Instead of $200,000, the loan would be for $204,000, with payments of $1,357.22. Over 360 payments, the extra $26.62 per month turns into $9,582.15.

Construction contractors, as noted previously, do not need long-term financing except for the purchase of equipment, which normally leads to loans of no longer than 5 years. If the purchase of real property is part of a company's long-term planning, term loans of 15 to 30 years may be of interest. In general, however, long-term financing is limited to equipment and similar purchases.

LOANS WITH END-OF-TERM BALLOON PAYMENTS

When purchasing an item on credit, a contractor may choose to pay a higher total loan interest if the terms of the load make the monthly installments more affordable during the first few years of loan repayment. That is the case when a contractor has a limited amount of cash to execute a project and needs to purchase a piece of equipment for the project. Instead of paying large monthly installments, it may be advantageous to pay a smaller monthly amount during the project execution and then, at the end of a relatively short period, pay a large lump sum to liquidate the loan principal that has not been repaid by the smaller installments. The unpaid principal at the end of the period is called the *balloon payment* amount and the type of loan that allows this arrangement is called a loan with an *end-of-term balloon payment*. The following example illustrates how this type of borrowing might be implemented.

A heavy construction company plans to purchase a front loader. The initial cost of the loader is $80,000. The company desires to borrow $90,000 to pay for taxes and other costs associated with the purchase. The bank offers a 9.00% APR and the loan must be repaid in no more than 5 years. The contractor wants to consider a balloon payment scheme to repay the loan.

First we will consider this purchase using a loan that stipulates uniform monthly payments at 9.00% APR over the 5-year term of the loan. This will be called Option A.

A second payment scheme will be called Option B. This payment sequence uses a balloon payment to reduce the monthly payments over the 5 year (60 month) period of the loan. In this case, the amount of monthly payments

TABLE 8.3 Fully Amortized and Balloon Payment Loans

A) Fully amortized loan	
Amount borrowed	\$90,000
APR (Compounded monthly)	9.00%
Years for payment computations	5
Monthly payments	\$1,868.25
Payment due at the end of 5 years	0
B) Loan with balloon payment after 5 years, amortized on a 20 year schedule	
Amount borrowed	\$90,000
APR (Compounded monthly)	9.00%
Years for payment computations	20
Monthly payments	\$809.75
Payment due at the end of 5 years	\$79,836.34

is based on uniform payments over a 20 year (240 month) term. However, the loan must be paid in full after 5 years. Therefore, at the end of 5 years, the remaining 15 years of the 20 year term will be paid out as a lump sum or balloon payment.

These two options (A & B), to include the monthly amounts and balloon payment to be paid are summarized in Table 8.3.

The calculation of uniform payments for Option A simply requires the application on the formula A = \$90,000 × (A/P, 0.75%, 60). This results in monthly payments of \$1,868.25.

The B Option involves reduced monthly payments (due to the 20 year term). The monthly payments are calculated as A = \$90,000 × (A/P, 0.75%, 240). The 240 month period leads to a reduced monthly payment of \$809.75. But, of course, the balloon payment must be paid at the end of the fifth year.

Therefore, we must calculate the present value of the remaining 15 years of payments comprising the balloon payment to be paid at the end of the fifth year. This involves taking the 15 years of monthly payments to a single PW value to be paid at the end of year 5. P = \$809.75 × (P/A, 0.75%, 180). N is 180 since there are 12 payments each year for 15 years (i.e.12 × 15) to be considered. This yields a balloon payment of \$79,836.34.

It can be observed that the balloon payment scheme results in payments of less than half of the uniform payments of Option A—\$809.75 versus \$1,868.25. The amount of the balloon payment is very large. After the five years, the company would owe nothing under the fully amortized plan (Option A), while \$79,836.34 would be still owed as a balloon payment under the alternative payment scheme (Option B).

The total payment using Option A is 60 months times \$1,868.25 or \$112,095. The total pay out for Option B is 60 months times \$809.75 plus the balloon payment of \$79,836.34. This amounts to \$128,421.34. Reducing the monthly payment and using a balloon payment will cost $16,326 more than Option A. Under certain circumstances, this may be acceptable. The cost versus the benefit must be carefully considered by the borrower. Lenders tend to be reluctant about such financing schemes due to the risk that the borrower may not be disciplined or reliable enough to escrow the required balloon payment. This where the Character—reliability and reputation—of the borrower must play a key role.

REVIEW QUESTIONS AND EXERCISES

1. What does it mean when it is said that a company is "excessively leveraged?" What could be the effects of excessive leverage? You may want to search the phrase "effect of excessive leverage" on the Internet.
2. An entrepreneur is considering a new project consisting of an apartment building with 50 units. Each unit can be rented for $1,000/month. It is expected to have an average vacancy rate of 10%. Net operating expenses are budgeted as 30% of the project's gross annual income. What would be the economic value of this project, using a cap rate of 8%?
3. How much would be the value of the long-term loan for the preceding project? How would this amount change if the entrepreneur offers to put a $100 million hotel as collateral for the loan?
4. A line of credit has a balance of $700,000 during the first 10 days of June, and $300,000 during the remaining 20 days. If the agreed-upon APR is 10% compounded daily, what is the total interest for the month?
5. A line of credit has a limit of $500,000 and had an average of $200,000 for July. How much is the commitment fee for the month, if it has an APR of 0.5% compounded daily?
6. Find the effective annualized discount rate offered by a supplier with payment terms of "1/15, net 30."
7. What is the monthly amortization of a 5-year term loan with a principal of $50,000 and an APR of 12% compounded monthly?
8. Find the monthly payment to amortize the balloon payment of $79,836.34 remaining after 5 years for the example developed in Table 8.1. Assume that the balloon amount must be repaid over 5 years, with an interest of 9.00%.
9. Are construction contractors used as financing sources by project owners? What would be the terms of a "perfect" contract between a project owner and a construction contractor, concerning financial resources?
10. Under what circumstances would you take the balloon payment scheme discussed in Table 8.3?

CHAPTER 9

THE IMPACT OF TAXES

INTRODUCTION

The profitability of any construction company is significantly affected by the amount of taxes it is required to pay. More than one third of a company's taxable income goes to pay its tax obligations to federal, state, and municipal entities. Understanding the rationale and conceptual framework of the current tax system is, therefore, of paramount importance to any contractor. Financial choices affect the amount of taxes paid and taxes affect the financial decisions that a construction company makes. This chapter provides an introductory treatment to this topic. It must be kept in mind, however, that tax law changes constantly and is continuously in a state of flux. Material presented in this text is based upon tax law and bulletins as of 2008. Keeping current on changes to this material is critical to managing the payment of tax obligations. The Internal Revenue Service (IRS) website at www.irs.gov contains excellent readable reference material that can help in keeping abreast of current events in this area. It is always, however, necessary to consult a tax expert before making important decisions involving taxes.

Taxes are primarily designed to finance government operations and to meet societal needs. In modern societies, taxes have also been implemented to influence taxpayers' behavior. For example, high taxes operate to increase the price of tobacco, which presumably causes fewer individuals to use tobacco products. Purchasing hybrid-powered cars results in significant tax credits meant to spur their use.

An unintended consequence of tax incentives and deterrents is that, once they are in place, they become extremely difficult to withdraw even when the issues that they originally addressed are no longer relevant. Obsolete tax breaks still abound and are the focus of the activities of countless lobbyists and politicians.

TYPES OF TAXES

A tax levied on the net income realized by a company or individual is called an income tax. Income taxes are referred to as *direct* taxes. *Indirect* taxes, in contrast, are levied on the cost, price or value of products or services.[1] The U.S. government relied almost exclusively on taxing resources other than income until the Sixteenth Amendment to the Constitution was passed in 1913 (previous attempts to impose income taxes were short-lived). Although the focus of this chapter is on income taxes, everyone also pays indirect taxes.

Some forms of indirect taxes relate to:

1. ***Excise taxes.*** These taxes are levied on the output (not the price) of a good. Although historically significant, this tax is currently applied to a small number of products (e.g., gasoline.). The excise tax on gasoline is applied to each gallon, independently of its selling price.
2. ***Import and export taxes.*** These taxes (or duties) are imposed on merchandise arriving or leaving a country, respectively. These taxes are popular in many countries with weak institutional administrative frameworks because they are easy to collect at their ports of entry and departure. Import duties are frequently used for political leverage with the exporting country and to protect local industries from cheaper imported goods.
3. ***Sales taxes.*** These taxes are computed as a percentage of the selling price of goods. In the United States, sales taxes are collected by merchants at the point of sale. In Europe, the Value Added Tax (VAT) is functionally similar from a consumer's viewpoint, but the tax is collected progressively up the production and distribution chain of the product.
4. ***Ad valorem taxes.*** These taxes are levied on the estimated value of a property. The most common form of ad valorem tax is the property tax imposed on real estate and its improvements. The semantics of the term *ad valorem* has been used to argue that all taxes based on value (including sales and import/export) fall into this classification.

Most indirect taxes can be viewed as location-related in that they depend on the area or location where the company or individual is based. This supports

[1] Some legal scholars define indirect taxes as those collected by an intermediary from the person who ultimately pays the tax. This is the familiar case of the sales tax collected by a retail store and paid by the customer.

the generation of local tax revenue and empowers citizens at the local level. Alternately, it can also result in a wide range of taxes that may put a brake on economic activity. Because of this, certain states and municipalities are viewed as tax friendly, while others are not.

The U.S. government entrusts the Internal Revenue Service (IRS) with the administration and collection of federal taxes. States and other levels of government have their own tax collection agencies. Outside of the United States, most countries have agencies similar to the IRS to administer tax collection. Failure to pay taxes normally results in harsh penalties for tax evasion.

INCOME TAX SYSTEMS

Business entities pay tax on taxable income, which is the revenue earned by a company minus the expense of doing business and any allowable deductions. Construction costs (direct costs plus field overhead) and general overhead costs associated with the operation of a construction company are deductible in principle, independent of the company's form of business organization. However, some expenses are difficult to classify and others are not fully deductible. For example, the IRS allows deduction of only half of the cost incurred for business meals and entertainment.

Revenue minus all costs of generating said revenue is the definition of profit, and it is tempting to conclude that income taxes are applied to a company's profit. However, deductions authorized by tax law invariably lead to a company paying taxes on a lower amount than the amount of net profit reported to stake and/or stockholders. The net amount used to compute a company's tax liability is called its tax base or *taxable income*.

Tax deductions constitute major administrative tools for influencing taxpayers to consider and conform to what government authorities view as a desired behavior. Charitable contributions, solar energy generation systems, extra costs resulting from locating in certain inner cities and other reputedly desirable deeds are rewarded by allowing costs associated with these actions to be deducted from a company's or individual's revenue. Deductions are, however, the source of much of the complexity of the current income tax system.

ALTERNATIVES FOR COMPANY LEGAL ORGANIZATION

The legal form of organization adopted by a company influences many management issues, including the level of effort required to start a company, minimum reporting requirements for its owners and government agencies, the risks faced by its owners in case of financial default, and the amount of taxes paid on its profits. This section offers a short summary of major organizational forms, emphasizing their tax implications.

Sole Proprietorships

A company organized as a sole proprietorship has a single owner, who is ultimately responsible for all its liabilities and enjoys all profit, if any, from its operation. The company's profit is treated as personal income for tax purposes. For instance, suppose that Elmer Fudd is the sole owner of Elmer Fudd Construction, a proprietorship whose profit was $500,000 for 2008. Additionally, he had income of $10,000 from interest generated by his personal bank savings.

His total personal income would be $510,000. Assuming that he had no other sources of income, he is married filing jointly with no dependent children, and he chooses the standard IRS deduction, the amount of his federal income tax for 2008 would be around $144,627. The details of this and the following tax computations are discussed later in this section.

Partnerships

A partnership is similar to a proprietorship in that the owners are ultimately responsible for all company liabilities and pay taxes on the profit received by each one. As in the case of proprietorships, a partnership is transparent for taxation purposes. The company is not independently taxed, nor can it retain any proportion of its profits. Consequently, the account Retained Earnings, discussed in the accounting chapters of this book, is meaningless for partnerships and proprietorships from a legal standpoint.

For comparison, suppose that Elmer Fudd and four nephews have equal ownership rights in a partnership called Fudd and Nephews Construction Company. The company had a profit of $500,000 for 2008. Each of the five partners is married filing jointly with no dependent children and received $2,000 in interest income. The total income of each one (their respective *gross adjusted income*, as discussed later) would be $102,000. Each one would pay approximately $13,713 of federal income tax, for a combined total of $68,565. The difference between this figure and the tax paid by Mr. Fudd as a sole proprietor is quite significant.

Corporations

From a legal perspective, a corporation is considered a unique entity, separate from its owners. As such, the owners' responsibility for its liabilities is limited to their share of company equity. In contrast, a regular partner or a sole proprietor can lose virtually all personal property if it is necessary to pay off their company debts. The tradeoff of this substantial advantage is that corporation profits are taxed before any dividends are distributed to owners. Owners, in turn, must pay tax on the dividends they receive, albeit at a lower rate than regular personal income. Most individuals paid a flat rate of 15% for dividends in 2008, as further discussed later.

Suppose that Elmer Fudd and his four nephews decide to incorporate their company as Fudd Associates, Inc. The corporation earnings before income taxes was $500,000 and all after-tax profit was distributed as dividends to the five shareholders, each one receiving the same amount. Tax computations in this case are more complicated than in the previous organizational forms.

The corporation would pay $170,000 in federal income taxes, and therefore, there would be $330,000 left over to distribute to owners. Each of the five owners would receive dividends of $66,000. The tax for this dividend amount (at 15%) would be $9,900. If each one had interest income of $2,000, paying a marginal tax rate of 28%, the tax levied would be $560. Each shareholder would pay $10,460 of federal income tax, for a combined total of $52,300. The total amount paid to the IRS (corporation + owners) would be $222,300, making this the most expensive of all alternatives.[2]

The preceding discussion applies to regular ("C") corporations. Shareholders can choose to incorporate as an S corporation (or "Subchapter S" corporation) and be treated as a partnership for tax purposes. S corporations, combine the limited liability of corporations without having double taxation (first on the corporation, then on the shareholder). However, this type of corporation has legal limitations in its type of shares, number of shareholders, and other aspects, while having essentially all the administrative burden of a regular corporation.

Limited Liability Partnerships and Companies

Recently, two types of increasingly popular legal organizations have come into fashion. They are regulated by each state, as opposed to by federal statute. The Limited Liability Partnership (LLP) works as a partnership for tax purposes, but each partner is responsible for any liability under his control and for a proportional part of all shared liabilities. Partners are not, however, responsible for liabilities arising exclusively from the actions of other partners. LLPs have other corporation-like stipulations, such as having a managing board. This type of organization is especially popular with professionals covered by malpractice insurance, such as medical doctors. This format inherently protects the other partners from the malpractice or failures of one of the partners. Many states limit LLPs to individually insured professionals and offer the more general alternative of LLCs, discussed below, for construction and other types of businesses.

The Limited Liability Company (LLC) works very similarly to an LLP, but has important differences. In general, LLCs are more closely regulated than LLPs and must pay an additional "franchise tax" or "capital values tax" to the state where they are incorporated. The advantages of an LLC include

[2] Assuming that three of the owners of the company are active in the operation of the firm and pay themselves a salary of $100,000 (i.e., $300,000 for the three), how would this impact the amount of tax paid?

the option to be taxed as a proprietorship, partnership, or corporation, while providing the limited liability protection of a corporation to its members. The main disadvantages common to LLCs and LLPs arise from the legal framework that varies significantly among states. Moreover, both organizational forms have limitations similar to those of partnerships. For example, the death of a member can trigger succession problems or even the termination of the company. Also, raising capital is harder for LLPs and LLCs than for corporations. Lenders and potential investors tend to distrust their long-term stability. The National Conference of Commissioners on Uniform State Laws adopted the Uniform Limited Liability Company Act in 1996 (revised in 2006) to address many of these problems for LLCs.

Other Options

A *joint venture* is a partnership between two or more participants—usually corporations—to reach an operational goal. They are usually, but not necessarily, temporary in the sense that they exist only for the duration of a given project or the time required to achieve a given objective. Joint ventures are common in the construction industry for large or specialized projects that require the pooling the resources of two or more contracting companies. Joint ventures between a large foreign company and a local company are frequently required by financing agencies for international construction contracts.

Another type of partnership is the *strategic alliance*. It is similar to a joint venture, but is geared toward a longer-term common goal not necessarily involving a high level of equity stake. The small and fractured nature of construction companies makes strategic alliances less popular than in other industries.

TAXATION OF BUSINESS

For business entities, income is defined by the equation:

$$\text{Income} = \text{Revenue} - \text{Expenses}$$

The concept of deductions, however, allows for this amount to be further reduced for tax purposes through the subtraction of deductions authorized by the Internal Revenue Service and existing tax law. Therefore, the taxable amount of income becomes:

$$\text{Taxable Income} = \text{Revenue} - \text{Expenses} - \text{Deductions}$$

In business organizations, the definitions of revenues and expenses must be scrutinized carefully and conform to guidelines established by tax law and the Internal Revenue Service (IRS) publications. The amount of expenses and the definition of deductions greatly impact the amount of taxable income. In other words, gross profit as reported in the Income Statement is not the same as taxable income because of the impact of deductions. As discussed earlier in this text, construction companies have many expenses that are directly incurred in the physical construction of projects. Other expenses are required to support construction and provide for the management of site activities and for company administration. These are categorized as direct and indirect expenses:

1. ***Direct Expenses.*** These expenses are incurred in operations that lead to the realization of the physical facility being constructed. Typical direct expenses are job payroll, project related equipment costs, materials used for construction of the project, and subcontractor payments.
2. ***Indirect Expenses.*** This term relates to costs incurred for the management support of the job site and for home office general and administrative costs (e.g., senior management and staff salaries, office equipment costs such as computer and copying charges, and costs related to the ownership or rental of office space). These types of indirect expenses are *overheads* or *overhead expenses.*

Expenses measure the expenditure of funds used to generate revenues. As such, they can be deducted from revenues when determining taxable income. As noted previously, expenses may be direct or indirect.

In general, most indirect expenses (e.g., field and home office overheads) are subtracted from revenues to establish the taxable income of a company. However, certain overhead expenses may not be fully deductible. For instance, the existing tax law limits travel and entertainment expenses as follows:

> "Generally, you can deduct only 50% or your business meal and entertainment expenses, including meals incurred while away from home or business."

This, of course, means that these indirect expenses are NOT fully deductible.

To illustrate this, consider the Income Statement shown in Figure 9.1.

This Income Statement lists the revenues and expenses referred to above in an organized fashion. In this case, revenues are referred to as "Net Sales." The Gross Profit is simply calculated as Sales minus Cost of Sales (i.e., Revenues minus Expenses). Gross Profit is reduced in the amount of "Job Related Indirect Charges" ($562,600) and "Fixed (Home Office) Overheads ($1,265,820). After subtracting these two Overhead Costs ($1,828,480), Net Profits before Taxes are shown as $1,454,810. At this point, taxes are subtracted yielding a Net Profit of $945,627. This amount will either be retained (i.e., held in the company as Retained Earnings) or distributed as dividends to the owners or stockholders of the company.

Two items in this report are of interest from a tax standpoint. Under job related costs, item 3 reports Transportation (Travel, Meals, etc.) costs. Assume that $60,000 of the reported "Transportation Costs" is related to expenses

Net Sales (Revenues)	$21,427,610
Cost of Sales (Direct Expenses)	
Materials	10,139,130
Labor (Includes all payroll taxes & (fringes)	5,482,710
Subcontracts	514,500
Total Cost of Sales	$18,159,320
Gross Profit	$ 3,283,290
Job Related Indirect Charges	
1. Salaries	276,640
2. Job Office Costs	190,430
3. Transportation (Travel, Meals, etc.)	80,710
4. Interest Charges (Job Carrying Costs)	12,750
5. Miscellaneous	2,130
Total Job Indirect Charges	$562,660
Fixed (Home Office) Overheads	
1. Salaries – Home Office Staff	234,180
2. Salaries – Officers	384,100
3. Home Office Administrative Cost (Equipment, etc.)	516,950
4. Legal and Audit	23,150
5. Rent	25,060
6. Depreciation	82,380
Total Fixed overheads	$1,265,820
Total Overheads	$1,828,480
Net Profit before Taxes	$1,454,810
State and Federal Taxes	509,183
Net Profit	$945,627

Figure 9.1 Income Statement, Fudd Construction Co. 3rd Qtr, 2XXX.

for meals and entertainment. As noted above, this amount cannot be fully deducted from revenues. Only half of meal and entertainment costs are available as a deduction. Therefore, before calculating federal tax, half of this amount (i.e., $30,000) must be added to net profit before taxes reported in Figure 9.1. That is, the taxable amount of profit becomes:

$$\text{Taxable Income} = \$1,454,810 + \$30,000 = \$1,484,810$$

since the total amount of the job related Item 3 is not fully deductible.

On the other hand, the amount shown as Item 6 under Fixed Overhead – Depreciation is a major deduction allowed by tax law as a way for each firm to recover the loss of value of equipment due to obsolescence over time.

Depreciation is one of the most significant deductions available to construction contractors. For instance, an off-highway truck or hauler will decline in value over time. After 4 years of use, a $100,000 hauler may have a market value of only $40,000. This decrease or decline in value is viewed as a cost of doing business (i.e., a cost of generating revenues). Therefore, the IRS allows this loss of value to be recovered through a deduction called *depreciation*. In Figure 9.1, Fudd Construction is claiming $82,380 for decline in the value of equipment in the firm used to generate revenues." The subtraction of this amount when calculating net profit provides a major way for a company to reduce its amount of taxable income. In effect, deductions cause:

$$\text{Income} = \text{Revenue} - \text{Expenses}$$

to be modified, yielding:

$$\text{Taxable Income} = \text{Revenue} - \text{Expenses} - \text{Deductions}$$

Obviously, the income and profit reported to stockholders or company owner is different from the taxable income reported for tax purposes.

BUSINESS DEDUCTIONS IN GENERAL

Deductions allow companies to reduce the amount of income subject to taxation. They can be viewed as a gift from the government recognizing both direct and indirect as well as latent costs of doing business. Some deductions, as noted previously, are designed to provide incentives to trigger certain actions on the part of business entities.

Construction costs (direct costs plus field overhead) and general (home office) overhead costs associated with the operation of a construction company are deductible independent of the company's form of business organization. However, some expenses are difficult to classify and others, as noted, are not fully deductible (e.g., meal and entertainment costs).

Therefore, the calculation of taxable income and the determination of what items can be deducted and to what degree must be carefully determined. Deductions such as depreciation typically lead to a company paying taxes on a lower amount than the net profit before income taxes appearing in a company's Income Statement. The net amount considering deductions is used to compute a company's tax liability and is referred to as its tax base or taxable income.

TAXABLE INCOME: INDIVIDUALS

Taxes levied on proprietorships and other forms of business organization are paid directly by their owners, and their computation is linked to individual tax considerations. The taxable income of individuals depends not only on the amount of money earned but also on their civil status and how they choose to report their deductions.

For tax purposes, individuals are classified into the following filing statuses:

- ***Single.*** The unmarried individual not falling into the head of household category below generally pays the highest amounts of tax compared to the other categories.
- ***Married filing separately.*** A married couple may decide to pay taxes separately, filing their tax returns as essentially two single individuals. The current tax code results in a so-called marriage penalty for married couples filing a joint return when both spouses work and earn higher than average, comparable incomes. In these cases, the couple pays more taxes by submitting a single joint return than by filing separately.
- ***Married filing jointly.*** This filing status comprises the traditional family of spouses and children, choosing to use a single tax return for the couple and dependents. Historically, tax codes protected this category, although such advantages have all but disappeared in the current code.
- ***Head of household.*** This category provides certain tax benefits. It includes persons contributing more than half of their household expenses. These persons are considered unmarried and have at least one dependent person. Single parents and single children supporting their parents and similar individuals are included in this category.

ITEMIZED DEDUCTIONS, STANDARD DEDUCTIONS, AND PERSONAL EXEMPTIONS

Every individual (or family, if filing jointly) can deduct certain items from their income, thus reducing the tax basis upon which their taxes will be computed. Contributions to charity, medical expenses over a certain threshold, and especially the interest paid on the mortgage loan for the individual's residence can add up to a substantial amount of money. This item-by-item reporting is called the itemized deduction approach. Alternatively, individuals can choose to subtract a *standard deduction* from their income instead of performing an itemized deduction calculation.

The amount of the standard deduction is specified by the IRS and varies according to the filing status. Furthermore, it is adjusted every year. An advantage of using the standard deduction is that no confirming documentation is needed compared to the sometimes excruciating exercise of backing each claimed itemized deduction with some form of proof (typically called a voucher). Both approaches cannot be taken simultaneously. If a person takes advantage of the standard deduction, no other deduction can be claimed

TABLE 9.1 Standard Deductions and Personal Exemptions, 2008

Filing Status	Standard Deduction
Single	\$5,450
Married filing jointly	\$10,900
Married filing separately	\$5,450
Heads of household	\$8,000
Personal Exemption	\$3,500

except for a few special cases, such as contributions to individual retirement accounts (IRAs). The standard deductions for 2008 are provided in Table 9.1.[3]

The IRS also allows a personal exemption amount to be deducted for the person filing and those reported as dependent (e.g., spouse and children). For example, if a married couple with two children files a joint return, they can claim four personal exemptions. The amount specified for each personal exemption does not vary with the filing status, but it is adjusted every year. As can be seen in Table 9.1, the amount for 2008 is \$3,500. The personal exemptions for the family of four, previously mentioned, would reduce taxable income in the amount of $4 \times \$3{,}500 = \$14{,}000$. If the taxpayer pays taxes at the rate of 25%, this yields a reduction in actual tax paid of $(.25) \times \$14{,}000$ or \$3,500.

There are several minor deductions applicable to businesses and/or individuals not discussed here, such as college tuition and fees and the interest paid on student loans. The taxable income of proprietorship owners and individuals in general results from applying all deductions and personal exemptions.

THE TAX SIGNIFICANCE OF DEPRECIATION

Deductions benefit the tax payer in that each dollar that can be deducted results in a reduction in the amount of tax paid. This reduction in tax amounts to a per dollar savings of $\$1.00 \times$ (Tax Rate). If a tax payer in the 28% marginal tax bracket[4] can deduct \$1,000 for whatever reason, this reduces the amount of tax by \$280.

Most tax agencies accept the concept that property declines in value over time and is a legitimate cost of doing business that can be claimed as *depreciation*. Depreciation reflects the loss of value and obsolescence of property (e.g., equipment) involved in the operation of a business and can constitute a

[3] Owners of proprietorships detail their business revenue and costs separately using Schedule C—a supplemental schedule of IRS Form1040. They can choose to use their standard deduction, independently of the costs deducted from their revenue to arrive to the business income (or loss) reported in their form 1040.

[4] The concept of marginal tax rates is discussed later in the Section entitled "Marginal Tax Rates."

very significant tax deduction. It results in a lowering of taxes (e.g., 34 cents on the dollar for businesses based on a tax rate of 34%).

Depreciation applies to all property that is required for the operation of a business. Computers, office furniture, heavy equipment, and real estate are all depreciable assets. Most construction companies, particularly heavy and highway constructors, own a substantial amount of equipment that can be depreciated over time. Contractors look at equipment as small "profit centers" and will apply any depreciation associated with the piece of equipment to offset the profit generated by that machine or equipment item.

It is important to align the declaration of depreciation so that it can be used to offset income that is taxable. If a company has $100,000 of depreciation that can be taken in a given year, but only $80,000 in taxable income in that year, $20,000 in tax savings are lost. That is, depreciation cannot be carried over for use in a future tax year. In general, depreciation must be applied in the tax year in which it occurs.

During the construction of the Alaska Pipeline, a midwestern contractor successfully received the contract to construct an access road running parallel to the pipeline. In order to prepare for this job, the contractor purchased a multimillion dollar fleet of equipment. The job was delayed several years because of impact assessments and studies triggered by environmental action groups. During this time, the equipment was placed in storage near the construction site and no depreciation was claimed. Since the fleet was not productive during this period, it was not generating taxable income. The strategy was to take depreciation on the equipment at the time the equipment was productive and generating income against which the depreciation could be used to reduce taxes.

CALCULATING DEPRECIATION

The method by which depreciation is calculated for tax purposes must conform to standards established by the Internal Revenue Service (IRS). The three bases available for calculation of depreciation are (1) Straight Line, or Linear, method, (2) the Accelerated method, and (3) a variant of the Straight Line method called the Production method. Federal law has introduced the use of fixed percentages as given in published tables to calculate the amount of depreciation for various classes of property when using the Accelerated method.

The major factors to be considered when calculating depreciation of an asset are shown in Figure 9.2. The three major factors form the three sides of the depreciation "box" that are linked by the method of depreciation selected. They are:

1. Initial cost or basis in dollars
2. Service life in years or hours
3. Salvage value in dollars

The amount that can be depreciated or claimed by way of a tax deduction is the difference between the initial net value of the asset and it's residual or

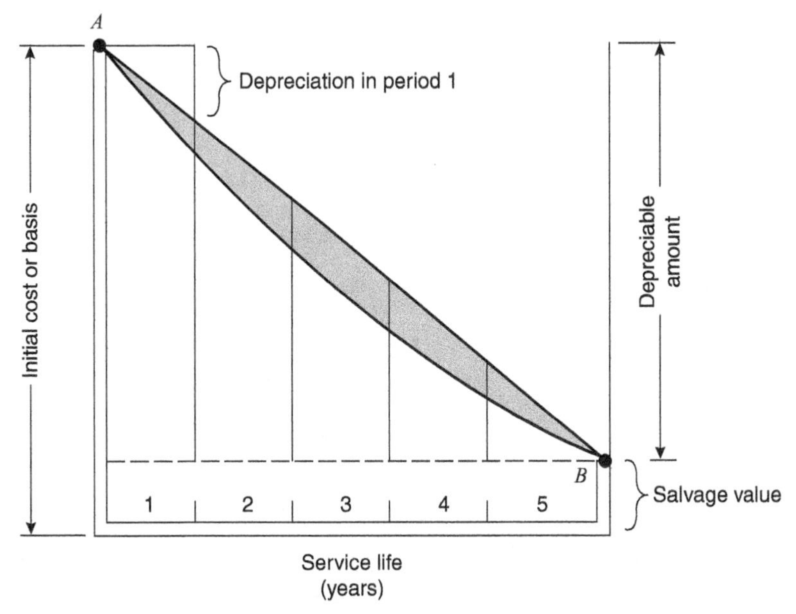

Figure 9.2 Factors involved in depreciation.

salvage value. This is referred to as the *depreciable amount* and establishes the maximum number of depreciation dollars available in the asset during its service life. U.S. tax law provides that salvage value less than 10% of the initial basis or cost of the item can be assumed to be zero (i.e., neglected). In such cases, the entire initial net value is considered to be depreciable.

STRAIGHT LINE METHOD

The Straight line method of calculating depreciation is based on the assumption that the depreciation, or loss of value through use, is uniform during the useful life of the property. This simply means that the net first cost or other basis for the calculation, less the estimated salvage value (if greater than 10% of the basis), is deductible in equal annual amounts over the estimated service life of the property. The depreciable amount is the difference between the purchase price or initial basis and the salvage value (if any).

Let us assume that we have a piece of equipment that has an initial cost or base value of $16,000 and a salvage value of $1,000. The service life of this equipment is 5 years. Since the salvage value of $1,000 is less than 10% of the initial cost, it is considered to be zero. The depreciable amount is $16,000. By prorating the depreciable amount over the service life, the amount of depreciation claimed each year is $16,000/5 years or $3200 per year. This is illustrated in Figure 9.3.

The remaining amount of depreciation of this asset can be determined by consulting the stepwise plot of declining value. During the 3rd year of the

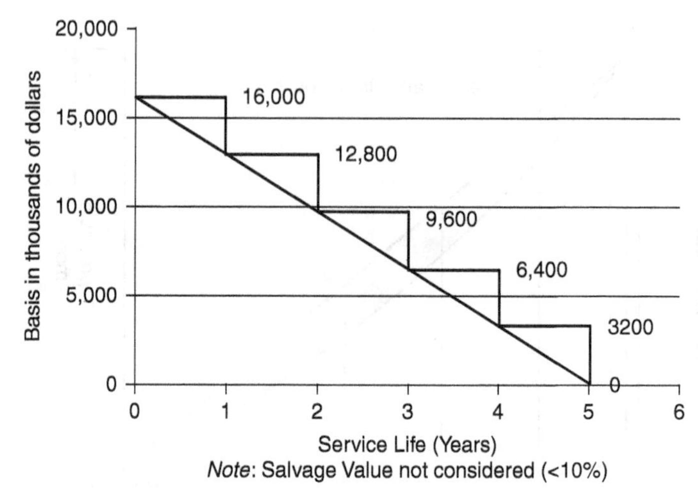

Figure 9.3 Straight line depreciation.

asset's service life, for example, the remaining base value (also called the book value) of the asset is $16,000 minus $6,400, or $9,600. If we connect the points representing the book value at the end of each year, we have the "straight line."

The concept of base value, or book value, has further tax implications. For instance, if we sell this asset in the third year for $13,000, we are receiving more from the buyer than the book value of $9,600. We are gaining $3,400 more than the depreciated book value of the asset. The $3,400 constitutes a *capital gain*. The reasoning is that we have claimed depreciation up to this point of $6,400, and we have declared that as part of the cost of doing business. Now the market has allowed us to sell at $3,400 over the previously declared value, and therefore, we have received taxable income.

THE PRODUCTION METHOD

As was stated previously, contractors try to claim depreciation on a given unit of equipment at the same time the equipment is generating profit. The production method allows the declaration of depreciation based on the number of hours the equipment unit was in production for a given year. The asset's cost is prorated and depreciation is recovered on a per-unit-of-output or hourly basis. If the $16,000 equipment unit we have been discussing has a 10,000 hour service life, the $16,000 is recovered on a per hour of use basis at the rate of $1.60 per hour. This method is popular with small contractors, since it is easy to calculate and ensures that the depreciation available from the asset will be recovered at the same time the unit is generating profit. A reasonable estimate of the total operating hours for a piece of equipment may be obtained by referring to the odometer on the unit together with the logbook or job time cards. The production method ensures that depreciation is available when the machine is productive and theoretically profitable or producing income.

DEPRECIATION BASED ON CURRENT LAW

As was noted above, current guidance from the IRS allows property to be depreciated using (1) the Straight Line method, (2) the Production method, or (3) the Accelerated method. A set of tables defines the acceptable accelerated depreciation amounts for equipment based on the appropriate service life categories. These tables establish rates of depreciation available using the Modified Accelerated Cost Recovery System, or MACRS approach presently in place. This system defines various classes of property ranging from small items (e.g., computers, file cabinets, etc.) with a 3-year service life up to a 20-year class for real property. Most construction equipment items fall into the 3-, 5-, or 7-year service life categories given in the IRS tables. For example, light trucks (less than 6.5 tons) are considered to be 3-year property. Most average-weight construction equipment is classified as 5-year equipment. Some heavy construction equipment such as dredging barges are depreciated over a 10-year life. The MACRS table is shown in Table 9.2.

The amount depreciated is calculated using prescribed depreciation methods for each class of property. For example 2-, 5-, 7- and 10-year property are depreciated based on the 200% declining-balance (DB) method with a switch to Straight Line method at a time that maximizes the deduction. The 15- and 20- classes use the 150% DB method.

When using the declining-balance (DB) method, the amount of depreciation claimed in the previous year is subtracted from the book value at the beginning of the previous year before computing the next year's depreciation. The rate is calculated by dividing either 200% or 150% by the total number of service life years associated with the property (e.g. 3, 5, 7, etc.). The rate for a 5-year class of property using the 200% DB method is 200%/5 or 40%. For a $10,000 piece of equipment with a service life of 5 years, the 200% and 150% DB depreciation for each year is shown below:

	200% DB Depreciation (Rate = 40%)*	150% DB Depreciation (Rate = 30%)*
Year 1	$10,000 × .40 = $4,000	$10,000 × .30 = $3,000
Year 2	$6,000 × .40 = $2,400	$7,000 × .30 = $2,100
Year 3	$3,600 × .40 = $1,440	$4,900 × .30 = $1,470
Year 4	$2,160 × .40 = $864, etc.	$3,430 × .30 = $1,029, etc

(*Note*: The 150% DB method is used only for 15- and 20-year assets, so this calculation is shown only as a sample calculation. The actual 15- and 20-year 150% DB rates to be used would be 150%/15 = 10% and 150%/20 =7.5%, respectively).

TABLE 9.2 MACRS Table for Accelerated Depreciation

	Annual recovery (Percent of original depreciable basis)					
Recovery year	3-year class (200 % d.b.)	5-year class (200 % d.b.)	7-year class (200 % d.b.)	10-year class (200% d.b.)	15-year class (150% d.b.)	20-year class (150% d.b.)
1	33.00	20.00	14.28	10.00	5.00	3.75
2	45.00	32.00	24.49	18.00	9.50	7.22
3	15.00	19.20	17.49	14.40	8.55	6.68
4	7.00	11.52	12.49	11.52	7.69	6.18
5		11.52	8.93	9.22	6.93	5.71
6		5.76	8.93	7.37	6.23	5.28
7			8.93	6.55	5.90	4.89
8			4.46	6,55	5.90	4.52
9				6.55	5.90	4.46
10				6.55	5.90	4.46
11				3.29	5.90	4.46
12					5.90	4.46
13					5.90	4.46
14					5.90	4,46
15					5.90	4.46
16					3.00	4.46
17						4.46
18						4.46
19						4.46
20						4.46
21						2.25

MACRS also uses the "half-year convention" to calculate the first-year depreciation. This means that half of the DB amount is taken in the first year.

Since the MACRS rate for a 5-year class property is 40%, only half of this percentage, that is, 20%, is taken in the first year. The rate for the second year then becomes $((1.0 - 0.2) \times 0.4) \times 100 = 32\%$. Similarly, the third year rate is $((0.8 - 0.32) \times 0.4) \times 100 = 19.2\%$, and so forth. For a 5-year class item with base value of $100,000, the depreciation amounts for years 1 to 3 are:

Year	Depreciation	Book Value
1	$20,000	$80,000
2	$32,000	$48,000
3	$19,200	$28,800

By consulting Table 9.2, it will be noted that although this is a 5-year property class, it is depreciated out across a 6-year period. Also, following the third year, MACRS switches from 200-DB to Straight Line method. The

6th year is assumed to be a half-year so that the depreciation for years 4 and 5 is equal to 28,800 divided by 2.5, or $11,520.

Year	Depreciation	Book Value
4	$11,520	$17,280
5	$11,520	$5,760

The remaining $5,760 is available for depreciation in year 6. In other words, the last year is considered to be a half-year. Verify the values for 10-year property, by using the "Half-Year Convention" and procedure explained above. The item is depreciated out over $10 + 1 = 11$ years. Note that all categories in Table 9.2 are depreciated over "$n + 1$" years. In what year, does the switch from 200 DB to straight line occur for the 10-year asset? All values in Table 9.2 can be confirmed using this approach.

IRS documents available on the Internet explain the calculation of depreciation. Moreover, the Internet has many websites with tax estimators (search income tax estimator). While it is perfectly legitimate to use these resources, they are "black boxes" that offer no guidance about the rationale behind the computations. The process discussed here provides a limited insight into the special topic of depreciation and its impact on taxable income.

MARGINAL TAX RATES

The income tax system is progressive. This term is used to indicate that tax rates do not stay "flat" or constant across all income levels. In a flat system, all taxpayers are subject to the same tax as a percentage of their income, regardless of the amount earned. That is, a flat or constant tax rate or percent is paid by all taxpayers regardless of the level of their income.

In a progressive system, the percentage rate of income tax paid by taxpayers with high income is greater than the percentage paid by individuals with lower income. There are a few instances of regressive taxes, which result in lower-income individuals paying more, as a percentage of their income, than higher income taxpayers. The excise tax on gas is an example: The tax levied on each gallon is more significant as a percentage of earnings for lower-income persons than to upper-income individuals (e.g., $3.00/gallon gasoline is a higher percent of the income of a person making $25,000/year than of a person making $180,000/year).

The income tax depends on two main factors: the filing status of the taxpayer (corporation, single individual, married filing jointly, etc.), and the taxable income on which the tax is computed. Each filing status has a corresponding tax table, similar to the one shown in Table 9.3. The table is divided into income brackets, and each income bracket has a different tax percentage

TABLE 9.3 Tax Table for Single Tax Payers, 2008

Bracket floor	Lump Sum	Marginal Tax Rate
8,025	0	10%
32,550	802.50	15%
78,850	4,481.25	25%
164,550	16,056.20	28%
357,700	40,052.20	33%
No limit	103,791.75	35%

called the marginal tax rate for the bracket. The column entitled Lump Sum is derived from other data, as discussed below. The brackets (i.e., lower and upper limits) are adjusted by the IRS every year. Changing the bracket percentages, however, is much more difficult, since percent changes require an Act of Congress.

Taxes are computed incrementally. Referring to Table 9.3, a single person with taxable income of \$10,000 will pay a 10% tax over the first \$8,025 and 15% on the amount over \$8,025.

$$\text{Tax for \$10,000 taxable income} = 10\% \times 8{,}025 + 15\% \times (10{,}000 - 8{,}025) = \$1{,}098.75$$

Similarly:

$$\text{Tax for \$50,000 taxable income} = 10\% \times 8{,}025 + 15\% \times (32{,}550 - 8{,}025) + 25\% \times (50{,}000 - 32{,}551) = \$8{,}843.50$$

Notice that for a taxable income of \$60,000, only the last term would be different from the above computations, that is, 25% × (60,000 − 32,550). In fact, only the last term of the tax computations for a taxable income differs from the computations for any other taxable income within the same bracket. To simplify the computations, most tax tables include a lump sum equal to the amount of tax paid up to the bracket base.

The lump sum for the first brackets of Table 9.3 can be found as follows:

$$\text{For the bracket \$0 to \$8,025, the Lump sum} = \$0, \text{ since there are no preceding brackets}$$

$$\text{For the bracket \$8,025 to \$32,550, Lump sum} = 10\% \times 8{,}025 = \$802.50$$

TABLE 9.4 Individual Taxpayer Income Brackets and Tax Rates, 2008

Single	Married Filing Jointly or Qualified Widow(er)	Married Filing Separately	Head of Household	Marginal Tax Rate
\$0	\$0	\$0	\$0	10%
\$8,025	\$16,050	\$8,025	\$11,450	15%
\$32,550	\$65,100	\$32,550	\$43,650	25%
\$78,850	\$131,450	\$65,725	\$112,650	28%
\$164,550	\$200,300	\$100,150	\$182,400	33%
\$357,700	\$357,700	\$178,850	\$357,700	35%

$$\text{For } \$32{,}550 \text{ to } \$78{,}850\text{: Lump sum} = 10\% \times 8{,}025 + 15\% \times (32{,}550 - 8{,}025) = \$4{,}481.25$$

For incomes in the \$78,850 to \$164,550 bracket:

$$\text{Lump sum} = 10\% \times 8{,}025 + 15\% \times (32{,}550 - 8{,}025) + 25\% \times (78{,}850 - 32{,}550) = \$16{,}056.20$$

Using the lump sum column, tax computations are straightforward:

$$\begin{aligned}
\text{Tax for } \$10{,}000 \text{ taxable income} &= 802.50 + 15\% \times (10{,}000 - 8{,}025) \\
&= \$1{,}098.75 \\
\text{Tax for } \$50{,}000 \text{ taxable income} &= 4{,}481.25 + 25\% \times (50{,}000 - 32{,}550) \\
&= \$8{,}843.50 \\
\text{Tax for } \$60{,}000 \text{ taxable income} &= 4{,}481.25 + 25\% \times (60{,}000 - 32{,}550) \\
&= \$11{,}343.75
\end{aligned}$$

A summary table for other individual taxpayer filing statuses is shown in Table 9.4. Notice that although lump sums are not included, the table is perfectly usable. Table 9.5 includes the brackets and rates for corporations.

TABLE 9.5 Corporate Income Brackets and Tax Rates, 2008

Bracket floor	Lump sum	Marginal tax rate
$0	$0	15%
$50,000	$7,500	25%
$75,000	$13,750	34%
$100,000	$22,250	39%
$335,000	$113,900	34%
$10,000,000	$3,400,000	35%
$15,000,000	$5,150,000	38%
$18,333,333	$6,416,667	35%

TAX CREDITS

A tax credit is an allowance that can be deducted directly from the actual tax to be paid as computed previously, instead of from the individual's or corporation's income. This is the case, for example, of the Hope credit and the Lifetime Learning credits, which can be claimed by college students when complying with given requisites. These credits cause a direct reduction in the amount of tax due for payment.

Other examples of tax credits are those given for child and dependent care expenses and the residential energy credit. As stated previously, these allowances usually subsidize a reputedly desirable behavior (college education and energy saving investments) or attempt to provide relief to individuals confronted with trying financial circumstances (child and dependent care, etc.). Similarly, corporations can claim credits. These credits tend to encourage business activities in certain regions (e.g., the American Samoa Economic Development Credit), reimburse taxes paid overseas, or foster a desirable behavior (e.g., the Rehabilitation Tax Credit).

It must be repeated that credits are subtracted directly from the computed tax. To emphasize the effect of a tax credit, consider the case of a person owing $10,000 in federal income tax who decided to purchase a hybrid car in 2008. At the time, the IRS allowed a credit of up to $3,000 to foster the use of this technology. By purchasing a model qualified for the full $3,000 credit the person would have lowered the tax to $7,000 instead of the original $10,000. In other words, a tax credit results in a direct reduction of the tax paid, whereas a deduction results in a reduction of the amount that is taxable. Therefore, a deduction of $3,000 for a person in the 28% tax bracket reduces the tax actually paid by only $3,000 × 0.28 = $840—not $3,000. Clearly, a tax credit is more desirable.

TAX PAYROLL WITHHOLDING

Employers are required to retain an amount from each paycheck that is roughly enough to cover the employee's tax liability at the end of the year. This withholding tax reduces the impact of the thousands of dollars levied in taxes to any average employee. The amount held back is essentially an escrow account held for the employee by the employer. At the time of payment, this escrow, or withheld, amount provides the employee with a basis for paying tax. As many taxpayers know, it often occurs that the total retention exceeds the total tax owed, in which case the IRS refunds the amount of over payment to the individual taxpayer.

TAX PAYMENT SCHEDULES

Corporations and some individuals must pay their taxes in quarterly installments, although their deadlines are slightly different. The requirement to make quarterly payments is similar, in concept, to withholding taxes for the individual employee discussed in the previous section. Installments are intended to avoid the situation of a company being confronted with a large payment at the end of the tax period. It ensures that companies and individuals "pay as they go" and spread tax payments over the tax period. This avoids the shock of paying a single large "balloon" payment at the end of the tax period.

Corporations can select a fiscal year, which does not need to coincide with the calendar year. While a calendar year ends on December 31st, a corporate fiscal year can end on any arbitrary date, usually at the end or the middle of a month. Many retailers, for example, do not like a December 31st fiscal year end because they are still occupied with the busy Christmas season. Nevertheless, about two thirds of American corporations choose to have their fiscal year end with the calendar year. A corporation with this calendar must estimate its profit and submit installments on the 15th of April, June, September, and December. Proprietorships and other organizational forms without an independent tax standing, as well as individuals with substantial income that is not subject to withholding, use the same calendar for the first three installments (15th of April, June, and September) but pay the last installment on January 15th of the following year. Oddly enough, this means that there are 2 months between the 1st and 2nd payments, 3 months between the 2nd and 3rd payment, and 4 months between the 3rd and last payment.

MARGINAL, AVERAGE, AND EFFECTIVE TAX RATES

The previous discussions of taxable income and tax computations yield insight about the tax paid by an individual or corporation compared to its income. As mentioned before, a progressive system imposes more taxes as a percentage of income on higher income. But, it is also important to view taxes as a percentage of the money earned by the taxpayer.

- The marginal tax rate is the percentage that an additional dollar earned by a taxpayer would have to pay. The marginal tax rate is relevant to correctly analyze the money saved or paid by relatively small changes in income. In the discussion of organizational forms, it was mentioned in the partnership form that Elmer Fudd (married filing jointly) would earn $102,000 and pay taxes for around $13,713 as a partner in a partnership. Table 9.3 shows that for a couple filing jointly, this income belongs to the bracket taxed at 25%, that is, its marginal tax rate would be 25%. This means that if he decides to give $1,000 to charity, he will be able to reduce his taxes by 25% × $1,000 = $250.
- The average tax rate is the ratio of total taxes paid to taxable income. The average tax rate paid by any individual with income over the minimum bracket will always be lower than the marginal rate, since the average includes the fraction of the income taxed at the lower tax rate of the low-income brackets. This is not true for corporations, as can be observed in its convoluted table. Returning to Mr. Fudd, his taxable income would be $102,000 (gross adjusted income) – $10,900 (std. deduction) – 2 × $3,500 = $84,100. Therefore, his average tax rate would be $13,713/$84,100 = 16.3%.
- The effective tax rate results from dividing the amount of paid taxes by the income without considering deductions or exemptions. Elmer Fudd had an effective rate of $13,713/$102,000 = 13.4%. The spread between average and effective income taxes is low in this case, compared to many large corporations with large tax breaks. The IRS reported that in 1999 the effective rate paid (in average) by the top 10,000 largest corporations was 20%, considerably less than their average tax rate of near 34%.

NET OPERATING LOSSES

A company has a net operating loss (NOL) when it reports a higher amount of allowable deductions than revenue. NOL is generally close, but not equal, to the loss (if any) reported in the company's Income Statement. As previously discussed, there are deductions for tax purposes that may be computed differently in the company's accounting system (e.g., depreciation) and may not appear in the Income Statement (e.g., incentives to locate in certain areas), or vice versa, may not be acceptable for tax purposes (e.g., some public relations and travel expenses). Individuals and all forms of business organizations can report a NOL.

No income taxes are levied in the year that an individual or company has an NOL. Furthermore, the loss can be used to revise the taxable income of previous years and to lower the taxable income of future years. The NOL applied to past years is called a carryback, and the amount applied to future years is a carryforward.

The rules applied to carryback and carryforward have changed with every major revision of the tax code, and depend on the type of loss, income, and even the institution (banks, for example, have their set of rules in this area). Being aware of the concepts of carryback and carryover should be enough for the intended level of this book. Carryback has been limited to the previous 3 years and carry forward to the next 15 years for most situations applicable to construction companies. The following example helps in the understanding of the process.

Suppose that a company reports a NOL of $1,000,000 for 2009. It had a taxable income of $200,000 in each of the previous four years. How would this company apply this NOL to revise previous tax computations and as a deduction for coming years?

The carryback process is usually performed before the carryforward, because it results in immediate tax relief (a refund) from the IRS. An exception happens when a company finds that the marginal tax rate of previous years is lower than the expected marginal rate for coming years. In such a (uncommon) case, the difference in rates may justify the wait. In this example, the company chooses to apply the carryback first. The taxable income of each of the three previous years is revised to zero:

- In 2008, the amount available to carry back is $1,000,000. After subtracting the $200,000 originally reported as taxable income, the revised taxable income is zero, and the amount available to further carryback is $800,000.
- In 2007, the revised taxable income is also zero, and the remaining carryback is $600,000.
- In 2006, the same process is applied. The revised taxable income is zero, and the remaining carryback is $400,000
- The taxable income for 2005 cannot be revised, despite the $400,000 of NOL still available, since this year is more than 3 years before the NOL.

The company will send the IRS revised tax returns showing zero taxable income for 2008, 2007, and 2006. It will get a refund for all taxes paid in those years (not $200,000 for each year but the tax amount paid in each year).

The unapplied portion of the NOL is saved as carryforward for the coming years. If the company has a taxable income before applying the NOL of $100,000 in 2010, it can set to zero the taxable income and still have $300,000 available. If it has a taxable income of $450,000 in 2011, it would apply all the remaining funds from the NOL to pay taxes on a taxable income of $150,000.

TAXES ON DIVIDENDS AND LONG-TERM CAPITAL GAINS

Dividends, as discussed in the accounting section of this book, are the payments made by a corporation to its shareholders. They are the portion of the corporation's profit not reinvested as retained earnings.

As noted in the section entitled "Straight Line Method," when a capital asset is held for more than a year and then sold at a higher price than its depreciated or base value, the resulting profit is called a long-term capital gain. An asset such as a piece of equipment or a building, typically appear in the Non-Current Assets section of a company's balance sheet. To qualify as a long-term capital gain, the capital asset cannot be purchased for the purpose of being resold.

Dividends and long-term capital gains are taxed under different rules. Currently, a flat tax of 15% is levied for dividends, as well as for long-term capital gains with the exception of real estate.[5] Since this part of a company's or individual's income is treated differently, dividends and long-term capital gains are not counted in the regular income tax computations discussed so far.

There are several types of capital gains that result in different rates. Long-term gains involving real estate are taxed at up to 25%, and collectibles at up to 28%. Short-term capital gains (those from capital assets held for less than a year) and stock in many small businesses are taxed at the marginal rate of regular income. Complicating this topic, current rates are scheduled to expire in 2010, making these specifics moot unless new legislation extends their term.

ALTERNATIVE MINIMUM TAX

The tax rules discussed so far can result in some wealthier individuals and corporations paying proportionally less taxes than lower-income taxpayers. The Alternative Minimum Tax (AMT) amounts to a parallel set of rules designed to avoid any tax loopholes in the federal tax system. This is accomplished by a simpler, flatter system, which sets the individual or corporation taxable income at a level much closer to their income as computed by accounting standards. In other words, it allows fewer exemptions to individuals and corporations falling under its scope. Its tax rates are nearly flat: 26–28% for individuals, 20% for corporations. If an individual or corporation has a high level of income, it must perform the AMT calculations. The amount owed will be the higher of the tax owed using the regular approach or the AMT.

The AMT was first applied in 1970 and revised in 1986. Its original scope made it applicable to literarily a few dozen very wealthy families and a few hundred large corporations. Unfortunately, its underlying legislation does not include a provision to adjust for inflation, and its thresholds have become

---

[5] Individuals with taxable income falling in the 10% or 15% brackets of their applicable tax table are subject to a long-term capital gain/dividend rate of 5% instead of 15%.

so low that an upper-middle income family may find itself dealing with this parallel system. A Faustian dilemma has developed. Although it is clear that the AMT is changing its fundamental purpose by including upper-middle-class families, cutting off the tax revenue generated from the AMT paid by these families would lead to a reduction of billions of dollars now paid into the coffers of the government.

The corporate AMT is more complex than the individual version. It is not applicable to small corporations and has its own set of deductions (preferences) and allows the equivalent of a "personal exemption" that is tapered off for higher alternative minimum taxable incomes (AMTI)

In order to deal with this and a host of other tax-related issues, even relatively small construction firms organized in the corporate form, require the service of experienced tax professionals.

SUMMARY

This chapter has discussed issues related to taxation that influence the bottom line of a construction company and impact financial decisions. Clearly, minimizing the amount of taxes paid increases the profit and retained earnings controlled by the company and ultimately distributed to the owners and stockholders. One area, which is often discussed in conjunction with taxation, is contributions made by companies to ensure the welfare of employees. Companies are involved in paying Social Security contributions required by federal law, Unemployment and Workers' Compensation insurance required by state governments, and so-called "fringe" benefits required of union contractors by union contracts. Although these are not strictly taxes, they do have a similar impact on the profitability of the company and must be carefully analyzed. A cursory introduction to these charges and how they impact labor costs is presented in *Construction Management* by Halpin published by John Wiley and Sons, Inc. (Halpin, 2006). The company's top management, and particularly the chief financial officer (CFO), must continuously study and analyze the impact of taxation on all financial and organizational decisions. Taxes are business costs which lead to a 34% reduction in before tax income and must be considered very seriously. These costs require at least the same level of professional and financial expertise that is required from a technical perspective to build innovative and successful construction projects.

REVIEW QUESTIONS AND EXERCISES

1. Explain in your own words the concept of marginal tax rate.
2. Does a single individual making $500,000 pay the same marginal tax rate as another making $1,000,000? Explain.
3. You decide to sell a house that you bought five years ago with the purpose of renting it. Its purchase price was $300,000, and now its market value

is about $500,000. How much should you expect to pay in taxes for the sale of this house?

4. What is the average tax rate paid by a single person with a $60,000 AGI for 2008?
5. What is the name of the amount of money that can be deducted directly from the computed tax, instead of from the gross adjusted income?
6. Which business enterprises pay untaxed profits to owners/partners who then pay the taxes on their individual returns?
7. Peachtree Construction Inc., which is incorporated as a corporation, reported a net operating loss (NOL) of $500,000 in 2008. Given the following data, how much of this NOL is likely to be available as a deduction from its profit in 2009?
   - The company was created in 2006.
   - It had a NOL of $100,000 in 2006.
   - It reported a profit before taxes of $400,000 in 2007.
8. A corporation had a net income before taxes of $1,200,000 for 2008. Find the tax liability for this company.
9. Pegg Construction, a partnership of three brothers, had a net income before taxes of $300,000. The brothers have agreed that all profit will be shared in equal parts. John Pegg is single, and has an income of $500,000 from other sources. How much will Mr. Pegg pay in taxes for the $100,000 from the partnership with his siblings?
10. Let us assume that we own our personal residences. One question of interest is, "If we depreciate real property (e.g. land, buildings, etc.), can we depreciate the house in which we live?" Depreciation allows for the recovery of the cost of doing business. Since, in most cases, we do not "do business" in our home, our residence is not a depreciable asset. You can, however, think of some instances in which a person conducts business at home. Can such a person working at home, depreciate the cost of his/her house?
11. Why are governments willing to agree to accept the concept of depreciation, which leads in many cases to a considerably lower level of taxation for business entities? The answer to this question provides an insight into how modern economies operate. How are economies impacted by depreciation?
12. Verify the 5- and 7-year property class percentages given in Table 9.2 by applying the 200% DB approach to a piece of equipment with a nominal value of $1,000. For the 7-year property class, in what year is the switch from 200% DB to Straight Line made based on the percentages in the Table 9.2. Only half of the first and last years are considered in the computations. This means that for a five-year class property, only half the 200% DB amount (40%/2 = 20%) is deducted.

13. Why do you think the 150% DB method is used as the basis for depreciation for 15- and 20-year classes of equipment, as given in Table 9.2?
14. Find the yearly depreciation of a $100,000 five-year class concrete mixer, first assuming that it is new when purchased, and then that it is purchased as used equipment.

APPENDIX A

TYPICAL CHART OF ACCOUNTS

TABLE A.1 General Ledger Accounts

Assets	
10.	Petty cash
11.	Bank deposits
.1	General bank account
.2	Payroll bank account
.3	Project bank accounts
.4	
12.	Accounts receivable
.1	
.2	Parent, associated, or affiliated companies
.3	Notes receivable
.4	Employees' accounts
.5	Sundry debtors
.6	
13.	Deferred receivables *All construction contracts are charged to this account, being diminished by progress payments as received. This account is offset by Account 48.0, Deferred Income.*
14.	Property, plant, and equipment
Property and General Plant	
.100	Real estate and improvements
.200	Leasehold improvements
.300	Shops and yards

(*continued*)

TABLE A.1 (*Continued*)

Mobile Equipment	
.400	Motor vehicles
.500	Tractors
.01	Repairs, parts, and labor
.05	Outside service
.12	Tire replacement
.15	Tire repair
.20	Fuel
.25	Oil, lubricants, filters
.30	Licenses, permits
.35	Depreciation
.40	Insurance
.45	Taxes
.510	Power shovels
.520	Bottom dumps
Stationary Equipment	
.530	Concrete mixing plant
.540	Concrete pavers
.550	Air compressors .560
Small Power Tools and Portable Equipment	
.600	Welders
.610	Concrete power buggies
.620	Electric drills .630
Marine Equipment	
.700	
Miscellaneous Construction Equipment	
.800	Scaffolding
.810	Concrete forms
.820	Wheelbarrows
.830	
Office and Engineering Equipment	
.900	Office equipment
.910	Office furniture
.920	Engineering instruments
15.	Reserve for depreciation
16.	Amortization for leasehold
17.	Inventory of materials and supplies
.1	Lumber
.2	Hand shovels

TABLE A.1

.3	Spare parts
.4	
	These accounts show the values of all expendable materials and supplies. Charges against these accounts are made by authenticated requisitions showing project where used.
18.	Returnable deposits
.1	Plan deposits
.2	Utilities
.3	
Liabilities	
40.	Accounts payable
41.	Subcontracts payable
42.	Notes payable
43.	Interest payable
44.	Contracts payable
45.	Taxes payable
.1	Old-age, survivors, and disability insurance (withheld from employees' pay)
.2	Federal income taxes (withheld from employees' pay)
.3	State income taxes (withheld from employees' pay)
.4	
46.	Accrued expenses
.1	Wages and salaries
.2	Old-age, survivors, and disability insurance (employer's portion)
.3	Federal unemployment tax
.4	State unemployment tax
.51	Payroll insurance (public liability and property damage)
.52	Payroll insurance (workmen's compensation)
.6	Interest
.7	
47.	Payrolls payable
48.	Deferred income
49.	Advances by clients
Net Worth	
50.	Capital stock
51.	Earned surplus
52.	Paid-in-surplus
53.	
Income	
70.	Income accounts
.101	Project income
.102	
.2	Cash discount earned

(continued)

TABLE A.1 (*Continued*)

.3	Profit or loss from sale of capital assets
.4	Equipment rental income
.5	Interest income
.6	Other income
Expense	
80.	Project expense
.100	Project work accounts
.700	Project overhead accounts
	These are control accounts for the detail project cost accounts that are maintained in the detail cost ledgers.
81.	Office expense
.10	Officer salaries
.11	Insurance on property and equipment
.20	Donations
.21	Utilities
.22	Telephone and telegraph
.23	Postage
.30	Repairs and maintenance
82.	Yard and warehouse expense (not assignable to a particular project)
.10	Yard salaries
.11	Yard supplies
83.	Estimating department expense accounts
.10	Estimating salaries
.11	Estimating supplies
.12	Estimating travel
84.	Engineering department expense accounts .10
85.	Cost of equipment ownership
.1	Depreciation
.2	Interest
.3	Taxes and licenses
.4	Insurance
.5	Storage
86.	Loss on bad debts
87.	Interest
90.	Expense on office employees
.1	Workmen's compensation insurance
.2	Old-age, survivors, and disability insurance
.3	Employees' insurance
.4	Other insurance
.5	Federal and state unemployment taxes
.6	
91.	Taxes and licenses
.1	Sales taxes
.2	Compensating taxes
.3	State income taxes
.4	Federal income taxes

APPENDIX B

FURTHER ILLUSTRATIONS OF TRANSACTIONS

TABLE B.1 Composite of Ledger Entries 1–20

Cash	
DR +	CR −
$ 50,000(1)	$330,000(5)
50,000(1)	5,000(5)
360,000(4)	85,000(5)
20,000(4)	2,000(5)
90,000(4)	11,000(9)
6,000(4)	2,250(10)
5,000(7)	33,500(11)
20,000(7)	3,160(13)
4,000(20)	

Accounts Receivable	
DR +	CR −
$414,000(2)	
112,500(2)	
	$360,000(4)
27,000(6)	90,000(4)
	20,000(7)

Retainage Receivable	
DR +	CR −
$46,000(2)	
12,500(2)	
	$20,000(4)
	6,000(4)
3,000(6)	5,000(7)

Accrued Interest Receivable	
DR +	CR −
$4,000(19)	
	$4,000(20)

(*continued*)

TABLE B.1 ***(Continued)***

Work-in-Progress Expense (Asset)*	
DR +	CR −
$370,000(3)	
100,000(3)	
11,000(8)	
2,250(10)	

Prepaid Insurance	
DR +	CR −
$7,200(14)	

Fixed Assets	
DR +	CR −
$90,000(16)	

Accumulated Depreciation	
DR +	CR −
	$7,500(18)

Accounts Payable	
DR −	CR +
	$340,000(3)
	92,000(3)
$330,000(5)	
85,000(5)	
	10,000(8)
10,000(9)	3,160(12)
3,160(13)	7,200(14)

Accrued Liability	
DR −	CR +
	$3000(15)
	4800(17)

Retainage Payable	
DR −	CR +
	$30,000(3)
	8,000(3)
$5,000(5)	
2,000(5)	
1,000(9)	1,000(8)

Billings in Excess of Revenues	
DR −	CR +

Notes Payable—Short-term	
DR −	CR +
	$30,000(16)

Notes Payable-Long-Term	
DR +	CR −
	$60,000(16)

*This expense account is a nominal account used to capture expenses on jobs in progress pending reconciliation at closing of accounts. These expenses are viewed as assets until closed to the real account, "contract expense." This occurs at the end of the accounting period.

TABLE B.1

Advanced Billings	
DR −	CR +
	$460,000(2)
	125,000(2)
	30,000(6)

Retained Earnings	
DR +	CR −

Capital-Common Stock	
DR −	CR +
	$50,000(1)
	50,000(1)

Project Revenue (Sales)	
DR −	CR +

Project Expense	
DR −	CR +

General Overhead	
DR −	CR +
$33,500(11)	
3,160(12)	
3,000(15)	

Interest Income	
DR −	CR +
	$4000(19)

Interest Expense	
DR +	CR −
$4800(17)	

Depreciation Expense	
DR +	CR −
$7500(18)	

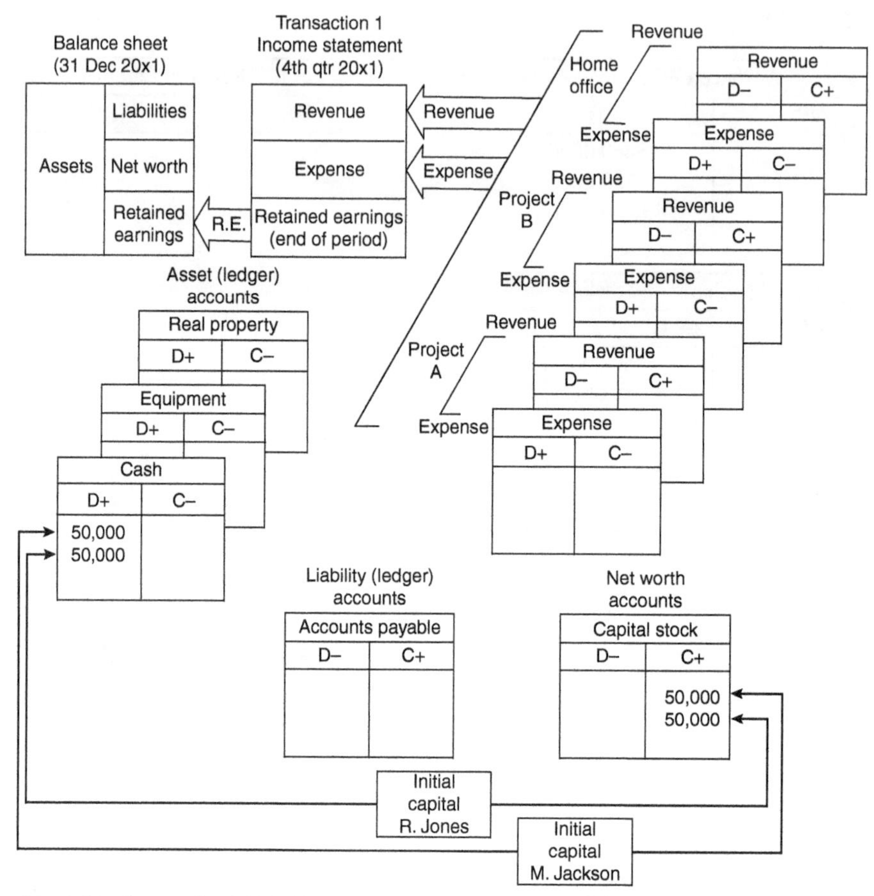

Description of transaction: Initial capital to form Apex Construction Company is contributed by R. Jones and M. Jackson. Each contributes $50,000. Stock in the amount of 10,000 shares is issued to each.

Journal entry:

Description	DR	CR
Cash	50,000	
Capital stock–R. Jones		50,000
Cash	50,000	
Capital stock–M. Jackson		50,000

Figure B.1

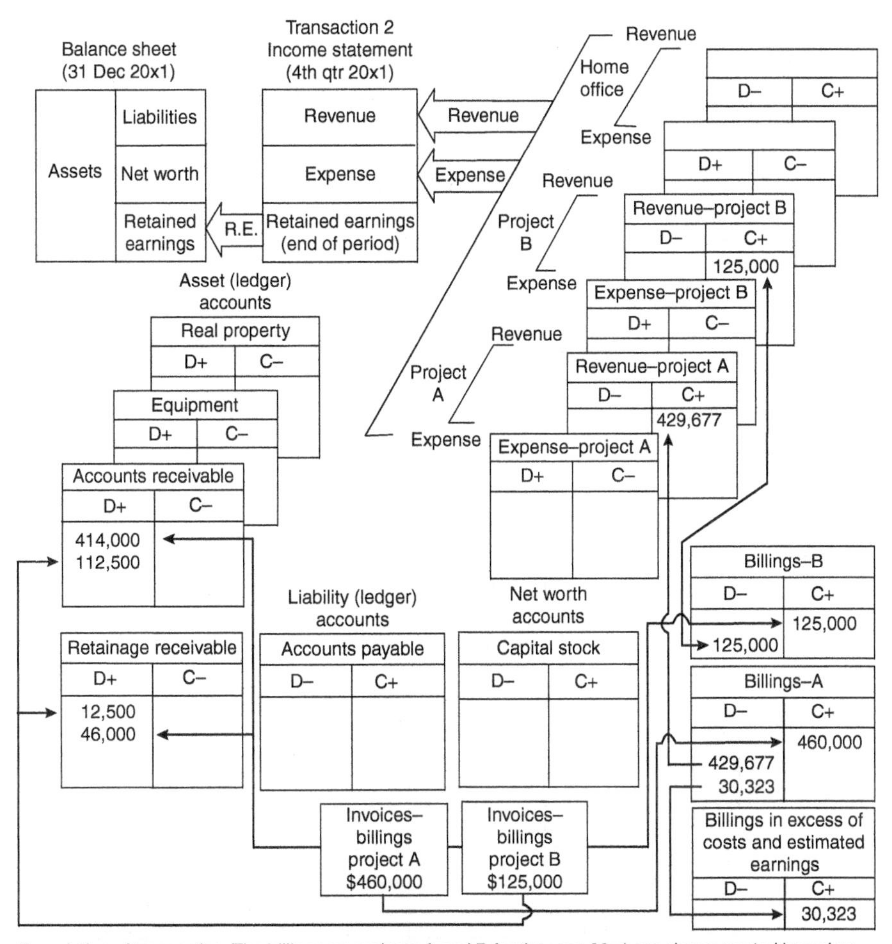

Description of transaction: The billings on projects A and B for the year 20x1 are shown posted based on the percentage-of-completion method. The actual billed amounts on A and B are $460,000 and $125,000, respectively. A retainage of 10 percent is held on each contract. Although many individual billings would be made over the year, they have been consolidated into single entries for the purpose of this illustration. The calculated revenues on A and B (with POC) are $429,677 and $125,000, respectively. The overbilling on A is accounted for in the liability account Billings in excess of costs and estimated earnings.

Description	DR	CR	
Accounts receivable	414,000		Accumulation of entries to billings through period
Retainage receivable	46,000		
Billings–project A		460,000	
Accounts receivable	112,500		
Retainage receivable	12,500		
Billings–project B		125,000	
Billings–project A	460,000		Closing entries
Revenue–project A		429,677	
Billings in excess of costs and estimated earnings		30,323	
Billings–project B	125,000		
Revenue–project B		125,000	

Figure B.2

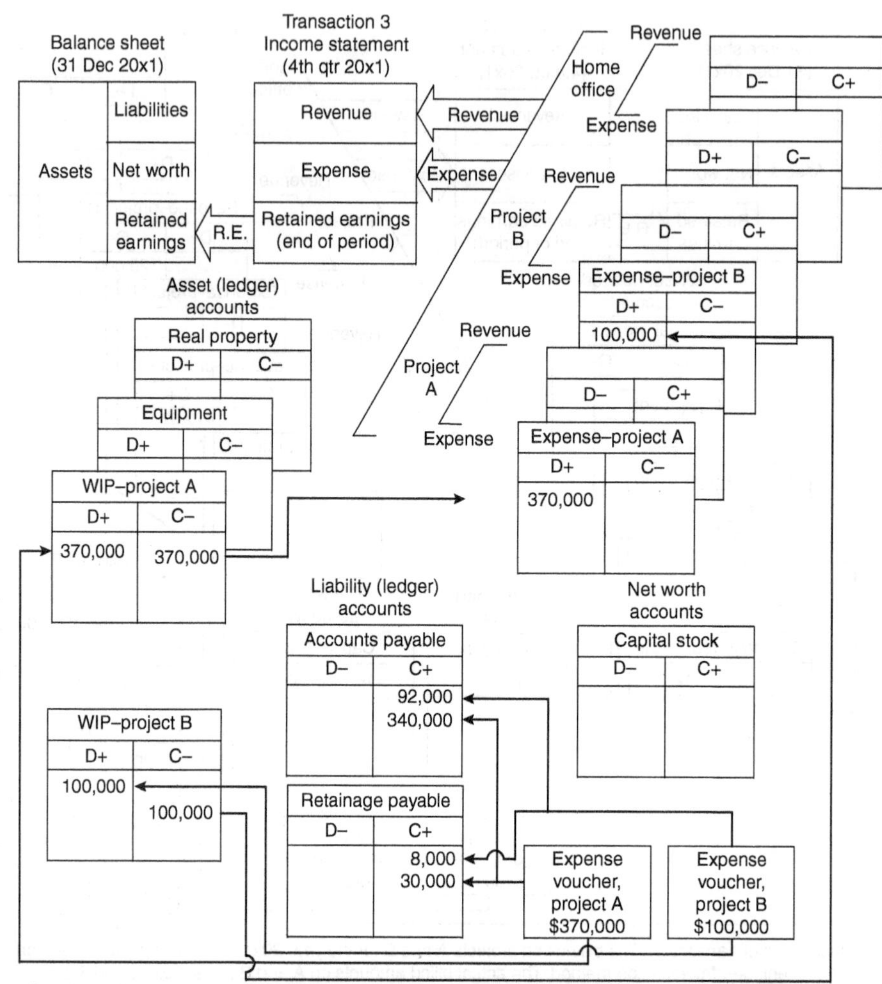

Description of transaction: The cost of projects A and B for the year 20x1 are $370,000 and $100,000, respectively. Retainage amounts of $30,000 and $8,000 respectively, on A and B are held by Apex from subcontract billings. These costs are carried as expenses under the appropriate project expense accounts. Although the costs would accumulate for many individual transactions through the year, they are shown as single lump sum entries in this example.

Journal entry:

Description	DR	CR	
WIP–expenses–project A	370,000		Accumulation of expense during the period
Accounts payable		340,000	
Retainage payable		30,000	
WIP–expenses–project B	100,000		
Accounts payable		92,000	
Retainage payable		8,00	
Project expenses–project A	370,000		Closing actions
Project expenses–project B	100,000		
WIP–expense–project A		370,000	
WIP–expense–project B		100,000	

Figure B.3

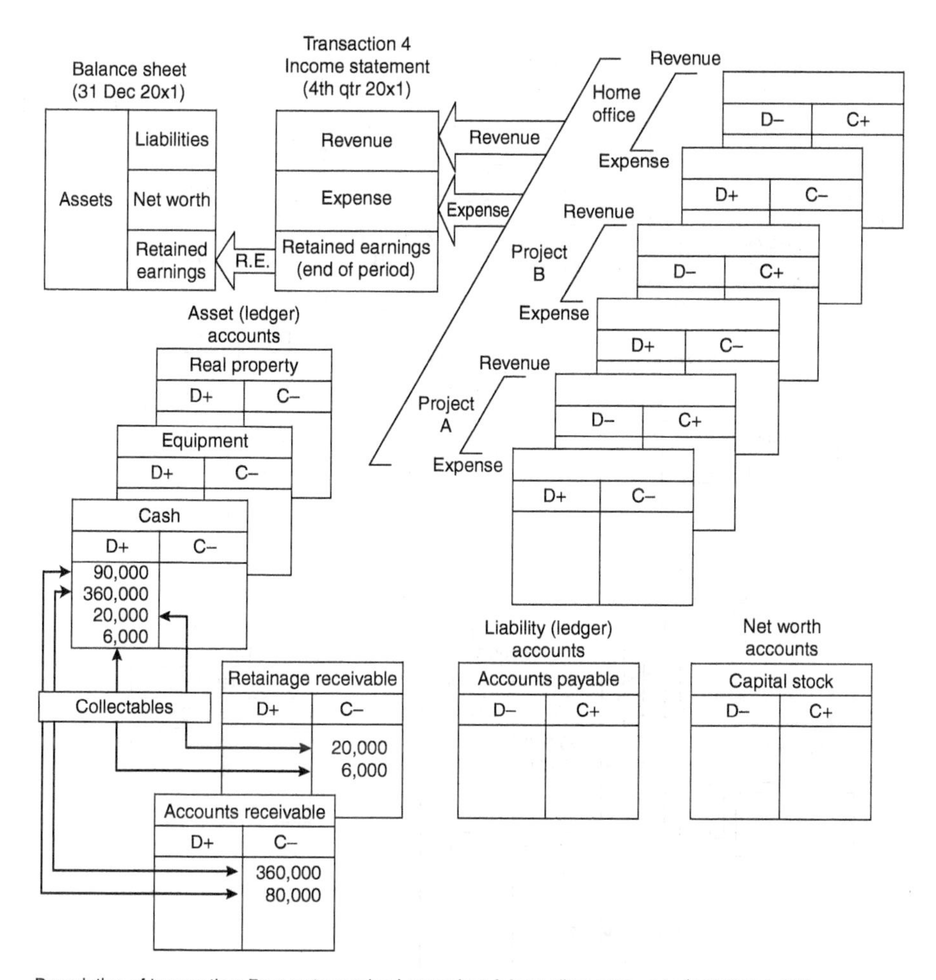

Description of transaction: Payments received on project A from client amount to $360,000 in billings and $20,000 in retainage. Payments received on project B amount to $90,000 in billings and $6,000 in retainage. These are accumulations of individual collections during the year.

Journal entry:

Description	DR	CR
Cash	360,000	
Accounts receivable		360,000
Cash	20,000	
Retainage receivable		20,000
Cash	90,000	
Accounts receivable		90,000
Cash	6,000	
Retainage receivable		6,000

Figure B.4

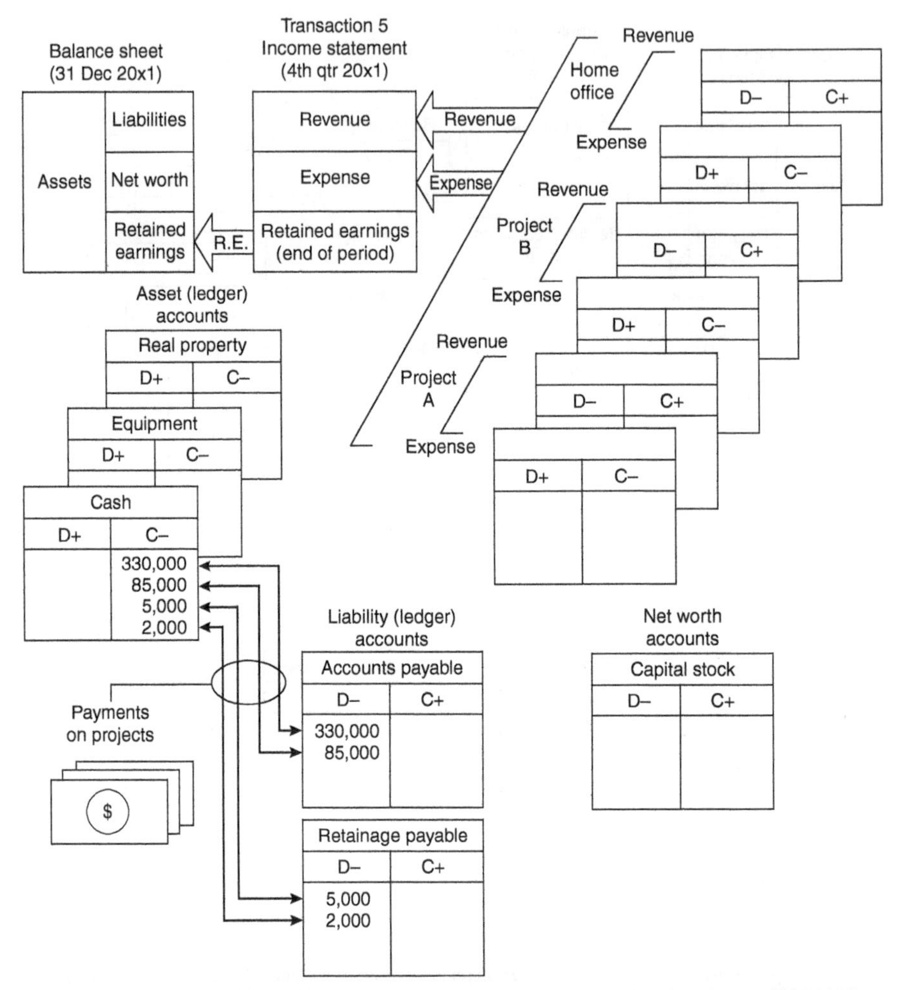

Description of transaction: Payments to vendors and subcontractors on project A amount to $330,000 in regular payments and $5,000 in retainage during period. Payments to vendors and subcontractors on Project B amount to $85,000 in regular payments and $2,000 in retainage during the period.

Journal entry:

Description	DR	CR
Accounts payable–A	330,000	
Cash		330,000
Retainage payable–A	5,000	
Cash		5,000
Accounts payable–B	85,000	
Cash		85,000
Retainage payable–B	2,000	
Cash		2,000

Figure B.5

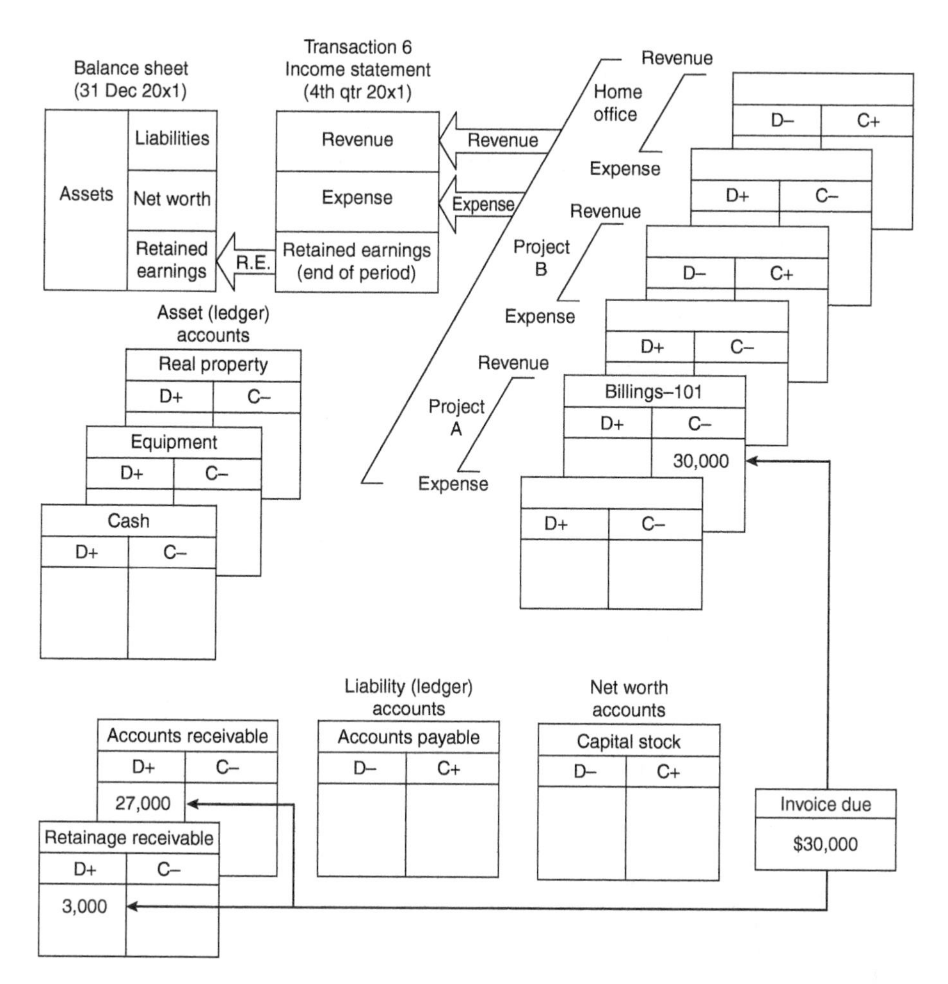

Description of transaction: Billing to a customer on project 101 of $30,000 including $3,000 retainage. Billing accounts are "work-in-progress–billings." Accounts during the period prior to clearing and closing.

Journal entry:

Description	DR	CR
Accounts receivable	27,000	
Retainage receivables	3,000	
Billings–project 101		30,000

Figure B.6

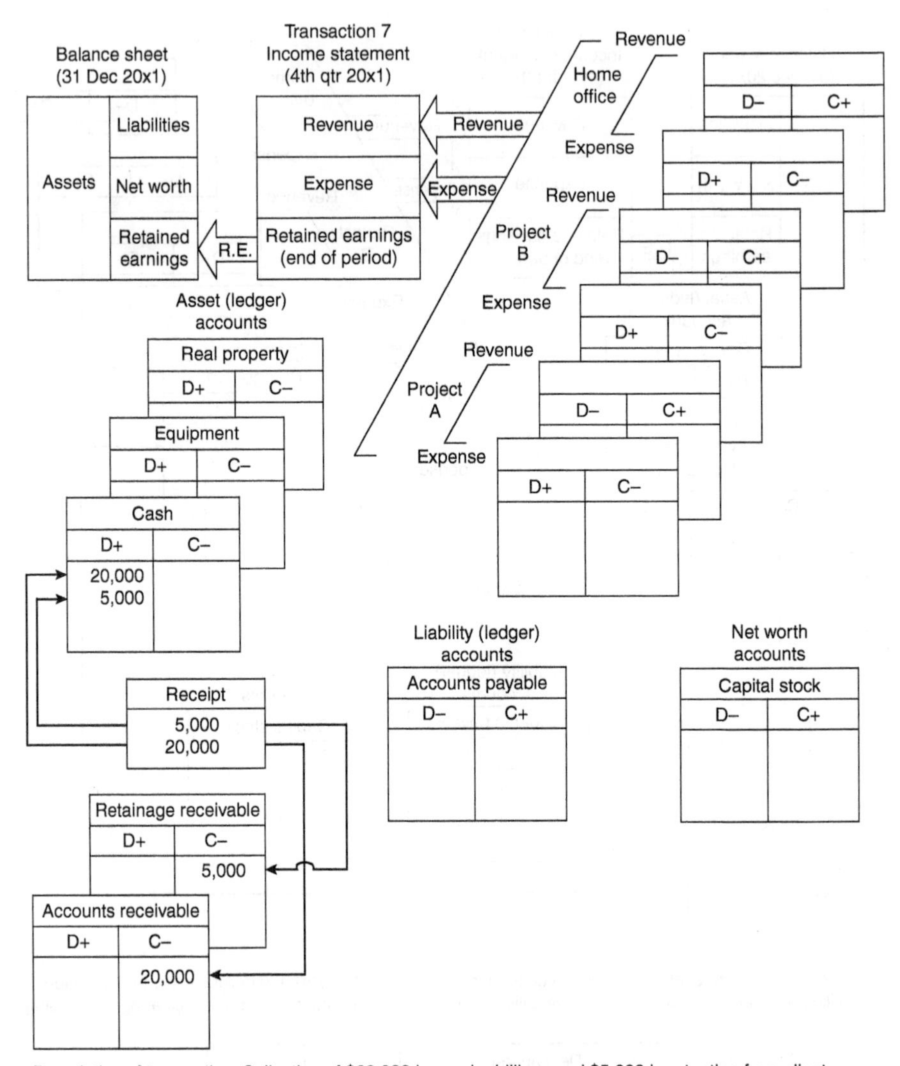

Description of transaction: Collection of $20,000 in regular billings and $5,000 in retention from client on project 101.

Journal entry:

Description	DR	CR
Cash	5,000	
Cash	20,000	
Retainage receivable		5,000
Accounts receivable		20,000

Figure B.7

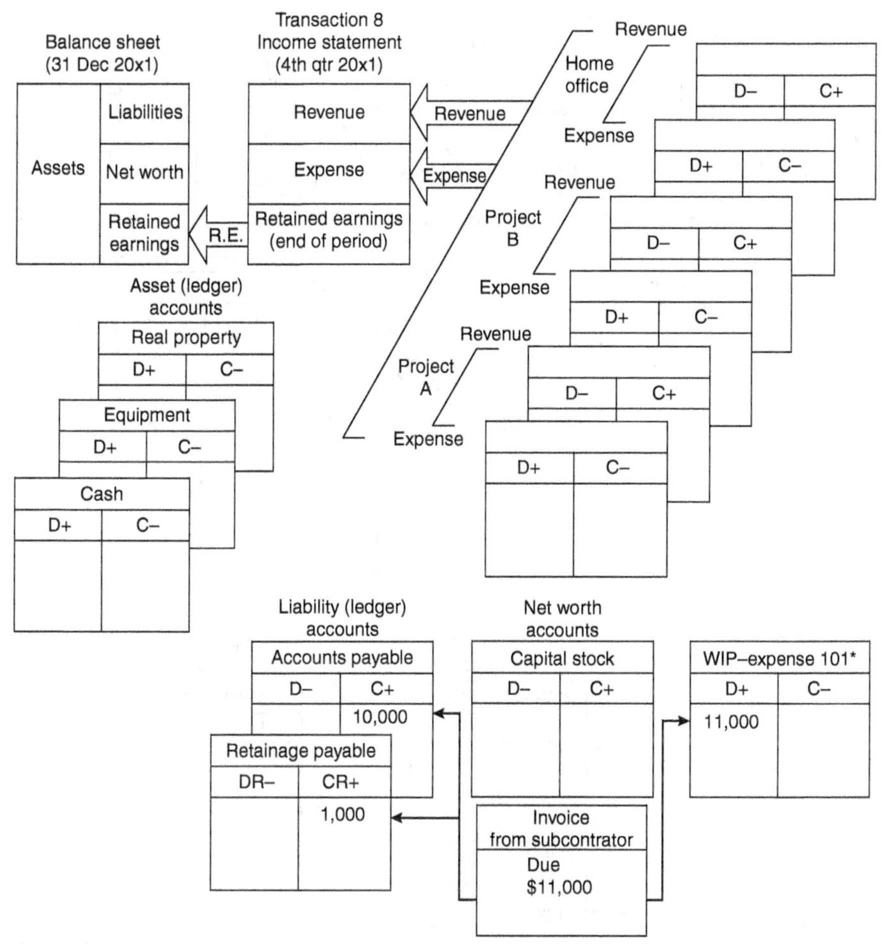

Description of transaction: Billing received from subcontractor in the amount of $10,000, including $1,000 retainage on project 101.

*Expenses are booked "work in progress–expense." Accounts prior to closing out of expense accounts at the end of the accounting period.

Journal entry:

Description	DR	CR
Work in progress– expense proj 101 (Jones Elec)	11,000	
Accounts payable		10,000
Retainage payable		1,000

Figure B.8

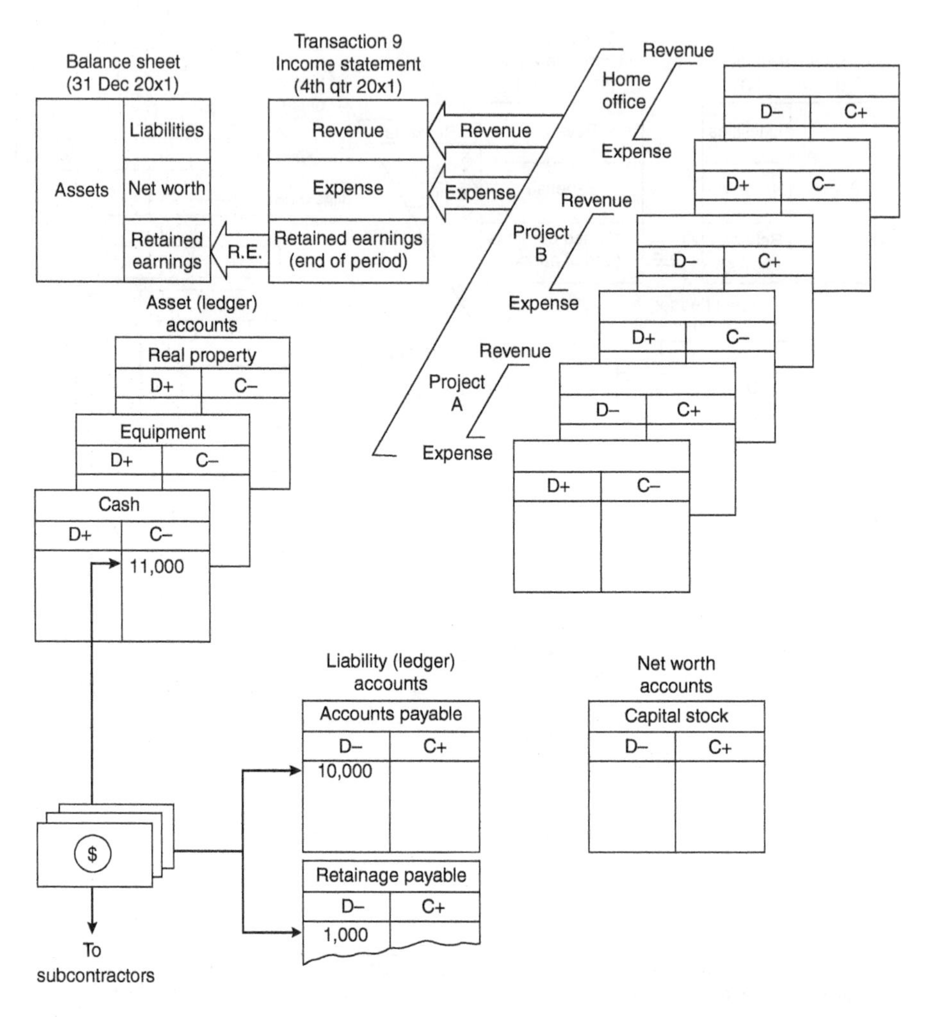

Description of transaction: Payment of $10,000 in regular billings and $1,000 in retainage on project 101. Subcontractor billings received previously.

Journal entry:

Description	DR	CR
Accounts payable	10,000	
Retainage payable	1,000	
Cash		11,000

Figure B.9

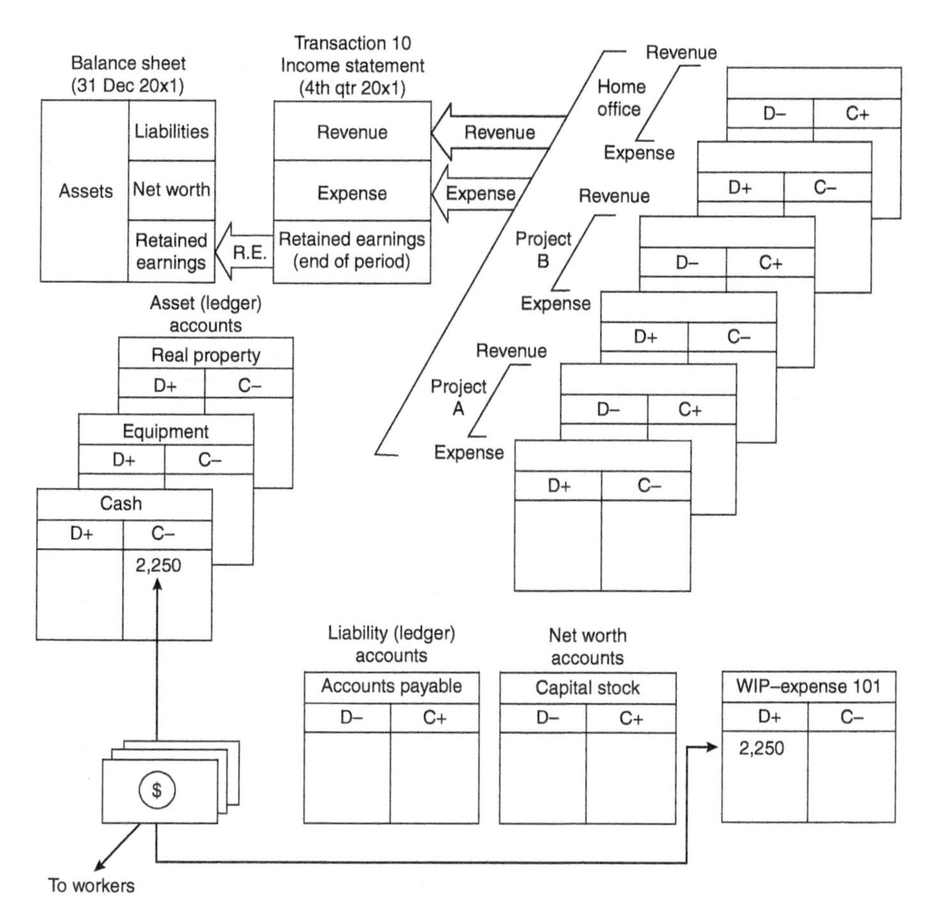

Description of transaction: Payment of weekly salary for craft workers on project 101 in the amount of $2,250.

Journal entry:

Description	DR	CR
WIP–expense Job 101 (payroll)	2,250	
Cash		2,250

Figure B.10

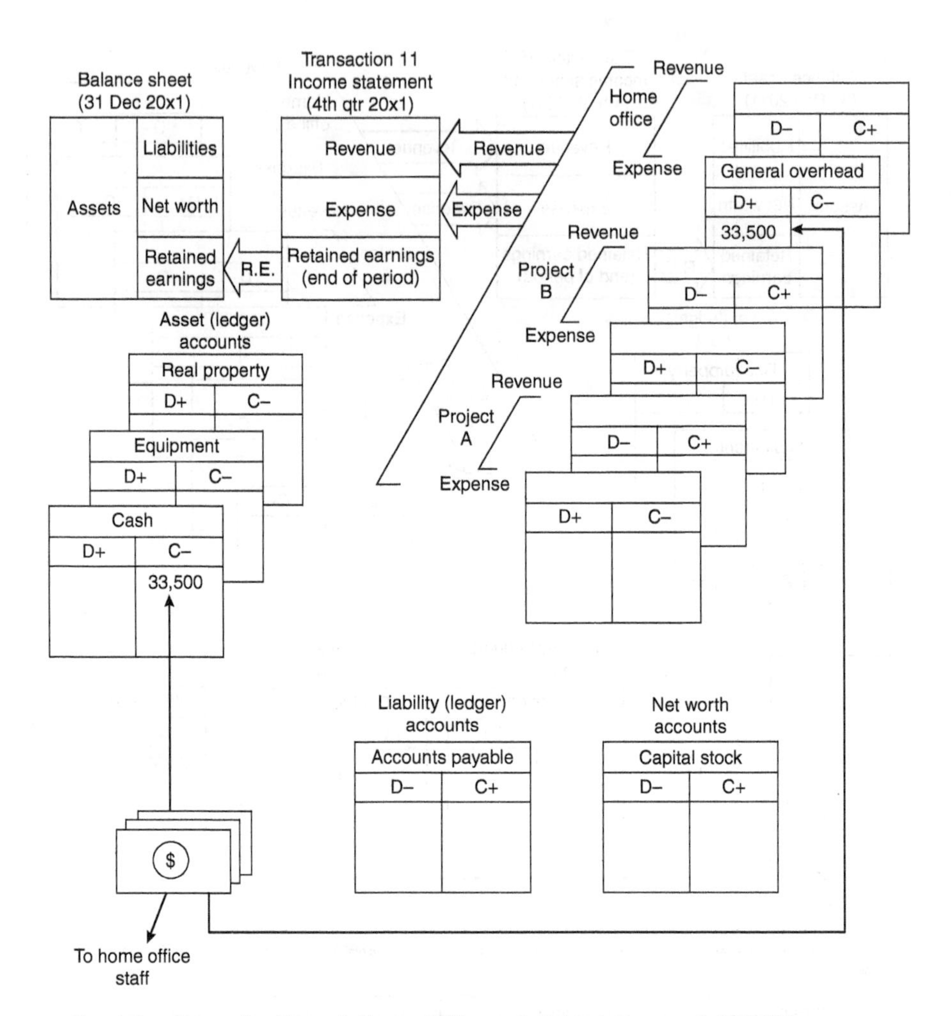

Description of transaction: Payment of home staff for month of June in the amount of $33,500.

Journal entry:

Description	DR	CR
General overhead (monthly home office salary)	33,500	
Cash		33,500

Figure B.11

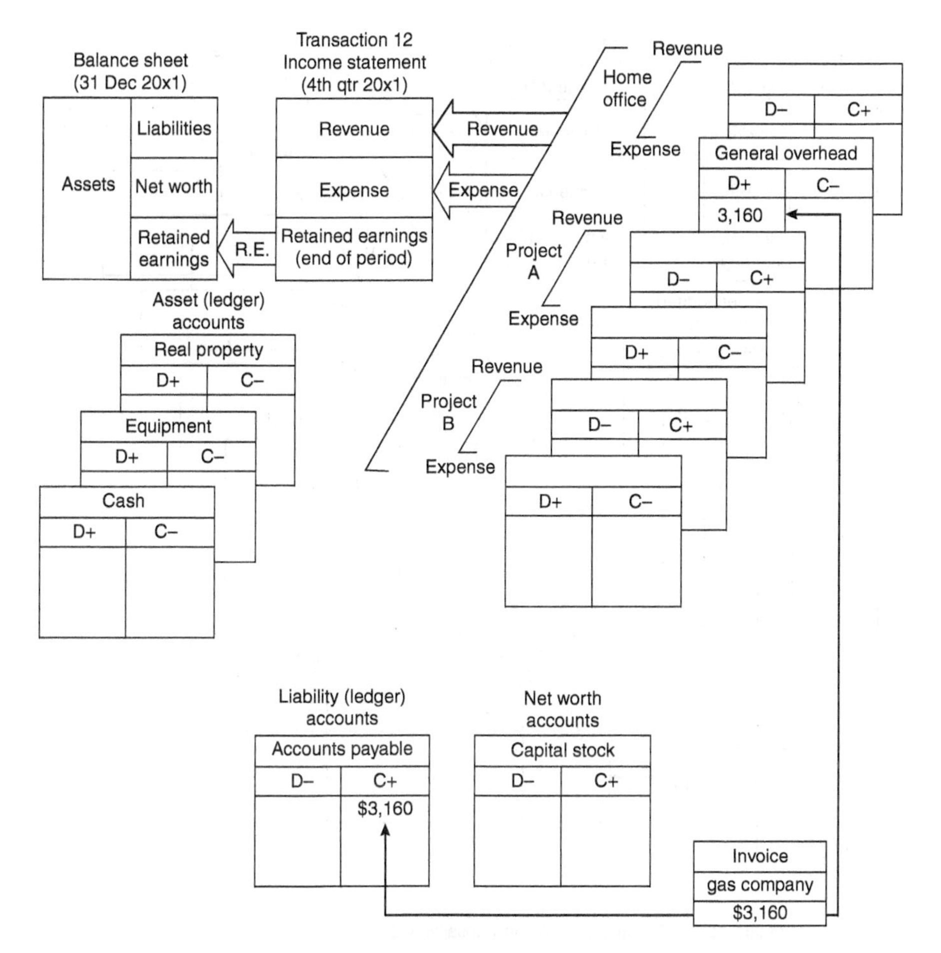

Description of transaction: Invoice for home office costs for heating headquarters building in the amount of $3,160.

Journal entry:

Description	DR	CR
General overhead (heating)	$3,160	
Accounts payable		$3,160

Figure B.12

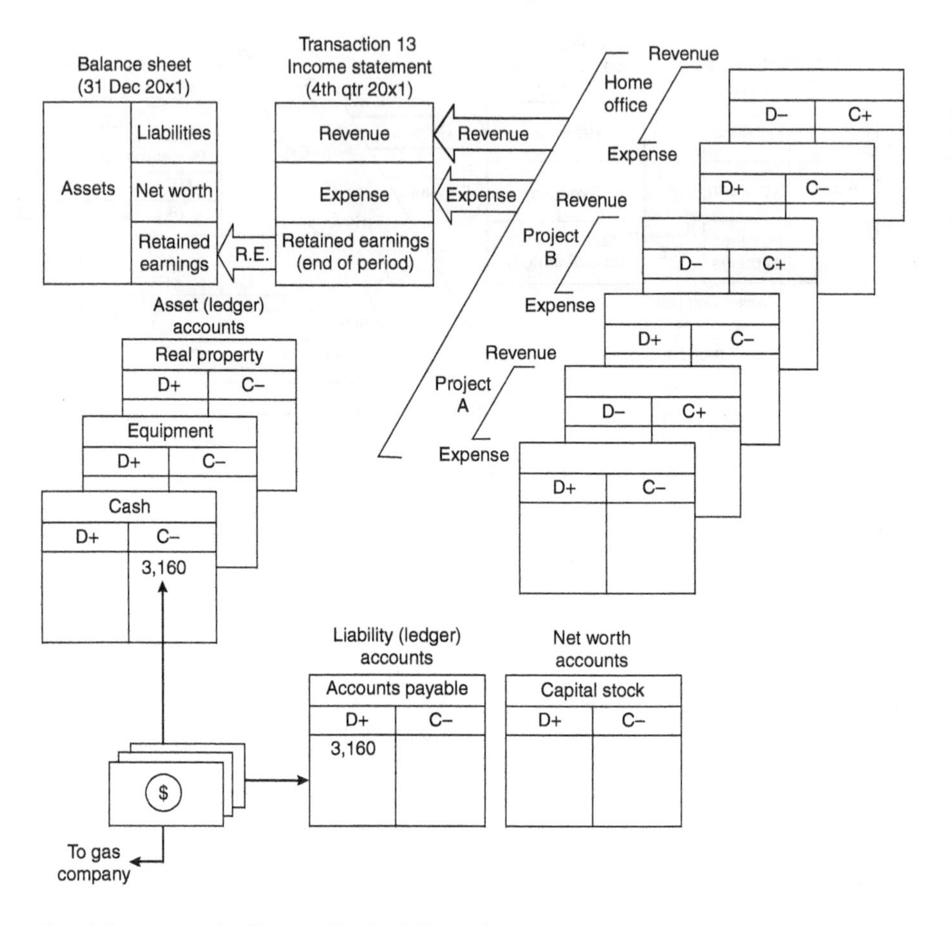

Description of transaction: Payment of invoice in Transaction 12.

Journal entry:

Description	DR	CR
Accounts payable (city gas company)	3,160	
Cash		3,160

Figure B.13

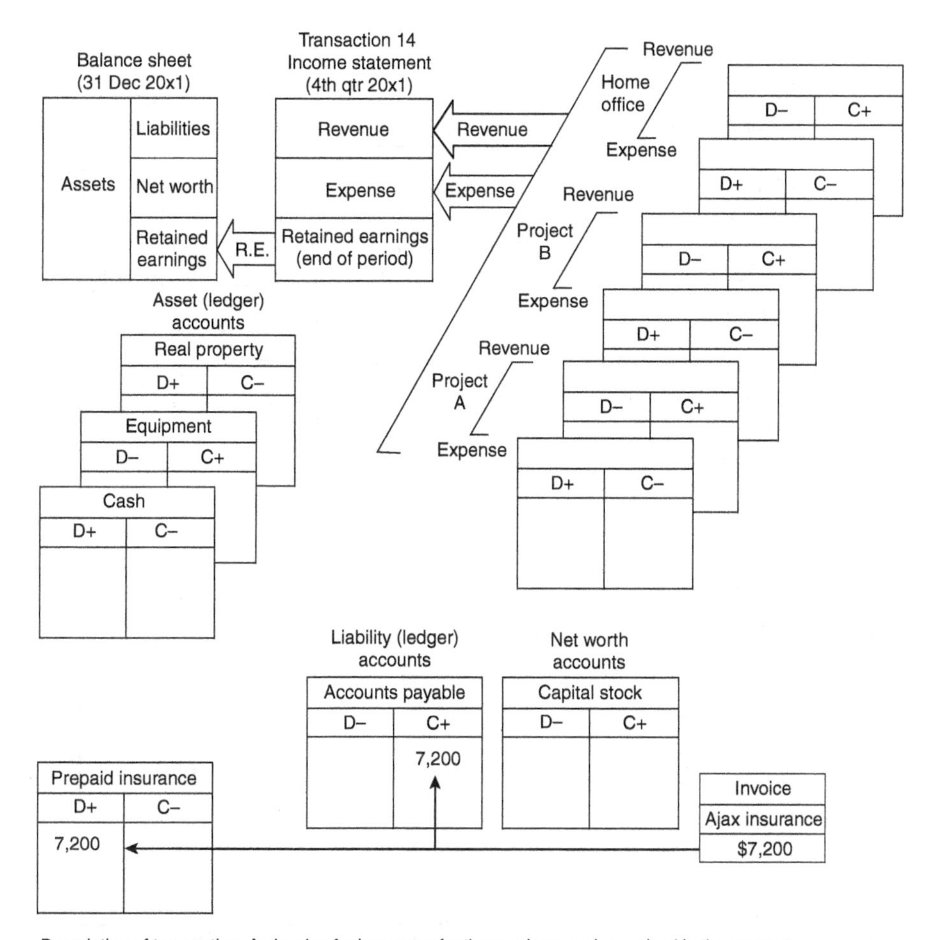

Decsription of transaction: An invoice for insurance for the coming year is received in the amount of $7,200. The payment represents a prepayment of insurance.

Journal entry:

Description	DR	CR
Prepaid insurance	7,200	
Accounts payable		7,200

Figure B.14

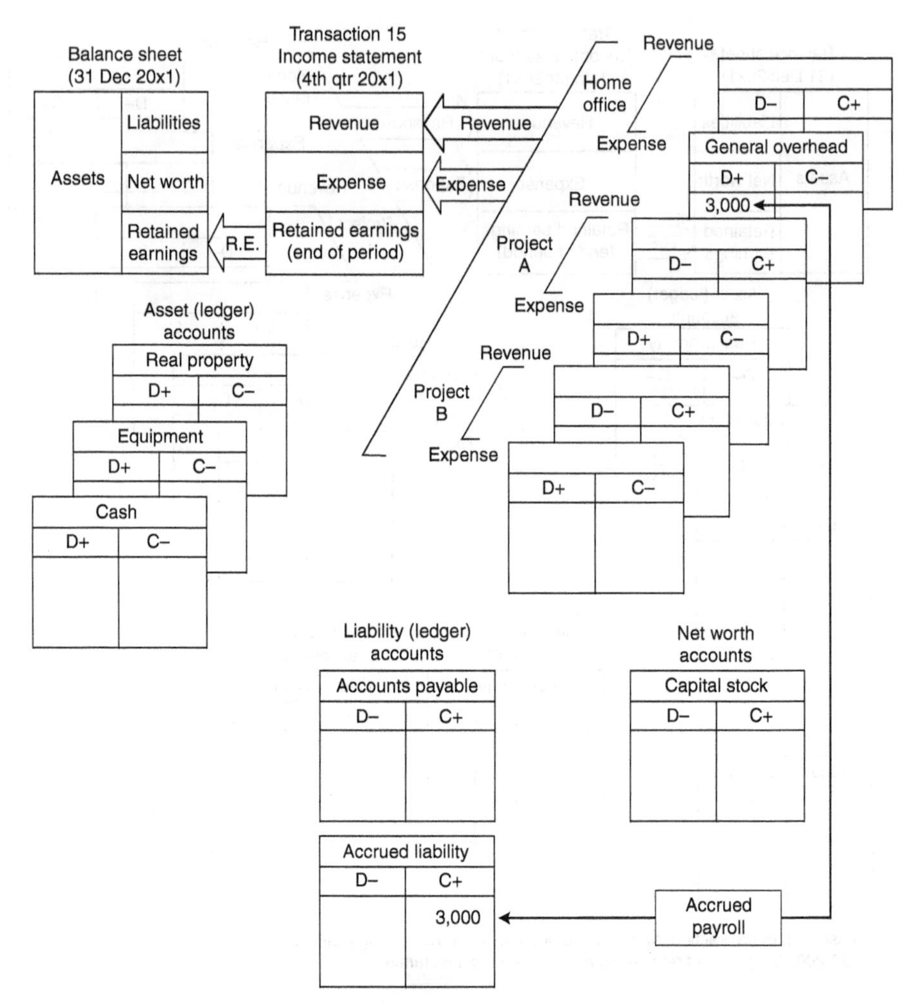

Description of transaction: Assume that company security personnel are paid on the fifteenth of the month. The books must be closed as of the end of December. Since 15 December, $3,000 in payroll to be paid (presumably on 15 January) has accrued. No specific invoice is received, but this obligation is incurred.

Journal entry:

Description	DR	CR
General overhead	3,000	
Accrued liability		3,000

Figure B.15

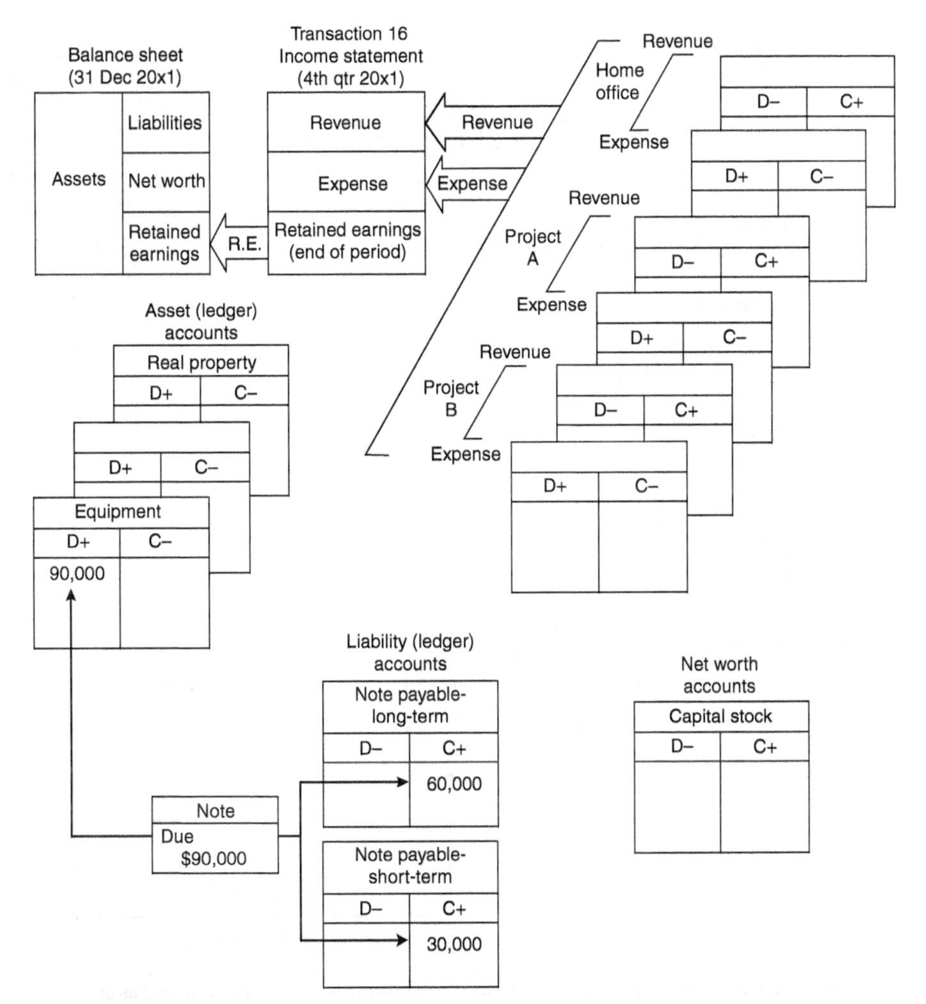

Description of transaction: A ditching machine is purchased on 1 July at a cost of $90,000. This company borrows this amount on a note at the First National Bank. $30,000 is due within one year. The remainder is due in years two and three of the three-year note.

Journal entry:

Description	DR	CR
Fixed assets–equipment	90,000	
Notes payable–short-term		30,000
Notes payable–long-term		60,000

Figure B.16

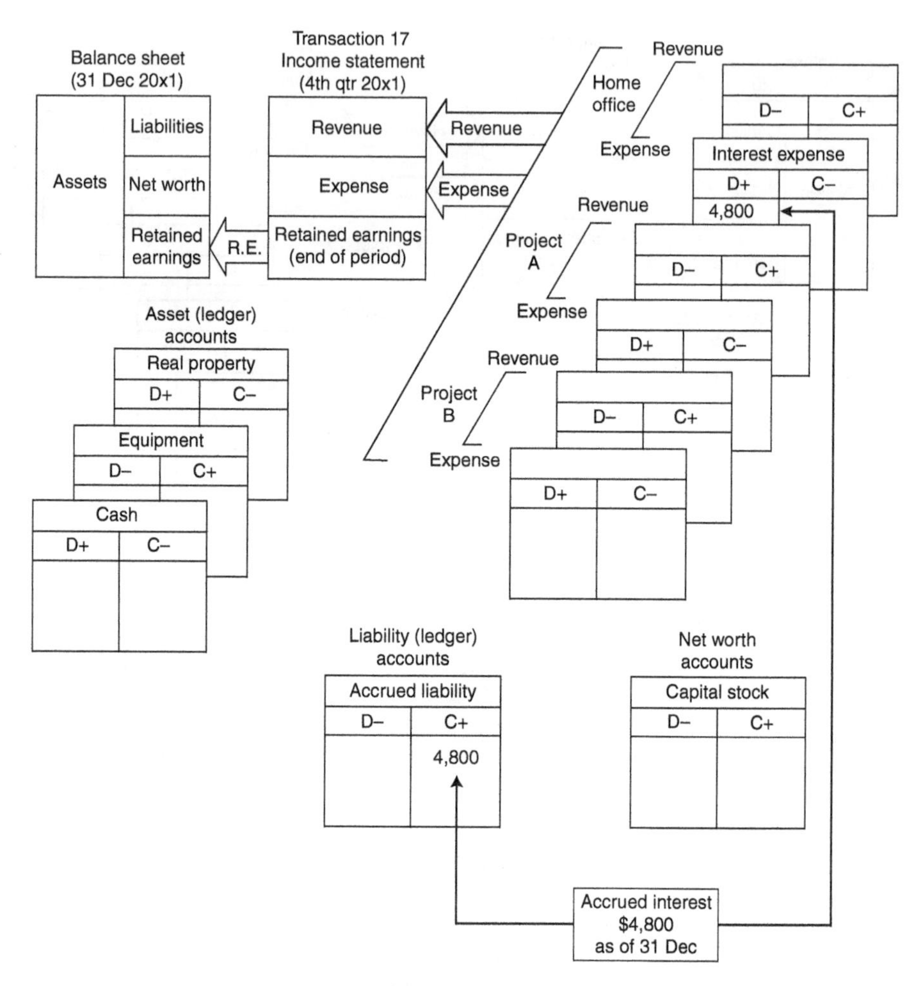

Description of transaction: Interest due on the note in Transaction 16 is computed to be $4,800 as of 31 December of the year of purchase. This is reflected as an accured liability for statement preparation purpose.

Journal entry:

Description	DR	CR
Interest expense	4,800	
Accrued liability		4,800

Figure B.17

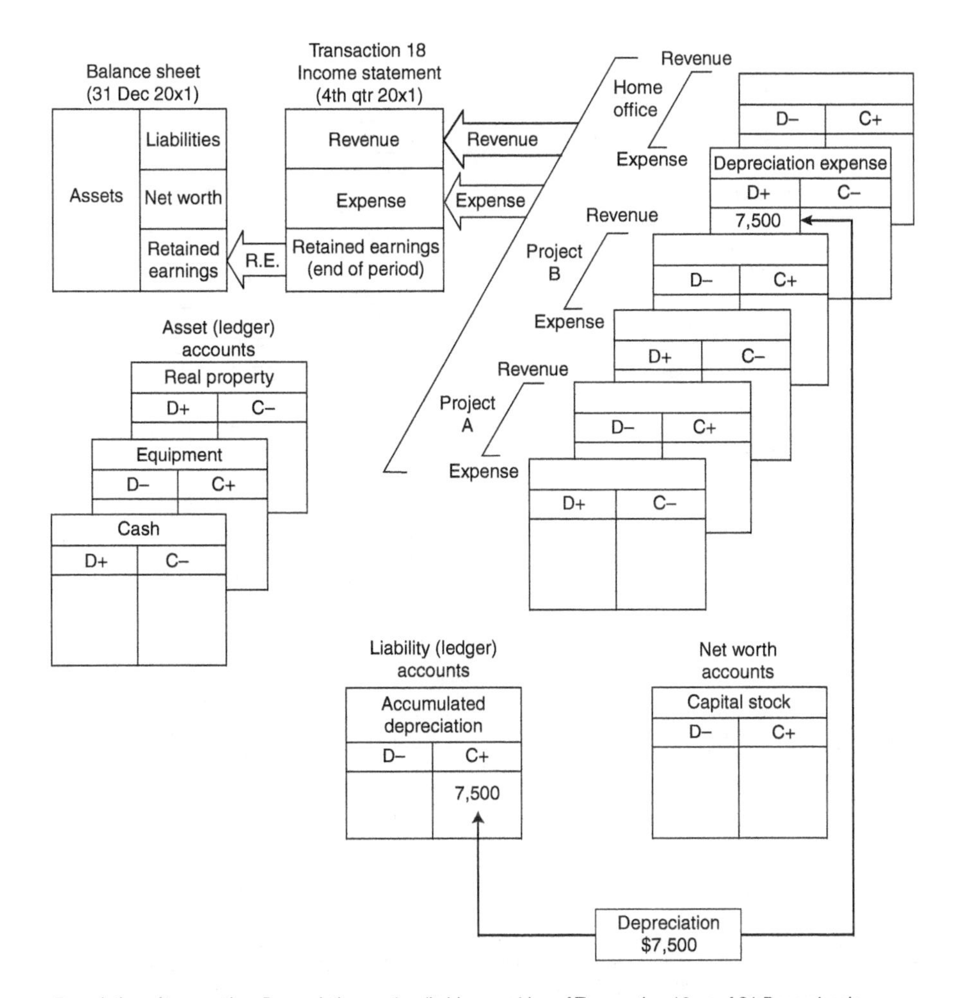

Description of transaction: Depreciation on the ditching machine of Transaction 16 as of 31 December is $7,500. An account is maintained to accumulate this expense. This type of account is called a contra account. It allows the original value of the asset to be maintained in the "fixed asset–equipment" account while placing the accumulated depreciation in the counterbalancing contra account. The contra account is considered a liability account.

Journal entry:

Description	DR	CR
Depreciation expense	7,500	
Accumulated depreciation		7,500

Figure B.18

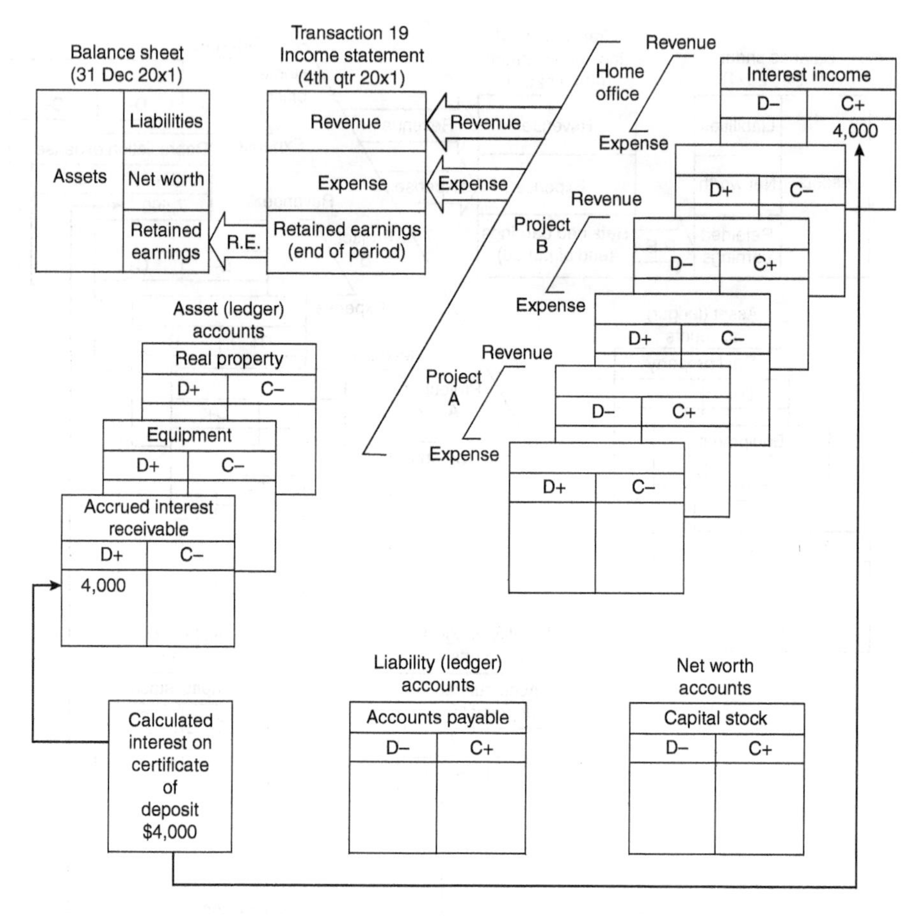

Description of transaction: Interest in the amount of $4,000 has accrued on a certificate of deposit (interest-producing-note) at the bank as of 31 December. This must be reflected in the year-end statements and is revenue to Apex Construction Company.

Journal entry:

Description	DR	CR
Accrued interest receivable	4,000	
Interest income		4,000

Figure B.19

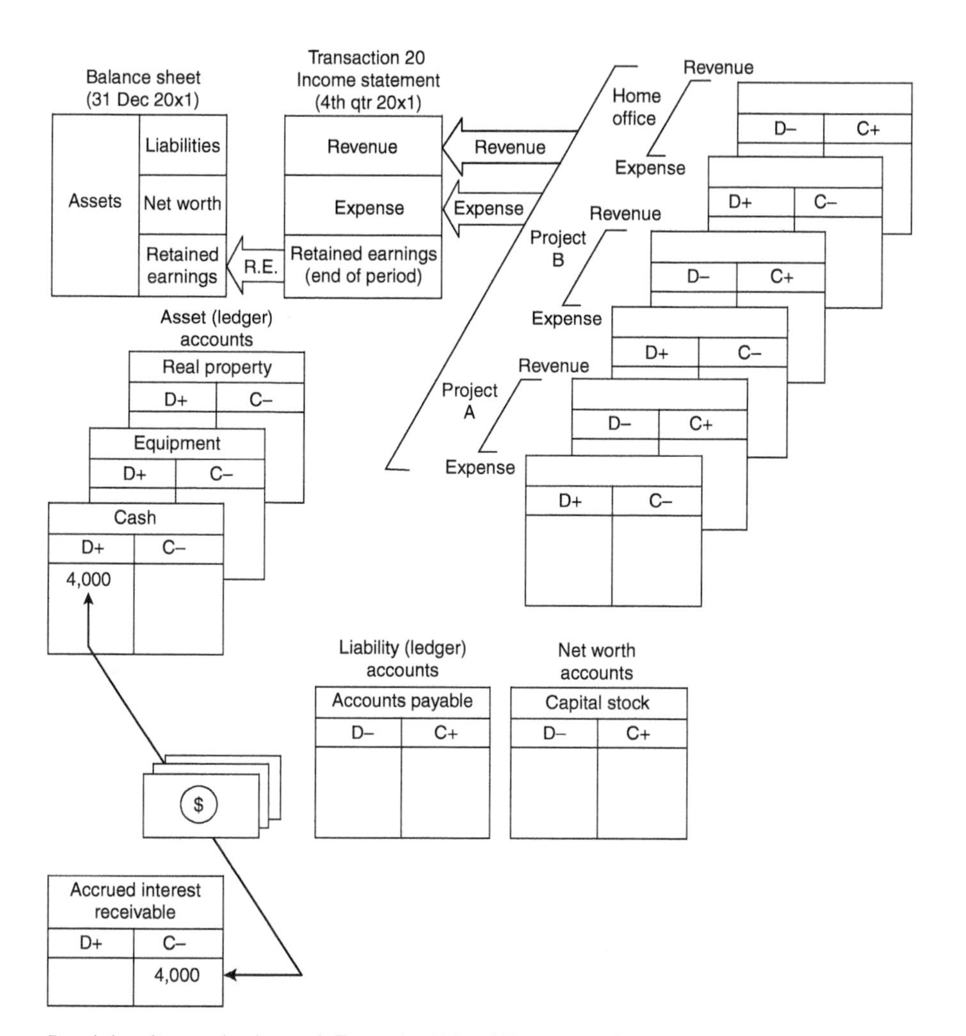

Description of transaction: Interest in Transaction 19 is paid into construction company account.

Journal entry:

Description	DR	CR
Cash	4,000	
Accrued interest receivable		4,000

Figure B.20

APPENDIX C

COMPOUND INTEREST TABLES

TABLE C.1

0.25%	Compound Interest Factors						0.25%
	Single Payment		Uniform Payment Series				
n	Compound Amount Factor	Present Worth Factor	Sinking Fund Factor	Capital Recovery Factor	Compound Amount Factor	Present Worth Factor	n
	Find F given P F/P	Find P given F P/F	Find A given F A/F	Find A given P A/P	Find F given A F/A	Find P given A P/A	
1	1.00250	0.99751	1.00000	1.00250	1.00000	0.99751	**1**
2	1.00501	0.99502	0.49938	0.50188	2.00250	1.99252	**2**
3	1.00752	0.99254	0.33250	0.33500	3.00751	2.98506	**3**
4	1.01004	0.99006	0.24906	0.25156	4.01503	3.97512	**4**
5	1.01256	0.98759	0.19900	0.20150	5.02506	4.96272	**5**
6	1.01509	0.98513	0.16563	0.16813	6.03763	5.94785	**6**
7	1.01763	0.98267	0.14179	0.14429	7.05272	6.93052	**7**
8	1.02018	0.98022	0.12391	0.12641	8.07035	7.91074	**8**
9	1.02273	0.97778	0.11000	0.11250	9.09053	8.88852	**9**
10	1.02528	0.97534	0.09888	0.10138	10.11325	9.86386	**10**
11	1.02785	0.97291	0.08978	0.09228	11.13854	10.83677	**11**
12	1.03042	0.97048	0.08219	0.08469	12.16638	11.80725	**12**
13	1.03299	0.96806	0.07578	0.07828	13.19680	12.77532	**13**
14	1.03557	0.96565	0.07028	0.07278	14.22979	13.74096	**14**
15	1.03816	0.96324	0.06551	0.06801	15.26537	14.70420	**15**
16	1.04076	0.96084	0.06134	0.06384	16.30353	15.66504	**16**
17	1.04336	0.95844	0.05766	0.06016	17.34429	16.62348	**17**
18	1.04597	0.95605	0.05438	0.05688	18.38765	17.57953	**18**
19	1.04858	0.95367	0.05146	0.05396	19.43362	18.53320	**19**
20	1.05121	0.95129	0.04882	0.05132	20.48220	19.48449	**20**
21	1.05383	0.94892	0.04644	0.04894	21.53341	20.43340	**21**
22	1.05647	0.94655	0.04427	0.04677	22.58724	21.37995	**22**
23	1.05911	0.94419	0.04229	0.04479	23.64371	22.32414	**23**
24	1.06176	0.94184	0.04048	0.04298	24.70282	23.26598	**24**
25	1.06441	0.93949	0.03881	0.04131	25.76457	24.20547	**25**
30	1.07778	0.92783	0.03214	0.03464	31.11331	28.86787	**30**
31	1.08048	0.92552	0.03106	0.03356	32.19109	29.79339	**31**
32	1.08318	0.92321	0.03006	0.03256	33.27157	30.71660	**32**
33	1.08589	0.92091	0.02911	0.03161	34.35475	31.63750	**33**
34	1.08860	0.91861	0.02822	0.03072	35.44064	32.55611	**34**
35	1.09132	0.91632	0.02738	0.02988	36.52924	33.47243	**35**
36	1.09405	0.91403	0.02658	0.02908	37.62056	34.38647	**36**
42	1.11057	0.90044	0.02261	0.02511	44.22603	39.82300	**42**
48	1.12733	0.88705	0.01963	0.02213	50.93121	45.17869	**48**
60	1.16162	0.86087	0.01547	0.01797	64.64671	55.65236	**60**
72	1.196948E+00	0.83546	0.01269	0.01519	7.877939E+01	65.81686	**72**
120	1.349354E+00	0.74110	0.00716	0.00966	1.397414E+02	103.56175	**120**
180	1.567432E+00	0.63799	0.00441	0.00691	2.269727E+02	144.80547	**180**
240	1.820755E+00	0.54922	0.00305	0.00555	3.283020E+02	180.31091	**240**
360	2.456842E+00	0.40703	0.00172	0.00422	5.827369E+02	237.18938	**360**

TABLE C.2

0.50%	Compound Interest Factors						0.50%
	Single Payment		Uniform Payment Series				
n	Compound Amount Factor	Present Worth Factor	Sinking Fund Factor	Capital Recovery Factor	Compound Amount Factor	Present Worth Factor	n
	Find F given P F/P	Find P given F P/F	Find A given F A/F	Find A given P A/P	Find F given A F/A	Find P given A P/A	
1	1.00500	0.99502	1.00000	1.00500	1.00000	0.99502	**1**
2	1.01003	0.99007	0.49875	0.50375	2.00500	1.98510	**2**
3	1.01508	0.98515	0.33167	0.33667	3.01502	2.97025	**3**
4	1.02015	0.98025	0.24813	0.25313	4.03010	3.95050	**4**
5	1.02525	0.97537	0.19801	0.20301	5.05025	4.92587	**5**
6	1.03038	0.97052	0.16460	0.16960	6.07550	5.89638	**6**
7	1.03553	0.96569	0.14073	0.14573	7.10588	6.86207	**7**
8	1.04071	0.96089	0.12283	0.12783	8.14141	7.82296	**8**
9	1.04591	0.95610	0.10891	0.11391	9.18212	8.77906	**9**
10	1.05114	0.95135	0.09777	0.10277	10.22803	9.73041	**10**
11	1.05640	0.94661	0.08866	0.09366	11.27917	10.67703	**11**
12	1.06168	0.94191	0.08107	0.08607	12.33556	11.61893	**12**
13	1.06699	0.93722	0.07464	0.07964	13.39724	12.55615	**13**
14	1.07232	0.93256	0.06914	0.07414	14.46423	13.48871	**14**
15	1.07768	0.92792	0.06436	0.06936	15.53655	14.41662	**15**
16	1.08307	0.92330	0.06019	0.06519	16.61423	15.33993	**16**
17	1.08849	0.91871	0.05651	0.06151	17.69730	16.25863	**17**
18	1.09393	0.91414	0.05323	0.05823	18.78579	17.17277	**18**
19	1.09940	0.90959	0.05030	0.05530	19.87972	18.08236	**19**
20	1.10490	0.90506	0.04767	0.05267	20.97912	18.98742	**20**
21	1.11042	0.90056	0.04528	0.05028	22.08401	19.88798	**21**
22	1.11597	0.89608	0.04311	0.04811	23.19443	20.78406	**22**
23	1.12155	0.89162	0.04113	0.04613	24.31040	21.67568	**23**
24	1.12716	0.88719	0.03932	0.04432	25.43196	22.56287	**24**
25	1.13280	0.88277	0.03765	0.04265	26.55912	23.44564	**25**
30	1.16140	0.86103	0.03098	0.03598	32.28002	27.79405	**30**
31	1.16721	0.85675	0.02990	0.03490	33.44142	28.65080	**31**
32	1.17304	0.85248	0.02889	0.03389	34.60862	29.50328	**32**
33	1.17891	0.84824	0.02795	0.03295	35.78167	30.35153	**33**
34	1.18480	0.84402	0.02706	0.03206	36.96058	31.19555	**34**
35	1.19073	0.83982	0.02622	0.03122	38.14538	32.03537	**35**
36	1.19668	0.83564	0.02542	0.03042	39.33610	32.87102	**36**
42	1.23303	0.81101	0.02146	0.02646	46.60654	37.79830	**42**
48	1.27049	0.78710	0.01849	0.02349	54.09783	42.58032	**48**
60	1.34885	0.74137	0.01433	0.01933	69.77003	51.72556	**60**
72	1.432044E+00	0.69830	0.01157	0.01657	8.640886E+01	60.33951	**72**
120	1.819397E+00	0.54963	0.00610	0.01110	1.638793E+02	90.07345	**120**
180	2.454094E+00	0.40748	0.00344	0.00844	2.908187E+02	118.50351	**180**
240	3.310204E+00	0.30210	0.00216	0.00716	4.620409E+02	139.58077	**240**
360	6.022575E+00	0.16604	0.00100	0.00600	1.004515E+03	166.79161	**360**

TABLE C.3

0.75%	Compound Interest Factors						0.75%
	Single Payment		Uniform Payment Series				
n	Compound Amount Factor	Present Worth Factor	Sinking Fund Factor	Capital Recovery Factor	Compound Amount Factor	Present Worth Factor	n
	Find F given P F/P	Find P given F P/F	Find A given F A/F	Find A given P A/P	Find F given A F/A	Find P given A P/A	
1	1.00750	0.99256	1.00000	1.00750	1.00000	0.99256	**1**
2	1.01506	0.98517	0.49813	0.50563	2.00750	1.97772	**2**
3	1.02267	0.97783	0.33085	0.33835	3.02256	2.95556	**3**
4	1.03034	0.97055	0.24721	0.25471	4.04523	3.92611	**4**
5	1.03807	0.96333	0.19702	0.20452	5.07556	4.88944	**5**
6	1.04585	0.95616	0.16357	0.17107	6.11363	5.84560	**6**
7	1.05370	0.94904	0.13967	0.14717	7.15948	6.79464	**7**
8	1.06160	0.94198	0.12176	0.12926	8.21318	7.73661	**8**
9	1.06956	0.93496	0.10782	0.11532	9.27478	8.67158	**9**
10	1.07758	0.92800	0.09667	0.10417	10.34434	9.59958	**10**
11	1.08566	0.92109	0.08755	0.09505	11.42192	10.52067	**11**
12	1.09381	0.91424	0.07995	0.08745	12.50759	11.43491	**12**
13	1.10201	0.90743	0.07352	0.08102	13.60139	12.34235	**13**
14	1.11028	0.90068	0.06801	0.07551	14.70340	13.24302	**14**
15	1.11860	0.89397	0.06324	0.07074	15.81368	14.13699	**15**
16	1.12699	0.88732	0.05906	0.06656	16.93228	15.02431	**16**
17	1.13544	0.88071	0.05537	0.06287	18.05927	15.90502	**17**
18	1.14396	0.87416	0.05210	0.05960	19.19472	16.77918	**18**
19	1.15254	0.86765	0.04917	0.05667	20.33868	17.64683	**19**
20	1.16118	0.86119	0.04653	0.05403	21.49122	18.50802	**20**
21	1.16989	0.85478	0.04415	0.05165	22.65240	19.36280	**21**
22	1.17867	0.84842	0.04198	0.04948	23.82230	20.21121	**22**
23	1.18751	0.84210	0.04000	0.04750	25.00096	21.05331	**23**
24	1.19641	0.83583	0.03818	0.04568	26.18847	21.88915	**24**
25	1.20539	0.82961	0.03652	0.04402	27.38488	22.71876	**25**
30	1.25127	0.79919	0.02985	0.03735	33.50290	26.77508	**30**
31	1.26066	0.79324	0.02877	0.03627	34.75417	27.56832	**31**
32	1.27011	0.78733	0.02777	0.03527	36.01483	28.35565	**32**
33	1.27964	0.78147	0.02682	0.03432	37.28494	29.13712	**33**
34	1.28923	0.77565	0.02593	0.03343	38.56458	29.91278	**34**
35	1.29890	0.76988	0.02509	0.03259	39.85381	30.68266	**35**
36	1.30865	0.76415	0.02430	0.03180	41.15272	31.44681	**36**
42	1.36865	0.73065	0.02034	0.02784	49.15329	35.91371	**42**
48	1.43141	0.69861	0.01739	0.02489	57.52071	40.18478	**48**
60	1.56568	0.63870	0.01326	0.02076	75.42414	48.17337	**60**
72	1.712553B+00	0.58392	0.01053	0.01803	9.500703E+01	55.47685	**72**
120	2.451357E+00	0.40794	0.00517	0.01267	1.935143E+02	78.94169	**120**
180	3.838043E+00	0.26055	0.00264	0.01014	3.784058E+02	98.59341	**180**
240	6.009152E+00	0.16641	0.00150	0.00900	6.678869E+02	111.14495	**240**
360	1.473058E+01	0.06789	0.00055	0.00805	1.830743E+03	124.28187	**360**

TABLE C.4

1.00%	Compound Interest Factors						1.00%
	Single Payment		Uniform Payment Series				
n	Compound Amount Factor	Present Worth Factor	Sinking Fund Factor	Capital Recovery Factor	Compound Amount Factor	Present Worth Factor	n
	Find F given P F/P	Find P given F P/F	Find A given F A/F	Find A given P A/P	Find F given A F/A	Find P given A P/A	
1	1.01000	0.99010	1.00000	1.01000	1.00000	0.99010	**1**
2	1.02010	0.98030	0.49751	0.50751	2.01000	1.97040	**2**
3	1.03030	0.97059	0.33002	0.34002	3.03010	2.94099	**3**
4	1.04060	0.96098	0.24628	0.25628	4.06040	3.90197	**4**
5	1.05101	0.95147	0.19604	0.20604	5.10101	4.85343	**5**
6	1.06152	0.94205	0.16255	0.17255	6.15202	5.79548	**6**
7	1.07214	0.93272	0.13863	0.14863	7.21354	6.72819	**7**
8	1.08286	0.92348	0.12069	0.13069	8.28567	7.65168	**8**
9	1.09369	0.91434	0.10674	0.11674	9.36853	8.56602	**9**
10	1.10462	0.90529	0.09558	0.10558	10.46221	9.47130	**10**
11	1.11567	0.89632	0.08645	0.09645	11.56683	10.36763	**11**
12	1.12683	0.88745	0.07885	0.08885	12.68250	11.25508	**12**
13	1.13809	0.87866	0.07241	0.08241	13.80933	12.13374	**13**
14	1.14947	0.86996	0.06690	0.07690	14.94742	13.00370	**14**
15	1.16097	0.86135	0.06212	0.07212	16.09690	13.86505	**15**
16	1.17258	0.85282	0.05794	0.06794	17.25786	14.71787	**16**
17	1.18430	0.84438	0.05426	0.06426	18.43044	15.56225	**17**
18	1.19615	0.83602	0.05098	0.06098	19.61475	16.39827	**18**
19	1.20811	0.82774	0.04805	0.05805	20.81090	17.22601	**19**
20	1.22019	0.81954	0.04542	0.05542	22.01900	18.04555	**20**
21	1.23239	0.81143	0.04303	0.05303	23.23919	18.85698	**21**
22	1.24472	0.80340	0.04086	0.05086	24.47159	19.66038	**22**
23	1.25716	0.79544	0.03889	0.04889	25.71630	20.45582	**23**
24	1.26973	0.78757	0.03707	0.04707	26.97346	21.24339	**24**
25	1.28243	0.77977	0.03541	0.04541	28.24320	22.02316	**25**
30	1.34785	0.74192	0.02875	0.03875	34.78489	25.80771	**30**
31	1.36133	0.73458	0.02768	0.03768	36.13274	26.54229	**31**
32	1.37494	0.72730	0.02667	0.03667	37.49407	27.26959	**32**
33	1.38869	0.72010	0.02573	0.03573	38.86901	27.98969	**33**
34	1.40258	0.71297	0.02484	0.03484	40.25770	28.70267	**34**
35	1.41660	0.70591	0.02400	0.03400	41.66028	29.40858	**35**
36	1.43077	0.69892	0.02321	0.03321	43.07688	30.10751	**36**
42	1.51879	0.65842	0.01928	0.02928	51.87899	34.15811	**42**
48	1.61223	0.62026	0.01633	0.02633	61.22261	37.97396	**48**
60	1.81670	0.55045	0.01224	0.02224	81.66967	44.95504	**60**
72	2.047099E+00	0.48850	0.00955	0.01955	1.047099E+02	51.15039	**72**
120	3.300387E+00	0.30299	0.00435	0.01435	2.300387E+02	69.70052	**120**
180	5.995802E+00	0.16678	0.00200	0.01200	4.995802E+02	83.32166	**180**
240	1.089255E+01	0.09181	0.00101	0.01101	9.892554E+02	90.81942	**240**
360	3.594964E+01	0.02782	0.00029	0.01029	3.494964E+03	97.21833	**360**

TABLE C.5

1.25%	Compound Interest Factors						1.25%
	Single Payment		Uniform Payment Series				
n	Compound Amount Factor	Present Worth Factor	Sinking Fund Factor	Capital Recovery Factor	Compound Amount Factor	Present Worth Factor	n
	Find F given P F/P	Find P given F P/F	Find A given F A/F	Find A given P A/P	Find F given A F/A	Find P given A P/A	
1	1.01250	0.98765	1.00000	1.01250	1.00000	0.98765	**1**
2	1.02516	0.97546	0.49689	0.50939	2.01250	1.96312	**2**
3	1.03797	0.96342	0.32920	0.34170	3.03766	2.92653	**3**
4	1.05095	0.95152	0.24536	0.25786	4.07563	3.87806	**4**
5	1.06408	0.93978	0.19506	0.20756	5.12657	4.81784	**5**
6	1.07738	0.92817	0.16153	0.17403	6.19065	5.74601	**6**
7	1.09085	0.91672	0.13759	0.15009	7.26804	6.66273	**7**
8	1.10449	0.90540	0.11963	0.13213	8.35889	7.56812	**8**
9	1.11829	0.89422	0.10567	0.11817	9.46337	8.46234	**9**
10	1.13227	0.88318	0.09450	0.10700	10.58167	9.34553	**10**
11	1.14642	0.87228	0.08537	0.09787	11.71394	10.21780	**11**
12	1.16075	0.86151	0.07776	0.09026	12.86036	11.07931	**12**
13	1.17526	0.85087	0.07132	0.08382	14.02112	11.93018	**13**
14	1.18995	0.84037	0.06581	0.07831	15.19638	12.77055	**14**
15	1.20483	0.82999	0.06103	0.07353	16.38633	13.60055	**15**
16	1.21989	0.81975	0.05685	0.06935	17.59116	14.42029	**16**
17	1.23514	0.80963	0.05316	0.06566	18.81105	15.22992	**17**
18	1.25058	0.79963	0.04988	0.06238	20.04619	16.02955	**18**
19	1.26621	0.78976	0.04696	0.05946	21.29677	16.81931	**19**
20	1.28204	0.78001	0.04432	0.05682	22.56298	17.59932	**20**
21	1.29806	0.77038	0.04194	0.05444	23.84502	18.36969	**21**
22	1.31429	0.76087	0.03977	0.05227	25.14308	19.13056	**22**
23	1.33072	0.75147	0.03780	0.05030	26.45737	19.88204	**23**
24	1.34735	0.74220	0.03599	0.04849	27.78808	20.62423	**24**
25	1.36419	0.73303	0.03432	0.04682	29.13544	21.35727	**25**
30	1.45161	0.68889	0.02768	0.04018	36.12907	24.88891	**30**
31	1.46976	0.68038	0.02661	0.03911	37.58068	25.56929	**31**
32	1.48813	0.67198	0.02561	0.03811	39.05044	26.24127	**32**
33	1.50673	0.66369	0.02467	0.03717	40.53857	26.90496	**33**
34	1.52557	0.65549	0.02378	0.03628	42.04530	27.56046	**34**
35	1.54464	0.64740	0.02295	0.03545	43.57087	28.20786	**35**
36	1.56394	0.63941	0.02217	0.03467	45.11551	28.84727	**36**
42	1.68497	0.59348	0.01825	0.03075	54.79734	32.52132	**42**
48	1.81535	0.55086	0.01533	0.02783	65.22839	35.93148	**48**
60	2.10718	0.47457	0.01129	0.02379	88.57451	42.03459	**60**
72	2.445920E+00	0.40884	0.00865	0.02115	1.156736E+02	47.29247	**72**
120	4.440213E+00	0.22521	0.00363	0.01613	2.752171E+02	61.98285	**120**
180	9.356334E+00	0.10688	0.00150	0.01400	6.685068E+02	71.44964	**180**
240	1.971549E+01	0.05072	0.00067	0.01317	1.497239E+03	75.94228	**240**
360	8.754100E+01	0.01142	0.00014	0.01264	6.923280E+03	79.08614	**360**

TABLE C.6

1.50%	Compound Interest Factors						1.50%
	Single Payment		Uniform Payment Series				
n	Compound Amount Factor	Present Worth Factor	Sinking Fund Factor	Capital Recovery Factor	Compound Amount Factor	Present Worth Factor	n
	Find F given P F/P	Find P given F P/F	Find A given F A/F	Find A given P A/P	Find F given A F/A	Find P given A P/A	
1	1.01500	0.98522	1.00000	1.01500	1.00000	0.98522	**1**
2	1.03023	0.97066	0.49628	0.51128	2.01500	1.95588	**2**
3	1.04568	0.95632	0.32838	0.34338	3.04522	2.91220	**3**
4	1.06136	0.94218	0.24444	0.25944	4.09090	3.85438	**4**
5	1.07728	0.92826	0.19409	0.20909	5.15227	4.78264	**5**
6	1.09344	0.91454	0.16053	0.17553	6.22955	5.69719	**6**
7	1.10984	0.90103	0.13656	0.15156	7.32299	6.59821	**7**
8	1.12649	0.88771	0.11858	0.13358	8.43284	7.48593	**8**
9	1.14339	0.87459	0.10461	0.11961	9.55933	8.36052	**9**
10	1.16054	0.86167	0.09343	0.10843	10.70272	9.22218	**10**
11	1.17795	0.84893	0.08429	0.09929	11.86326	10.07112	**11**
12	1.19562	0.83639	0.07668	0.09168	13.04121	10.90751	**12**
13	1.21355	0.82403	0.07024	0.08524	14.23683	11.73153	**13**
14	1.23176	0.81185	0.06472	0.07972	15.45038	12.54338	**14**
15	1.25023	0.79985	0.05994	0.07494	16.68214	13.34323	**15**
16	1.26899	0.78803	0.05577	0.07077	17.93237	14.13126	**16**
17	1.28802	0.77639	0.05208	0.06708	19.20136	14.90765	**17**
18	1.30734	0.76491	0.04881	0.06381	20.48938	15.67256	**18**
19	1.32695	0.75361	0.04588	0.06088	21.79672	16.42617	**19**
20	1.34686	0.74247	0.04325	0.05825	23.12367	17.16864	**20**
21	1.36706	0.73150	0.04087	0.05587	24.47052	17.90014	**21**
22	1.38756	0.72069	0.03870	0.05370	25.83758	18.62082	**22**
23	1.40838	0.71004	0.03673	0.05173	27.22514	19.33086	**23**
24	1.42950	0.69954	0.03492	0.04992	28.63352	20.03041	**24**
25	1.45095	0.68921	0.03326	0.04826	30.06302	20.71961	**25**
30	1.56308	0.63976	0.02664	0.04164	37.53868	24.01584	**30**
31	1.58653	0.63031	0.02557	0.04057	39.10176	24.64615	**31**
32	1.61032	0.62099	0.02458	0.03958	40.68829	25.26714	**32**
33	1.63448	0.61182	0.02364	0.03864	42.29861	25.87895	**33**
34	1.65900	0.60277	0.02276	0.03776	43.93309	26.48173	**34**
35	1.68388	0.59387	0.02193	0.03693	45.59209	27.07559	**35**
36	1.70914	0.58509	0.02115	0.03615	47.27597	27.66068	**36**
42	1.86885	0.53509	0.01726	0.03226	57.92314	30.99405	**42**
48	2.04348	0.48936	0.01437	0.02937	69.56522	34.04255	**48**
60	2.44322	0.40930	0.01039	0.02539	96.21465	39.38027	**60**
72	2.921158E+00	0.34233	0.00781	0.02281	1.280772E+02	43.84467	**72**
120	5.969323E+00	0.16752	0.00302	0.01802	3.312882E+02	55.49845	**120**
180	1.458437E+01	0.06857	0.00110	0.01610	9.056245E+02	62.09556	**180**
240	3.563282E+01	0.02806	0.00043	0.01543	2.308854E+03	64.79573	**240**
360	2.127038E+02	0.00470	0.00007	0.01507	1.411359E+04	66.35324	**360**

TABLE C.7

1.75%	Compound Interest Factors						1.75%
	Single Payment		Uniform Payment Series				
n	Compound Amount Factor	Present Worth Factor	Sinking Fund Factor	Capital Recovery Factor	Compound Amount Factor	Present Worth Factor	n
	Find F given P F/P	Find P given F P/F	Find A given F A/F	Find A given P A/P	Find F given A F/A	Find P given A P/A	
1	1.01750	0.98280	1.00000	1.01750	1.00000	0.98280	1
2	1.03531	0.96590	0.49566	0.51316	2.01750	1.94870	2
3	1.05342	0.94929	0.32757	0.34507	3.05281	2.89798	3
4	1.07186	0.93296	0.24353	0.26103	4.10623	3.83094	4
5	1.09062	0.91691	0.19312	0.21062	5.17809	4.74786	5
6	1.10970	0.90114	0.15952	0.17702	6.26871	5.64900	6
7	1.12912	0.88564	0.13553	0.15303	7.37841	6.53464	7
8	1.14888	0.87041	0.11754	0.13504	8.50753	7.40505	8
9	1.16899	0.85544	0.10356	0.12106	9.65641	8.26049	9
10	1.18944	0.84073	0.09238	0.10988	10.82540	9.10122	10
11	1.21026	0.82627	0.08323	0.10073	12.01484	9.92749	11
12	1.23144	0.81206	0.07561	0.09311	13.22510	10.73955	12
13	1.25299	0.79809	0.06917	0.08667	14.45654	11.53764	13
14	1.27492	0.78436	0.06366	0.08116	15.70953	12.32201	14
15	1.29723	0.77087	0.05888	0.07638	16.98445	13.09288	15
16	1.31993	0.75762	0.05470	0.07220	18.28168	13.85050	16
17	1.34303	0.74459	0.05102	0.06852	19.60161	14.59508	17
18	1.36653	0.73178	0.04774	0.06524	20.94463	15.32686	18
19	1.39045	0.71919	0.04482	0.06232	22.31117	16.04606	19
20	1.41478	0.70682	0.04219	0.05969	23.70161	16.75288	20
21	1.43954	0.69467	0.03981	0.05731	25.11639	17.44755	21
22	1.46473	0.68272	0.03766	0.05516	26.55593	18.13027	22
23	1.49036	0.67098	0.03569	0.05319	28.02065	18.80125	23
24	1.51644	0.65944	0.03389	0.05139	29.51102	19.46069	24
25	1.54298	0.64810	0.03223	0.04973	31.02746	20.10878	25
30	1.68280	0.59425	0.02563	0.04313	39.01715	23.18585	30
31	1.71225	0.58403	0.02457	0.04207	40.69995	23.76988	31
32	1.74221	0.57398	0.02358	0.04108	42.41220	24.34386	32
33	1.77270	0.56411	0.02265	0.04015	44.15441	24.90797	33
34	1.80372	0.55441	0.02177	0.03927	45.92712	25.46238	34
35	1.83529	0.54487	0.02095	0.03845	47.73084	26.00725	35
36	1.86741	0.53550	0.02018	0.03768	49.56613	26.54275	36
42	2.07227	0.48256	0.01632	0.03382	61.27236	29.56780	42
48	2.29960	0.43486	0.01347	0.03097	74.26278	32.29380	48
60	2.83182	0.35313	0.00955	0.02705	104.67522	36.96399	60
72	3.487210E+00	0.28676	0.00704	0.02454	1.421263E+02	40.75645	72
120	8.019183E+00	0.12470	0.00249	0.01999	4.010962E+02	50.01709	120
180	2.270885E+01	0.04404	0.00081	0.01831	1.240506E+03	54.62653	180
240	6.430730E+01	0.01555	0.00028	0.01778	3.617560E+03	56.25427	240
360	5.156921E+02	0.00194	0.00003	0.01753	2.941097E+04	57.03205	360

TABLE C.8

2.00%	Compound Interest Factors						2.00%
	Single Payment		Uniform Payment Series				
n	Compound Amount Factor	Present Worth Factor	Sinking Fund Factor	Capital Recovery Factor	Compound Amount Factor	Present Worth Factor	n
	Find F given P F/P	Find P given F P/F	Find A given F A/F	Find A given P A/P	Find F given A F/A	Find P given A P/A	
1	1.02000	0.98039	1.00000	1.02000	1.00000	0.98039	**1**
2	1.04040	0.96117	0.49505	0.51505	2.02000	1.94156	**2**
3	1.06121	0.94232	0.32675	0.34675	3.06040	2.88388	**3**
4	1.08243	0.92385	0.24262	0.26262	4.12161	3.80773	**4**
5	1.10408	0.90573	0.19216	0.21216	5.20404	4.71346	**5**
6	1.12616	0.88797	0.15853	0.17853	6.30812	5.60143	**6**
7	1.14869	0.87056	0.13451	0.15451	7.43428	6.47199	**7**
8	1.17166	0.85349	0.11651	0.13651	8.58297	7.32548	**8**
9	1.19509	0.83676	0.10252	0.12252	9.75463	8.16224	**9**
10	1.21899	0.82035	0.09133	0.11133	10.94972	8.98259	**10**
;11	1.24337	0.80426	0.08218	0.10218	12.16872	9.78685	**11**
12	1.26824	0.78849	0.07456	0.09456	13.41209	10.57534	**12**
13	1.29361	0.77303	0.06812	0.08812	14.68033	11.34837	**13**
14	1.31948	0.75788	0.06260	0.08260	15.97394	12.10625	**14**
15	1.34587	0.74301	0.05783	0.07783	17.29342	12.84926	**15**
16	1.37279	0.72845	0.05365	0.07365	18.63929	13.57771	**16**
17	1.40024	0.71416	0.04997	0.06997	20.01207	14.29187	**17**
18	1.42825	0.70016	0.04670	0.06670	21.41231	14.99203	**18**
19	1.45681	0.68643	0.04378	0.06378	22.84056	15.67846	**19**
20	1.48595	0.67297	0.04116	0.06116	24.29737	16.35143	**20**
21	1.51567	0.65978	0.03878	0.05878	25.78332	17.01121	**21**
22	1.54598	0.64684	0.03663	0.05663	27.29898	17.65805	**22**
23	1.57690	0.63416	0.03467	0.05467	28.84496	18.29220	**23**
24	1.60844	0.62172	0.03287	0.05287	30.42186	18.91393	**24**
25	1.64061	0.60953	0.03122	0.05122	32.03030	19.52346	**25**
30	1.81136	0.55207	0.02465	0.04465	40.56808	22.39646	**30**
31	1.84759	0.54125	0.02360	0.04360	42.37944	22.93770	**31**
32	1.88454	0.53063	0.02261	0.04261	44.22703	23.46833	**32**
33	1.92223	0.52023	0.02169	0.04169	46.11157	23.98856	**33**
34	1.96068	0.51003	0.02082	0.04082	48.03380	24.49859	**34**
35	1.99989	0.50003	0.02000	0.04000	49.99448	24.99862	**35**
36	2.03989	0.49022	0.01923	0.03923	51.99437	25.48884	**36**
42	2.29724	0.43530	0.01542	0.03542	64.86222	28.23479	**42**
48	2.58707	0.38654	0.01260	0.03260	79.35352	30.67312	**48**
60	3.28103	0.30478	0.00877	0.02877	114.05154	34.76089	**60**
72	4.161140E+00	0.24032	0.00633	0.02633	1.580570E+02	37.98406	**72**
120	1.076516E+01	0.09289	0.00205	0.02205	4.882582E+02	45.35539	**120**
180	3.532083E+01	0.02831	0.00058	0.02058	1.716042 E+03	48.58440	**180**
240	1.158887E+02	0.00863	0.00017	0.02017	5.744437E+03	49.56855	**240**
360	1.247561E+03	0.00080	0.00002	0.02002	6.232806E+04	49.95992	**360**

TABLE C.9

2.50%	Compound Interest Factors						2.50%
	Single Payment		Uniform Payment Series				
n	Compound Amount Factor	Present Worth Factor	Sinking Fund Factor	Capital Recovery Factor	Compound Amount Factor	Present Worth Factor	n
	Find F given P F/P	Find P given F P/F	Find A given F A/F	Find A given P A/P	Find F given A F/A	Find P given A P/A	
1	1.02500	0.97561	1.00000	1.02500	1.00000	0.97561	1
2	1.05063	0.95181	0.49383	0.51883	2.02500	1.92742	2
3	1.07689	0.92860	0.32514	0.35014	3.07563	2.85602	3
4	1.10381	0.90595	0.24082	0.26582	4.15252	3.76197	4
5	1.13141	0.88385	0.19025	0.21525	5.25633	4.64583	5
6	1.15969	0.86230	0.15655	0.18155	6.38774	5.50813	6
7	1.18869	0.84127	0.13250	0.15750	7.54743	6.34939	7
8	1.21840	0.82075	0.11447	0.13947	8.73612	7.17014	8
9	1.24886	0.80073	0.10046	0.12546	9.95452	7.97087	9
10	1.28008	0.78120	0.08926	0.11426	11.20338	8.75206	10
11	1.31209	0.76214	0.08011	0.10511	12.48347	9.51421	11
12	1.34489	0.74356	0.07249	0.09749	13.79555	10.25776	12
13	1.37851	0.72542	0.06605	0.09105	15.14044	10.98318	13
14	1.41297	0.70773	0.06054	0.08554	16.51895	11.69091	14
15	1.44830	0.69047	0.05577	0.08077	17.93193	12.38138	15
16	1.48451	0.67362	0.05160	0.07660	19.38022	13.05500	16
17	1.52162	0.65720	0.04793	0.07293	20.86473	13.71220	17
18	1.55966	0.64117	0.04467	0.06967	22.38635	14.35336	18
19	1.59865	0.62553	0.04176	0.06676	23.94601	14.97889	19
20	1.63862	0.61027	0.03915	0.06415	25.54466	15.58916	20
21	1.67958	0.59539	0.03679	0.06179	27.18327	16.18455	21
22	1.72157	0.58086	0.03465	0.05965	28.86286	16.76541	22
23	1.76461	0.56670	0.03270	0.05770	30.58443	17.33211	23
24	1.80873	0.55288	0.03091	0.05591	32.34904	17.88499	24
25	1.85394	0.53939	0.02928	0.05428	34.15776	18.42438	25
30	2.09757	0.47674	0.02278	0.04778	43.90270	20.93029	30
31	2.15001	0.46511	0.02174	0.04674	46.00027	21.39541	31
32	2.20376	0.45377	0.02077	0.04577	48.15028	21.84918	32
33	2.25885	0.44270	0.01986	0.04486	50.35403	22.29188	33
34	2.31532	0.43191	0.01901	0.04401	52.61289	22.72379	34
35	2.37321	0.42137	0.01821	0.04321	54.92821	23.14516	35
36	2.43254	0.41109	0.01745	0.04245	57.30141	23.55625	36
42	2.82100	0.35448	0.01373	0.03873	72.83981	25.82061	42
48	3.27149	0.30567	0.01101	0.03601	90.85958	27.77315	48
60	4.39979	0.22728	0.00735	0.03235	135.99159	30.90866	60
72	5.917228E+00	0.16900	0.00508	0.03008	1.966891E+02	33.24008	72
120	1.935815E+01	0.05166	0.00136	0.02636	7.343260E+02	37.93369	120
180	8.517179E+01	0.01174	0.00030	0.02530	3.366872E+03	39.53036	180
240	3.747380E+02	0.00267	0.00007	0.02507	1.494952E+04	39.89326	240
360	7.254234E+03	0.00014	0.00000	0.02500	2.901293E+05	39.99449	360

TABLE C.10

3.00%	Compound Interest Factors						3.00%
	Single Payment		Uniform Payment Series				
n	Compound Amount Factor	Present Worth Factor	Sinking Fund Factor	Capital Recovery Factor	Compound Amount Factor	Present Worth Factor	n
	Find F given P F/P	Find P given F P/F	Find A given F A/F	Find A given P A/P	Find F given A F/A	Find P given A P/A	
1	1.03000	0.97087	1.00000	1.03000	1.00000	0.97087	**1**
2	1.06090	0.94260	0.49261	0.52261	2.03000	1.91347	**2**
3	1.09273	0.91514	0.32353	0.35353	3.09090	2.82861	**3**
4	1.12551	0.88849	0.23903	0.26903	4.18363	3.71710	**4**
5	1.15927	0.86261	0.18835	0.21835	5.30914	4.57971	**5**
6	1.19405	0.83748	0.15460	0.18460	6.46841	5.41719	**6**
7	1.22987	0.81309	0.13051	0.16051	7.66246	6.23028	**7**
8	1.26677	0.78941	0.11246	0.14246	8.89234	7.01969	**8**
9	1.30477	0.76642	0.09843	0.12843	10.15911	7.78611	**9**
10	1.34392	0.74409	0.08723	0.11723	11.46388	8.53020	**10**
11	1.38423	0.72242	0.07808	0.10808	12.80780	9.25262	**11**
12	1.42576	0.70138	0.07046	0.10046	14.19203	9.95400	**12**
13	1.46853	0.68095	0.06403	0.09403	15.61779	10.63496	**13**
14	1.51259	0.66112	0.05853	0.08853	17.08632	11.29607	**14**
15	1.55797	0.64186	0.05377	0.08377	18.59891	11.93794	**15**
16	1.60471	0.62317	0.04961	0.07961	20.15688	12.56110	**16**
17	1.65285	0.60502	0.04595	0.07595	21.76159	13.16612	**17**
18	1.70243	0.58739	0.04271	0.07271	23.41444	13.75351	**18**
19	1.75351	0.57029	0.03981	0.06981	25.11687	14.32380	**19**
20	1.80611	0.55368	0.03722	0.06722	26.87037	14.87747	**20**
21	1.86029	0.53755	0.03487	0.06487	28.67649	15.41502	**21**
22	1.91610	0.52189	0.03275	0.06275	30.53678	15.93692	**22**
23	1.97359	0.50669	0.03081	0.06081	32.45288	16.44361	**23**
24	2.03279	0.49193	0.02905	0.05905	34.42647	16.93554	**24**
25	2.09378	0.47761	0.02743	0.05743	36.45926	17.41315	**25**
30	2.42726	0.41199	0.02102	0.05102	47.57542	19.60044	**30**
31	2.50008	0.39999	0.02000	0.05000	50.00268	20.00043	**31**
32	2.57508	0.38834	0.01905	0.04905	52.50276	20.38877	**32**
33	2.65234	0.37703	0.01816	0.04816	55.07784	20.76579	**33**
34	2.73191	0.36604	0.01732	0.04732	57.73018	21.13184	**34**
35	2.81386	0.35538	0.01654	0.04654	60.46208	21.48722	**35**
36	2.89828	0.34503	0.01580	0.04580	63.27594	21.83225	**36**
42	3.46070	0.28896	0.01219	0.04219	82.02320	23.70136	**42**
48	4.13225	0.24200	0.00958	0.03958	104.40840	25.26671	**48**
60	5.89160	0.16973	0.00613	0.03613	163.05344	27.67556	**60**
72	8.400017E+00	0.11905	0.00405	0.03405	2.466672E+02	29.36509	**72**
120	3.471099E+01	0.02881	0.00089	0.03089	1.123700E+03	32.37302	**120**
180	2.045034E+02	0.00489	0.00015	0.03015	6.783445E+03	33.17034	**180**
240	1.204853E+03	0.00083	0.00002	0.03002	4.012842E+04	33.30567	**240**
360	4.182162E+04	0.00002	0.00000	0.03000	1.394021E+06	33.33254	**360**

TABLE C.11

4.00%	Compound Interest Factors						4.00%
	Single Payment		Uniform Payment Series				
n	Compound Amount Factor	Present Worth Factor	Sinking Fund Factor	Capital Recovery Factor	Compound Amount Factor	Present Worth Factor	n
	Find F given P F/P	Find P given F P/F	Find A given F A/F	Find A given P A/P	Find F given A F/A	Find P given A P/A	
1	1.04000	0.96154	1.00000	1.04000	1.00000	0.96154	1
2	1.08160	0.92456	0.49020	0.53020	2.04000	1.88609	2
3	1.12486	0.88900	0.32035	0.36035	3.12160	2.77509	3
4	1.16986	0.85480	0.23549	0.27549	4.24646	3.62990	4
5	1.21665	0.82193	0.18463	0.22463	5.41632	4.45182	5
6	1.26532	0.79031	0.15076	0.19076	6.63298	5.24214	6
7	1.31593	0.75992	0.12661	0.16661	7.89829	6.00205	7
8	1.36857	0.73069	0.10853	0.14853	9.21423	6.73274	8
9	1.42331	0.70259	0.09449	0.13449	10.58280	7.43533	9
10	1.48024	0.67556	0.08329	0.12329	12.00611	8.11090	10
11	1.53945	0.64958	0.07415	0.11415	13.48635	8.76048	11
12	1.60103	0.62460	0.06655	0.10655	15.02581	9.38507	12
13	1.66507	0.60057	0.06014	0.10014	16.62684	9.98565	13
14	1.73168	0.57748	0.05467	0.09467	18.29191	10.56312	14
15	1.80094	0.55526	0.04994	0.08994	20.02359	11.11839	15
16	1.87298	0.53391	0.04582	0.08582	21.82453	11.65230	16
17	1.94790	0.51337	0.04220	0.08220	23.69751	12.16567	17
18	2.02582	0.49363	0.03899	0.07899	25.64541	12.65930	18
19	2.10685	0.47464	0.03614	0.07614	27.67123	13.13394	19
20	2.19112	0.45639	0.03358	0.07358	29.77808	13.59033	20
21	2.27877	0.43883	0.03128	0.07128	31.96920	14.02916	21
22	2.36992	0.42196	0.02920	0.06920	34.24797	14.45112	22
23	2.46472	0.40573	0.02731	0.06731	36.61789	14.85684	23
24	2.56330	0.39012	0.02559	0.06559	39.08260	15.24696	24
25	2.66584	0.37512	0.02401	0.06401	41.64591	15.62208	25
30	3.24340	0.30832	0.01783	0.05783	56.08494	17.29203	30
31	3.37313	0.29646	0.01686	0.05686	59.32834	17.58849	31
32	3.50806	0.28506	0.01595	0.05595	62.70147	17.87355	32
33	3.64838	0.27409	0.01510	0.05510	66.20953	18.14765	33
34	3.79432	0.26355	0.01431	0.05431	69.85791	18.41120	34
35	3.94609	0.25342	0.01358	0.05358	73.65222	18.66461	35
36	4.10393	0.24367	0.01289	0.05289	77.59831	18.90828	36
42	5.19278	0.19257	0.00954	0.04954	104.81960	20.18563	42
48	6.57053	0.15219	0.00718	0.04718	139.26321	21.19513	48
60	10.51963	0.09506	0.00420	0.04420	237.99069	22.62349	60
72	1.684226E+01	0.05937	0.00252	0.04252	3.960566E+02	23.51564	72
120	1.106626E+02	0.00904	0.00036	0.04036	2.741564E+03	24.77409	120
180	1.164129E+03	0.00086	0.00003	0.04003	2.907822E+04	24.97852	180
240	1.224620E+04	0.00008	0.00000	0.04000	3.061301E+05	24.99796	240
360	1.355196E+06	0.00000	0.00000	0.04000	3.387988E+07	24.99998	360

TABLE C.12

5.00%	Compound Interest Factors						5.00%
	Single Payment		Uniform Payment Series				
n	Compound Amount Factor	Present Worth Factor	Sinking Fund Factor	Capital Recovery Factor	Compound Amount Factor	Present Worth Factor	n
	Find F given P F/P	Find P given F P/F	Find A given F A/F	Find A given P A/P	Find F given A F/A	Find P given A P/A	
1	1.05000	0.95238	1.00000	1.05000	1.00000	0.95238	**1**
2	1.10250	0.90703	0.48780	0.53780	2.05000	1.85941	**2**
3	1.15763	0.86384	0.31721	0.36721	3.15250	2.72325	**3**
4	1.21551	0.82270	0.23201	0.28201	4.31013	3.54595	**4**
5	1.27628	0.78353	0.18097	0.23097	5.52563	4.32948	**5**
6	1.34010	0.74622	0.14702	0.19702	6.80191	5.07569	**6**
7	1.40710	0.71068	0.12282	0.17282	8.14201	5.78637	**7**
8	1.47746	0.67684	0.10472	0.15472	9.54911	6.46321	**8**
9	1.55133	0.64461	0.09069	0.14069	11.02656	7.10782	**9**
10	1.62889	0.61391	0.07950	0.12950	12.57789	7.72173	**10**
11	1.71034	0.58468	0.07039	0.12039	14.20679	8.30641	**11**
12	1.79586	0.55684	0.06283	0.11283	15.91713	8.86325	**12**
13	1.88565	0.53032	0.05646	0.10646	17.71298	9.39357	**13**
14	1.97993	0.50507	0.05102	0.10102	19.59863	9.89864	**14**
15	2.07893	0.48102	0.04634	0.09634	21.57856	10.37966	**15**
16	2.18287	0.45811	0.04227	0.09227	23.65749	10.83777	**16**
17	2.29202	0.43630	0.03870	0.08870	25.84037	11.27407	**17**
18	2.40662	0.41552	0.03555	0.08555	28.13238	11.68959	**18**
19	2.52695	0.39573	0.03275	0.08275	30.53900	12.08532	**19**
20	2.65330	0.37689	0.03024	0.08024	33.06595	12.46221	**20**
21	2.78596	0.35894	0.02800	0.07800	35.71925	12.82115	**21**
22	2.92526	0.34185	0.02597	0.07597	38.50521	13.16300	**22**
23	3.07152	0.32557	0.02414	0.07414	41.43048	13.48857	**23**
24	3.22510	0.31007	0.02247	0.07247	44.50200	13.79864	**24**
25	3.38635	0.29530	0.02095	0.07095	47.72710	14.09394	**25**
30	4.32194	0.23138	0.01505	0.06505	66.43885	15.37245	**30**
31	4.53804	0.22036	0.01413	0.06413	70.76079	15.59281	**31**
32	4.76494	0.20987	0.01328	0.06328	75.29883	15.80268	**32**
33	5.00319	0.19987	0.01249	0.06249	80.06377	16.00255	**33**
34	5.25335	0.19035	0.01176	0.06176	85.06696	16.19290	**34**
35	5.51602	0.18129	0.01107	0.06107	90.32031	16.37419	**35**
36	5.79182	0.17266	0.01043	0.06043	95.83632	16.54685	**36**
42	7.76159	0.12884	0.00739	0.05739	135.23175	17.42321	**42**
48	10.40127	0.09614	0.00532	0.05532	188.02539	18.07716	**48**
60	18.67919	0.05354	0.00283	0.05283	353.58372	18.92929	**60**
72	3.354513E+01	0.02981	0.00154	0.05154	6.509027E+02	19.40379	**72**
120	3.489120E+02	0.00287	0.00014	0.05014	6.958240E+03	19.94268	**120**
180	6.517392E+03	0.00015	0.00001	0.05001	1.303278E+05	19.99693	**180**
240	1.217396E+05	0.00001	0.00000	0.05000	2.434771E+06	19.99984	**240**
360	4.247640E+07	0.00000	0.00000	0.05000	8.495279E+08	20.00000	**360**

TABLE C.13

6.00%	Compound Interest Factors						6.00%
	Single Payment		Uniform Payment Series				
n	Compound Amount Factor	Present Worth Factor	Sinking Fund Factor	Capital Recovery Factor	Compound Amount Factor	Present Worth Factor	n
	Find F given P F/P	Find P given F P/F	Find A given F A/F	Find A given P A/P	Find F given A F/A	Find P given A P/A	
1	1.06000	0.94340	1.00000	1.06000	1.00000	0.94340	1
2	1.12360	0.89000	0.48544	0.54544	2.06000	1.83339	2
3	1.19102	0.83962	0.31411	0.37411	3.18360	2.67301	3
4	1.26248	0.79209	0.22859	0.28859	4.37462	3.46511	4
5	1.33823	0.74726	0.17740	0.23740	5.63709	4.21236	5
6	1.41852	0.70496	0.14336	0.20336	6.97532	4.91732	6
7	1.50363	0.66506	0.11914	0.17914	8.39384	5.58238	7
8	1.59385	0.62741	0.10104	0.16104	9.89747	6.20979	8
9	1.68948	0.59190	0.08702	0.14702	11.49132	6.80169	9
10	1.79085	0.55839	0.07587	0.13587	13.18079	7.36009	10
11	1.89830	0.52679	0.06679	0.12679	14.97164	7.88687	11
12	2.01220	0.49697	0.05928	0.11928	16.86994	8.38384	12
13	2.13293	0.46884	0.05296	0.11296	18.88214	8.85268	13
14	2.26090	0.44230	0.04758	0.10758	21.01507	9.29498	14
15	2.39656	0.41727	0.04296	0.10296	23.27597	9.71225	15
16	2.54035	0.39365	0.03895	0.09895	25.67253	10.10590	16
17	2.69277	0.37136	0.03544	0.09544	28.21288	10.47726	17
18	2.85434	0.35034	0.03236	0.09236	30.90565	10.82760	18
19	3.02560	0.33051	0.02962	0.08962	33.75999	11.15812	19
20	3.20714	0.31180	0.02718	0.08718	36.78559	11.46992	20
21	3.39956	0.29416	0.02500	0.08500	39.99273	11.76408	21
22	3.60354	0.27751	0.02305	0.08305	43.39229	12.04158	22
23	3.81975	0.26180	0.02128	0.08128	46.99583	12.30338	23
24	4.04893	0.24698	0.01968	0.07968	50.81558	12.55036	24
25	4.29187	0.23300	0.01823	0.07823	54.86451	12.78336	25
30	5.74349	0.17411	0.01265	0.07265	79.05819	13.76483	30
31	6.08810	0.16425	0.01179	0.07179	84.80168	13.92909	31
32	6.45339	0.15496	0.01100	0.07100	90.88978	14.08404	32
33	6.84059	0.14619	0.01027	0.07027	97.34316	14.23023	33
34	7.25103	0.13791	0.00960	0.06960	104.18375	14.36814	34
35	7.68609	0.13011	0.00897	0.06897	111.43478	14.49825	35
36	8.14725	0.12274	0.00839	0.06839	119.12087	14.62099	36
42	11.55703	0.08653	0.00568	0.06568	175.95054	15.22454	42
48	16.39387	0.06100	0.00390	0.06390	256.56453	15.65003	48
60	32.98769	0.03031	0.00188	0.06188	533.12818	16.16143	60
72	6.637772E+01	0.01507	0.00092	0.06092	1.089629E+03	16.41558	72
120	1.088188E+03	0.00092	0.00006	0.06006	1.811980E+04	16.65135	120
180	3.589680E+04	0.00003	0.00000	0.06000	5.982634E+05	16.66620	180
240	1.184153E+06	0.00000	0.00000	0.06000	1.973586E+07	16.66665	240
360	1.288580E+09	0.00000	0.00000	0.06000	2.147634E+10	16.66667	360

TABLE C.14

7.00%	Compound Interest Factors						7.00%
	Single Payment		Uniform Payment Series				
n	Compound Amount Factor	Present Worth Factor	Sinking Fund Factor	Capital Recovery Factor	Compound Amount Factor	Present Worth Factor	n
	Find F given P F/P	Find P given F P/F	Find A given F A/F	Find A given P A/P	Find F given A F/A	Find P given A P/A	
1	1.07000	0.93458	1.00000	1.07000	1.00000	0.93458	**1**
2	1.14490	0.87344	0.48309	0.55309	2.07000	1.80802	**2**
3	1.22504	0.81630	0.31105	0.38105	3.21490	2.62432	**3**
4	1.31080	0.76290	0.22523	0.29523	4.43994	3.38721	**4**
5	1.40255	0.71299	0.17389	0.24389	5.75074	4.10020	**5**
6	1.50073	0.66634	0.13980	0.20980	7.15329	4.76654	**6**
7	1.60578	0.62275	0.11555	0.18555	8.65402	5.38929	**7**
8	1.71819	0.58201	0.09747	0.16747	10.25980	5.97130	**8**
9	1.83846	0.54393	0.08349	0.15349	11.97799	6.51523	**9**
10	1.96715	0.50835	0.07238	0.14238	13.81645	7.02358	**10**
11	2.10485	0.47509	0.06336	0.13336	15.78360	7.49867	**11**
12	2.25219	0.44401	0.05590	0.12590	17.88845	7.94269	**12**
13	2.40985	0.41496	0.04965	0.11965	20.14064	8.35765	**13**
14	2.57853	0.38782	0.04434	0.11434	22.55049	8.74547	**14**
15	2.75903	0.36245	0.03979	0.10979	25.12902	9.10791	**15**
16	2.95216	0.33873	0.03586	0.10586	27.88805	9.44665	**16**
17	3.15882	0.31657	0.03243	0.10243	30.84022	9.76322	**17**
18	3.37993	0.29586	0.02941	0.09941	33.99903	10.05909	**18**
19	3.61653	0.27651	0.02675	0.09675	37.37896	10.33560	**19**
20	3.86968	0.25842	0.02439	0.09439	40.99549	10.59401	**20**
21	4.14056	0.24151	0.02229	0.09229	44.86518	10.83553	**21**
22	4.43040	0.22571	0.02041	0.09041	49.00574	11.06124	**22**
23	4.74053	0.21095	0.01871	0.08871	53.43614	11.27219	**23**
24	5.07237	0.19715	0.01719	0.08719	58.17667	11.46933	**24**
25	5.42743	0.18425	0.01581	0.08581	63.24904	11.65358	**25**
30	7.61226	0.13137	0.01059	0.08059	94.46079	12.40904	**30**
31	8.14511	0.12277	0.00980	0.07980	102.07304	12.53181	**31**
32	8.71527	0.11474	0.00907	0.07907	110.21815	12.64656	**32**
33	9.32534	0.10723	0.00841	0.07841	118.93343	12.75379	**33**
34	9.97811	0.10022	0.00780	0.07780	128.25876	12.85401	**34**
35	10.67658	0.09366	0.00723	0.07723	138.23688	12.94767	**35**
36	11.42394	0.08754	0.00672	0.07672	148.91346	13.03521	**36**
42	17.14426	0.05833	0.00434	0.07434	230.63224	13.45245	**42**
48	25.72891	0.03887	0.00283	0.07283	353.27009	13.73047	**48**
60	57.94643	0.01726	0.00123	0.07123	813.52038	14.03918	**60**
72	1.305065E+02	0.00766	0.00054	0.07054	1.850092E+03	14.17625	**72**
120	3.357788E+03	0.00030	0.00002	0.07002	4.795412E+04	14.28146	**120**
180	1.945718E+05	0.00001	0.00000	0.07000	2.779583E+06	14.28564	**180**
240	1.127474E+07	0.00000	0.00000	0.07000	1.610677E+08	14.28571	**240**
360	3.785820E+10	0.00000	0.00000	0.07000	5.408314E+11	14.28571	**360**

TABLE C.15

8.00%	Compound Interest Factors						8.00%
	Single Payment		Uniform Payment Series				
n	Compound Amount Factor	Present Worth Factor	Sinking Fund Factor	Capital Recovery Factor	Compound Amount Factor	Present Worth Factor	n
	Find F given P F/P	Find P given F P/F	Find A given F A/F	Find A given P A/P	Find F given A F/A	Find P given A P/A	
1	1.08000	0.92593	1.00000	1.08000	1.00000	0.92593	**1**
2	1.16640	0.85734	0.48077	0.56077	2.08000	1.78326	**2**
3	1.25971	0.79383	0.30803	0.38803	3.24640	2.57710	**3**
4	1.36049	0.73503	0.22192	0.30192	4.50611	3.31213	**4**
5	1.46933	0.68058	0.17046	0.25046	5.86660	3.99271	**5**
6	1.58687	0.63017	0.13632	0.21632	7.33593	4.62288	**6**
7	1.71382	0.58349	0.11207	0.19207	8.92280	5.20637	**7**
8	1.85093	0.54027	0.09401	0.17401	10.63663	5.74664	**8**
9	1.99900	0.50025	0.08008	0.16008	12.48756	6.24689	**9**
10	2.15892	0.46319	0.06903	0.14903	14.48656	6.71008	**10**
11	2.33164	0.42888	0.06008	0.14008	16.64549	7.13896	**11**
12	2.51817	0.39711	0.05270	0.13270	18.97713	7.53608	**12**
13	2.71962	0.36770	0.04652	0.12652	21.49530	7.90378	**13**
14	2.93719	0.34046	0.04130	0.12130	24.21492	8.24424	**14**
15	3.17217	0.31524	0.03683	0.11683	27.15211	8.55948	**15**
16	3.42594	0.29189	0.03298	0.11298	30.32428	8.85137	**16**
17	3.70002	0.27027	0.02963	0.10963	33.75023	9.12164	**17**
18	3.99602	0.25025	0.02670	0.10670	37.45024	9.37189	**18**
19	4.31570	0.23171	0.02413	0.10413	41.44626	9.60360	**19**
20	4.66096	0.21455	0.02185	0.10185	45.76196	9.81815	**20**
21	5.03383	0.19866	0.01983	0.09983	50.42292	10.01680	**21**
22	5.43654	0.18394	0.01803	0.09803	55.45676	10.20074	**22**
23	5.87146	0.17032	0.01642	0.09642	60.89330	10.37106	**23**
24	6.34118	0.15770	0.01498	0.09498	66.76476	10.52876	**24**
25	6.84848	0.14602	0.01368	0.09368	73.10594	10.67478	**25**
30	10.06266	0.09938	0.00883	0.08883	113.28321	11.25778	**30**
31	10.86767	0.09202	0.00811	0.08811	123.34587	11.34980	**31**
32	11.73708	0.08520	0.00745	0.08745	134.21354	11.43500	**32**
33	12.67605	0.07889	0.00685	0.08685	145.95062	11.51389	**33**
34	13.69013	0.07305	0.00630	0.08630	158.62667	11.58693	**34**
35	14.78534	0.06763	0.00580	0.08580	172.31680	11.65457	**35**
36	15.96817	0.06262	0.00534	0.08534	187.10215	11.71719	**36**
42	25.33948	0.03946	0.00329	0.08329	304.24352	12.00670	**42**
48	40.21057	0.02487	0.00204	0.08204	490.13216	12.18914	**48**
60	101.25706	0.00988	0.00080	0.08080	1253.21330	12.37655	**60**
72	2.549825E+02	0.00392	0.00031	0.08031	3.174781E+03	12.45098	**72**
120	1.025299E+04	0.00010	0.00001	0.08001	1.281499E+05	12.49878	**120**
180	1.038188E+06	0.00000	0.00000	0.08000	1.297734E+07	12.49999	**180**
240	1.051239E+08	0.00000	0.00000	0.08000	1.314048E+09	12.50000	**240**
360	1.077834E+12	0.00000	0.00000	0.08000	1.347293E+13	12.50000	**360**

TABLE C.16

9.00%	Compound Interest Factors						9.00%
	Single Payment		Uniform Payment Series				
n	Compound Amount Factor	Present Worth Factor	Sinking Fund Factor	Capital Recovery Factor	Compound Amount Factor	Present Worth Factor	n
	Find F given P F/P	Find P given F P/F	Find A given F A/F	Find A given P A/P	Find F given A F/A	Find P given A P/A	
1	1.09000	0.91743	1.00000	1.09000	1.00000	0.91743	**1**
2	1.18810	0.84168	0.47847	0.56847	2.09000	1.75911	**2**
3	1.29503	0.77218	0.30505	0.39505	3.27810	2.53129	**3**
4	1.41158	0.70843	0.21867	0.30867	4.57313	3.23972	**4**
5	1.53862	0.64993	0.16709	0.25709	5.98471	3.88965	**5**
6	1.67710	0.59627	0.13292	0.22292	7.52333	4.48592	**6**
7	1.82804	0.54703	0.10869	0.19869	9.20043	5.03295	**7**
8	1.99256	0.50187	0.09067	0.18067	11.02847	5.53482	**8**
9	2.17189	0.46043	0.07680	0.16680	13.02104	5.99525	**9**
10	2.36736	0.42241	0.06582	0.15582	15.19293	6.41766	**10**
11	2.58043	0.38753	0.05695	0.14695	17.56029	6.80519	**11**
12	2.81266	0.35553	0.04965	0.13965	20.14072	7.16073	**12**
13	3.06580	0.32618	0.04357	0.13357	22.95338	7.48690	**13**
14	3.34173	0.29925	0.03843	0.12843	26.01919	7.78615	**14**
15	3.64248	0.27454	0.03406	0.12406	29.36092	8.06069	**15**
16	3.97031	0.25187	0.03030	0.12030	33.00340	8.31256	**16**
17	4.32763	0.23107	0.02705	0.11705	36.97370	8.54363	**17**
18	4.71712	0.21199	0.02421	0.11421	41.30134	8.75563	**18**
19	5.14166	0.19449	0.02173	0.11173	46.01846	8.95011	**19**
20	5.60441	0.17843	0.01955	0.10955	51.16012	9.12855	**20**
21	6.10881	0.16370	0.01762	0.10762	56.76453	9.29224	**21**
22	6.65860	0.15018	0.01590	0.10590	62.87334	9.44243	**22**
23	7.25787	0.13778	0.01438	0.10438	69.53194	9.58021	**23**
24	7.91108	0.12640	0.01302	0.10302	76.78981	9.70661	**24**
25	8.62308	0.11597	0.01181	0.10181	84.70090	9.82258	**25**
30	13.26768	0.07537	0.00734	0.09734	136.30754	10.27365	**30**
31	14.46177	0.06915	0.00669	0.09669	149.57522	10.34280	**31**
32	15.76333	0.06344	0.00610	0.09610	164.03699	10.40624	**32**
33	17.18203	0.05820	0.00556	0.09556	179.80032	10.46444	**33**
34	18.72841	0.05339	0.00508	0.09508	196.98234	10.51784	**34**
35	20.41397	0.04899	0.00464	0.09464	215.71075	10.56682	**35**
36	22.25123	0.04494	0.00424	0.09424	236.12472	10.61176	**36**
42	37.31753	0.02680	0.00248	0.09248	403.52813	10.81337	**42**
48	62.58524	0.01598	0.00146	0.09146	684.28041	10.93358	**48**
60	176.03129	0.00568	0.00051	0.09051	1944.79213	11.04799	**60**
72	4.951170E+02	0.00202	0.00018	0.09018	5.490189E+03	11.08867	**72**
120	3.098702E+04	0.00003	0.00000	0.09000	3.442891E+05	11.11075	**120**
180	5.454684E+06	0.00000	0.00000	0.09000	6.060759E+07	11.11111	**180**
240	9.601951E+08	0.00000	0.00000	0.09000	1.066883E+10	11.11111	**240**
360	2.975358E+13	0.00000	0.00000	0.09000	3.305954E+14	11.11111	**360**

TABLE C.17

10.00%	Compound Interest Factors						10.00%
	Single Payment		Uniform Payment Series				
n	Compound Amount Factor	Present Worth Factor	Sinking Fund Factor	Capital Recovery Factor	Compound Amount Factor	Present Worth Factor	n
	Find F given P F/P	Find P given F P/F	Find A given F A/F	Find A given P A/P	Find F given A F/A	Find P given A P/A	
1	1.10000	0.90909	1.00000	1.10000	1.00000	0.90909	**1**
2	1.21000	0.82645	0.47619	0.57619	2.10000	1.73554	**2**
3	1.33100	0.75131	0.30211	0.40211	3.31000	2.48685	**3**
4	1.46410	0.68301	0.21547	0.31547	4.64100	3.16987	**4**
5	1.61051	0.62092	0.16380	0.26380	6.10510	3.79079	**5**
6	1.77156	0.56447	0.12961	0.22961	7.71561	4.35526	**6**
7	1.94872	0.51316	0.10541	0.20541	9.48717	4.86842	**7**
8	2.14359	0.46651	0.08744	0.18744	11.43589	5.33493	**8**
9	2.35795	0.42410	0.07364	0.17364	13.57948	5.75902	**9**
10	2.59374	0.38554	0.06275	0.16275	15.93742	6.14457	**10**
11	2.85312	0.35049	0.05396	0.15396	18.53117	6.49506	**11**
12	3.13843	0.31863	0.04676	0.14676	21.38428	6.81369	**12**
13	3.45227	0.28966	0.04078	0.14078	24.52271	7.10336	**13**
14	3.79750	0.26333	0.03575	0.13575	27.97498	7.36669	**14**
15	4.17725	0.23939	0.03147	0.13147	31.77248	7.60608	**15**
16	4.59497	0.21763	0.02782	0.12782	35.94973	7.82371	**16**
17	5.05447	0.19784	0.02466	0.12466	40.54470	8.02155	**17**
18	5.55992	0.17986	0.02193	0.12193	45.59917	8.20141	**18**
19	6.11591	0.16351	0.01955	0.11955	51.15909	8.36492	**19**
20	6.72750	0.14864	0.01746	0.11746	57.27500	8.51356	**20**
21	7.40025	0.13513	0.01562	0.11562	64.00250	8.64869	**21**
22	8.14027	0.12285	0.01401	.0.11401	71.40275	8.77154	**22**
23	8.95430	0.11168	0.01257	0.11257	79.54302	8.88322	**23**
24	9.84973	0.10153	0.01130	0.11130	88.49733	8.98474	**24**
25	10.83471	0.09230	0.01017	0.11017	98.34706	9.07704	**25**
30	17.44940	0.05731	0.00608	0.10608	164.49402	9.42691	**30**
31	19.19434	0.05210	0.00550	0.10550	181.94342	9.47901	**31**
32	21.11378	0.04736	0.00497	0.10497	201.13777	9.52638	**32**
33	23.22515	0.04306	0.00450	0.10450	222.25154	9.56943	**33**
34	25.54767	0.03914	0.00407	0.10407	245.47670	9.60857	**34**
35	28.10244	0.03558	0.00369	0.10369	271.02437	9.64416	**35**
36	30.91268	0.03235	0.00334	0.10334	299.12681	9.67651	**36**
42	54.76370	0.01826	0.00186	0.10186	537.63699	9.81740	**42**
48	97.01723	0.01031	0.00104	0.10104	960.17234	9.89693	**48**
60	304.48164	0.00328	0.00033	0.10033	3034.81640	9.96716	**60**
72	9.555938E+02	0.00105	0.00010	0.10010	9.545938E+03	9.98954	**72**
120	9.270907E+04	0.00001	0.00000	0.10000	9.270807E+05	9.99989	**120**
180	2.822821E+07	0.00000	0.00000	0.10000	2.822821E+08	10.00000	**180**
240	8.594971E+09	0.00000	0.00000	0.10000	8.594971E+10	10.00000	**240**
360	7.968318E+14	0.00000	0.00000	0.10000	7.968318E+15	10.00000	**360**

TABLE C.18

11.00%	Compound Interest Factors						11.00%
	Single Payment		Uniform Payment Series				
n	Compound Amount Factor	Present Worth Factor	Sinking Fund Factor	Capital Recovery Factor	Compound Amount Factor	Present Worth Factor	n
	Find F given P F/P	Find P given F P/F	Find A given F A/F	Find A given P A/P	Find F given A F/A	Find P given A P/A	
1	1.11000	0.90090	1.00000	1.11000	1.00000	0.90090	**1**
2	1.23210	0.81162	0.47393	0.58393	2.11000	1.71252	**2**
3	1.36763	0.73119	0.29921	0.40921	3.34210	2.44371	**3**
4	1.51807	0.65873	0.21233	0.32233	4.70973	3.10245	**4**
5	1.68506	0.59345	0.16057	0.27057	6.22780	3.69590	**5**
6	1.87041	0.53464	0.12638	0.23638	7.91286	4.23054	**6**
7	2.07616	0.48166	0.10222	0.21222	9.78327	4.71220	**7**
8	2.30454	0.43393	0.08432	0.19432	11.85943	5.14612	**8**
9	2.55804	0.39092	0.07060	0.18060	14.16397	5.53705	**9**
10	2.83942	0.35218	0.05980	0.16980	16.72201	5.88923	**10**
11	3.15176	0.31728	0.05112	0.16112	19.56143	6.20652	**11**
12	3.49845	0.28584	0.04403	0.15403	22.71319	6.49236	**12**
13	3.88328	0.25751	0.03815	0.14815	26.21164	6.74987	**13**
14	4.31044	0.23199	0.03323	0.14323	30.09492	6.98187	**14**
15	4.78459	0.20900	0.02907	0.13907	34.40536	7.19087	**15**
16	5.31089	0.18829	0.02552	0.13552	39.18995	7.37916	**16**
17	5.89509	0.16963	0.02247	0.13247	44.50084	7.54879	**17**
18	6.54355	0.15282	0.01984	0.12984	50.39594	7.70162	**18**
19	7.26334	0.13768	0.01756	0.12756	56.93949	7.83929	**19**
20	8.06231	0.12403	0.01558	0.12558	64.20283	7.96333	**20**
21	8.94917	0.11174	0.01384	0.12384	72.26514	8.07507	**21**
22	9.93357	0.10067	0.01231	0.12231	81.21431	8.17574	**22**
23	11.02627	0.09069	0.01097	0.12097	91.14788	8.26643	**23**
24	12.23916	0.08170	0.00979	0.11979	102.17415	8.34814	**24**
25	13.58546	0.07361	0.00874	0.11874	114.41331	8.42174	**25**
30	22.89230	0.04368	0.00502	0.11502	199.02088	8.69379	**30**
31	25.41045	0.03935	0.00451	0.11451	221.91317	8.73315	**31**
32	28.20560	0.03545	0.00404	0.11404	247.32362	8.76860	**32**
33	31.30821	0.03194	0.00363	0.11363	275.52922	8.80054	**33**
34	34.75212	0.02878	0.00326	0.11326	306.83744	8.82932	**34**
35	38.57485	0.02592	0.00293	0.11293	341.58955	8.85524	**35**
36	42.81808	0.02335	0.00263	0.11263	380.16441	8.87859	**36**
42	80.08757	0.01249	0.00139	0.11139	718.97790	8.97740	**42**
48	149.79695	0.00668	0.00074	0.11074	1352.69958	9.03022	**48**
60	524.05724	0.00191	0.00021	0.11021	4755.06584	9.07356	**60**
72	1.833388E+03	0.00055	0.00006	0.11006	1.665808E+04	9.08595	**72**
120	2.746360E+05	0.00000	0.00000	0.11000	2.496682E+06	9.09088	**120**
180	1.439250E+08	0.00000	0.00000	0.11000	1.308409E+09	9.09091	**180**
240	7.542493E+10	0.00000	0.00000	0.11000	6.856812E+11	9.09091	**240**
360	2.071440E+16	0.00000	0.00000	0.11000	1.883127E+17	9.09091	**360**

TABLE C.19

12.00%	Compound Interest Factors						12.00%
	Single Payment		Uniform Payment Series				
n	Compound Amount Factor	Present Worth Factor	Sinking Fund Factor	Capital Recovery Factor	Compound Amount Factor	Present Worth Factor	n
	Find F given P F/P	Find P given F P/F	Find A given F A/F	Find A given P A/P	Find F given A F/A	Find P given A P/A	
1	1.12000	0.89286	1.00000	1.12000	1.00000	0.89286	**1**
2	1.25440	0.79719	0.47170	0.59170	2.12000	1.69005	**2**
3	1.40493	0.71178	0.29635	0.41635	3.37440	2.40183	**3**
4	1.57352	0.63552	0.20923	0.32923	4.77933	3.03735	**4**
5	1.76234	0.56743	0:15741	0.27741	6.35285	3.60478	**5**
6	1.97382	0.50663	0.12323	0.24323	8.11519	4.11141	**6**
7	2.21068	0.45235	0.09912	0.21912	10.08901	4.56376	**7**
8	2.47596	0.40388	0.08130	0.20130	12.29969	4.96764	**8**
9	2.77308	0.36061	0.06768	0.18768	14.77566	5.32825	**9**
10	3.10585	0.32197	0.05698	0.17698	17.54874	5.65022	**10**
11	3.47855	0.28748	0.04842	0.16842	20.65458	5.93770	**11**
12	3.89598	0.25668	0.04144	0.16144	24.13313	6.19437	**12**
13	4.36349	0.22917	0.03568	0.15568	28.02911	6.42355	**13**
14	4.88711	0.20462	0.03087	0.15087	32.39260	6.62817	**14**
15	5.47357	0.18270	0.02682	0.14682	37.27971	6.81086	**15**
16	6.13039	0.16312	0.02339	0.14339	42.75328	6.97399	**16**
17	6.86604	0.14564	0.02046	0.14046	48.88367	7.11963	**17**
18	7.68997	0.13004	0.01794	0.13794	55.74971	7.24967	**18**
19	8.61276	0.11611	0.01576	0.13576	63.43968	7.36578	**19**
20	9.64629	0.10367	0.01388	0.13388	72.05244	7.46944	**20**
21	10.80385	0.09256	0.01224	0.13224	81.69874	7.56200	**21**
22	12.10031	0.08264	0.01081	0.13081	92.50258	7.64465	**22**
23	13.55235	0.07379	0.00956	0.12956	104.60289	7.71843	**23**
24	15.17863	0.06588	0.00846	0.12846	118.15524	7.78432	**24**
25	17.00006	0.05882	0.00750	0.12750	133.33387	7.84314	**25**
30	29.95992	0.03338	0.00414	0.12414	241.33268	8.05518	**30**
31	33.55511	0.02980	0.00369	0.12369	271.29261	8.08499	**31**
32	37.58173	0.02661	0.00328	0.12328	304.84772	8.11159	**32**
33	42.09153	0.02376	0.00292	0.12292	342.42945	8.13535	**33**
34	47.14252	0.02121	0.00260	0.12260	384.52098	8.15656	**34**
35	52.79962	0.01894	0.00232	0.12232	431.66350	8.17550	**35**
36	59.13557	0.01691	0.00206	0.12206	484.46312	8.19241	**36**
42	116.72314	0.00857	0.00104	0.12104	964.35948	8.26194	**42**
48	230.39078	0.00434	0.00052	0.12052	1911.58980	8.29716	**48**
60	897.59693	0.00111	0.00013	0.12013	7471.64111	8.32405	**60**
72	3.497016E+03	0.00029	0.00003	0.12003	2.913347E+04	8.33095	**72**
120	8.056803E+05	0.00000	0.00000	0.12000	6.713994E+06	8.33332	**120**
180	7.231761E+08	0.00000	0.00000	0.12000	6.026468E+09	8.33333	**180**
240	6.491207E+11	0.00000	0.00000	0.12000	5.409339E+12	8.33333	**240**
360	5.229837E+17	0.00000	0.00000	0.12000	4.358198E+18	8.33333	**360**

TABLE C.20

13.00%	Compound Interest Factors						13.00%
	Single Payment		Uniform Payment Series				
n	Compound Amount Factor	Present Worth Factor	Sinking Fund Factor	Capital Recovery Factor	Compound Amount Factor	Present Worth Factor	n
	Find F given P F/P	Find P given F P/F	Find A given F A/F	Find A given P A/P	Find F given A F/A	Find P given A P/A	
1	1.13000	0.88496	1.00000	1.13000	1.00000	0.88496	**1**
2	1.27690	0.78315	0.46948	0.59948	2.13000	1.66810	**2**
3	1.44290	0.69305	0.29352	0.42352	3.40690	2.36115	**3**
4	1.63047	0.61332	0.20619	0.33619	4.84980	2.97447	**4**
5	1.84244	0.54276	0.15431	0.28431	6.48027	3.51723	**5**
6	2.08195	0.48032	0.12015	0.25015	8.32271	3.99755	**6**
7	2.35261	0.42506	0.09611	0.22611	10.40466	4.42261	**7**
8	2.65844	0.37616	0.07839	0.20839	12.75726	4.79877	**8**
9	3.00404	0.33288	0.06487	0.19487	15.41571	5.13166	**9**
10	3.39457	0.29459	0.05429	0.18429	18.41975	5.42624	**10**
11	3.83586	0.26070	0.04584	0.17584	21.81432	5.68694	**11**
12	4.33452	0.23071	0.03899	0.16899	25.65018	5.91765	**12**
13	4.89801	0.20416	0.03335	0.16335	29.98470	6.12181	**13**
14	5.53475	0.18068	0.02867	0.15867	34.88271	6.30249	**14**
15	6.25427	0.15989	0.02474	0.15474	40.41746	6.46238	**15**
16	7.06733	0.14150	0.02143	0.15143	46.67173	6.60388	**16**
17	7.98608	0.12522	0.01861	0.14861	53.73906	6.72909	**17**
18	9.02427	0.11081	0.01620	0.14620	61.72514	6.83991	**18**
19	10.19742	0.09806	0.01413	0.14413	70.74941	6.93797	**19**
20	11.52309	0.08678	0.01235	0.14235	80.94683	7.02475	**20**
21	13.02109	0.07680	0.01081	0.14081	92.46992	7.10155	**21**
22	14.71383	0.06796	0.00948	0.13948	105.49101	7.16951	**22**
23	16.62663	0.06014	0.00832	0.13832	120.20484	7.22966	**23**
24	18.78809	0.05323	0.00731	0.13731	136.83147	7.28288	**24**
25	21.23054	0.04710	0.00643	0.13643	155.61956	7.32998	**25**
30	39.11590	0.02557	0.00341	0.13341	293.19922	7.49565	**30**
31	44.20096	0.02262	0.00301	0.13301	332.31511	7.51828	**31**
32	49.94709	0.02002	0.00266	0.13266	376.51608	7.53830	**32**
33	56.44021	0.01772	0.00234	0.13234	426.46317	7.55602	**33**
34	63.77744	0.01568	0.00207	0.13207	482.90338	7.57170	**34**
35	72.06851	0.01388	0.00183	0.13183	546.68082	7.58557	**35**
36	81.43741	0.01228	0.00162	0.13162	618.74933	7.59785	**36**
42	169.54876	0.00590	0.00077	0.13077	1296.52895	7.64694	**42**
48	352.99234	0.00283	0.00037	0.13037	2707.63342	7.67052	**48**
60	1530.05347	0.00065	0.00009	0.13009	11761.94979	7.68728	**60**
72	6.632052E+03	0.00015	0.00002	0.13002	5.100809E+04	7.69115	**72**
120	2.341064E+06	0.00000	0.00000	0.13000	1.800817E+07	7.69230	**120**
180	3.581953E+09	0.00000	0.00000	0.13000	2.755348E+10	7.69231	**180**
240	5.480579E+12	0.00000	0.00000	0.13000	4.215830E+13	7.69231	**240**
360	1.283038E+19	0.00000	0.00000	0.13000	9.869526E+19	7.69231	**360**

TABLE C.21

14.00%	Compound Interest Factors						14.00%
	Single Payment		Uniform Payment Series				
n	Compound Amount Factor	Present Worth Factor	Sinking Fund Factor	Capital Recovery Factor	Compound Amount Factor	Present Worth Factor	n
	Find F given P F/P	Find P given F P/F	Find A given F A/F	Find A given P A/P	Find F given A F/A	Find P given A P/A	
1	1.14000	0.87719	1.00000	1.14000	1.00000	0.87719	**1**
2	1.29960	0.76947	0.46729	0.60729	2.14000	1.64666	**2**
3	1.48154	0.67497	0.29073	0.43073	3.43960	2.32163	**3**
4	1.68896	0.59208	0.20320	0.34320	4.92114	2.91371	**4**
5	1.92541	0.51937	0.15128	0.29128	6.61010	3.43308	**5**
6	2.19497	0.45559	0.11716	0.25716	8.53552	3.88867	**6**
7	2.50227	0.39964	0.09319	0.23319	10.73049	4.28830	**7**
8	2.85259	0.35056	0.07557	0.21557	13.23276	4.63886	**8**
9	3.25195	0.30751	0.06217	0.20217	16.08535	4.94637	**9**
10	3.70722	0.26974	0.05171	0.19171	19.33730	5.21612	**10**
11	4.22623	0.23662	0.04339	0.18339	23.04452	5.45273	**11**
12	4.81790	0.20756	0.03667	0.17667	27.27075	5.66029	**12**
13	5.49241	0.18207	0.03116	0.17116	32.08865	5.84236	**13**
14	6.26135	0.15971	0.02661	0.16661	37.58107	6.00207	**14**
15	7.13794	0.14010	0.02281	0.16281	43.84241	6.14217	**15**
16	8.13725	0.12289	0.01962	0.15962	50.98035	6.26506	**16**
17	9.27646	0.10780	0.01692	0.15692	59.11760	6.37286	**17**
18	10.57517	0.09456	0.01462	0.15462	68.39407	6.46742	**18**
19	12.05569	0.08295	0.01266	0.15266	78.96923	6.55037	**19**
20	13.74349	0.07276	0.01099	0.15099	91.02493	6.62313	**20**
21	15.66758	0.06383	0.00954	0.14954	104.76842	6.68696	**21**
22	17.86104	0.05599	0.00830	0.14830	120.43600	6.74294	**22**
23	20.36158	0.04911	0.00723	0.14723	138.29704	6.79206	**23**
24	23.21221	0.04308	0.00630	0.14630	158.65862	6.83514	**24**
25	26.46192	0.03779	0.00550	0.14550	181.87083	6.87293	**25**
30	50.95016	0.01963	0.00280	0.14280	356.78685	7.00266	**30**
31	58.08318	0.01722	0.00245	0.14245	407.73701	7.01988	**31**
32	66.21483	0.01510	0.00215	0.14215	465.82019	7.03498	**32**
33	75.48490	0.01325	0.00188	0.14188	532.03501	7.04823	**33**
34	86.05279	0.01162	0.00165	0.14165	607.51991	7.05985	**34**
35	98.10018	0.01019	0.00144	0.14144	693.57270	7.07005	**35**
36	111.83420	0.00894	0.00126	0.14126	791.67288	7.07899	**36**
42	245.47301	0.00407	0.00057	0.14057	1746.23582	7.11376	**42**
48	538.80655	0.00186	0.00026	0.14026	3841.47534	7.12960	**48**
60	2595.91866	0.00039	0.00005	0.14005	18535.13328	7.14011	**60**
72	1.250689E+04	0.00008	0.00001	0.14001	8.932778E+04	7.14229	**72**
120	6.738794E+06	0.00000	0.00000	0.14000	4.813423E+07	7.14286	**120**
180	1.749336E+10	0.00000	0.00000	0.14000	1.249526E+11	7.14286	**180**
240	4.541134E+13	0.00000	0.00000	0.14000	3.243667E+14	7.14286	**240**
360	3.060177E+20	0.00000	0.00000	0.14000	2.185840E+21	7.14286	**360**

TABLE C.22

15.00%	Compound Interest Factors						15.00%
	Single Payment		Uniform Payment Series				
n	Compound Amount Factor	Present Worth Factor	Sinking Fund Factor	Capital Recovery Factor	Compound Amount Factor	Present Worth Factor	n
	Find F given P F/P	Find P given F P/F	Find A given F A/F	Find A given P A/P	Find F given A F/A	Find P given A P/A	
1	1.15000	0.86957	1.00000	1.15000	1.00000	0.86957	**1**
2	1.32250	0.75614	0.46512	0.61512	2.15000	1.62571	**2**
3	1.52088	0.65752	0.28798	0.43798	3.47250	2.28323	**3**
4	1.74901	0.57175	0.20027	0.35027	4.99338	2.85498	**4**
5	2.01136	0.49718	0.14832	0.29832	6.74238	3.35216	**5**
6	2.31306	0.43233	0.11424	0.26424	8.75374	3.78448	**6**
7	2.66002	0.37594	0.09036	0.24036	11.06680	4.16042	**7**
8	3.05902	0.32690	0.07285	0.22285	13.72682	4.48732	**8**
9	3.51788	0.28426	0.05957	0.20957	16.78584	4.77158	**9**
10	4.04556	0.24718	0.04925	0.19925	20.30372	5.01877	**10**
11	4.65239	0.21494	0.04107	0.19107	24.34928	5.23371	**11**
12	5.35025	0.18691	0.03448	0.18448	29.00167	5.42062	**12**
13	6.15279	0.16253	0.02911	0.17911	34.35192	5.58315	**13**
14	7.07571	0.14133	0.02469	0.17469	40.50471	5.72448	**14**
15	8.13706	0.12289	0.02102	0.17102	47.58041	5.84737	**15**
16	9.35762	0.10686	0.01795	0.16795	55.71747	5.95423	**16**
17	10.76126	0.09293	0.01537	0.16537	65.07509	6.04716	**17**
18	12.37545	0.08081	0.01319	0.16319	75.83636	6.12797	**18**
19	14.23177	0.07027	0.01134	0.16134	88.21181	6.19823	**19**
20	16.36654	0.06110	0.00976	0.15976	102.44358	6.25933	**20**
21	18.82152	0.05313	0.00842	0.15842	118.81012	6.31246	**21**
22	21.64475	0.04620	0.00727	0.15727	137.63164	6.35866	**22**
23	24.89146	0.04017	0.00628	0.15628	159.27638	6.39884	**23**
24	28.62518	0.03493	0.00543	0.15543	184.16784	6.43377	**24**
25	32.91895	0.03038	0.00470	0.15470	212.79302	6.46415	**25**
30	66.21177	0.01510	0.00230	0.15230	434.74515	6.56598	**30**
31	76.14354	0.01313	0.00200	0.15200	500.95692	6.57911	**31**
32	87.56507	0.01142	0.00173	0.15173	577.10046	6.59053	**32**
33	100.69983	0.00993	0.00150	0.15150	664.66552	6.60046	**33**
34	115.80480	0.00864	0.00131	0.15131	765.36535	6.60910	**34**
35	133.17552	0.00751	0.00113	0.15113	881.17016	6.61661	**35**
36	153.15185	0.00653	0.00099	0.15099	1014.34568	6.62314	**36**
42	354.24954	0.00282	0.00042	0.15042	2354.99693	6.64785	**42**
48	819.40071	0.00122	0.00018	0.15018	5456.00475	6.65853	**48**
60	4383.99875	0.00023	0.00003	0.15003	29219.99164	6.66515	**60**
72	2.345549E+04	0.00004	0.00001	0.15001	1.563633E+05	6.66638	**72**
120	1.921945E+07	0.00000	0.00000	0.15000	1.281296E+08	6.66667	**120**
180	8.425802E+10	0.00000	0.00000	0.15000	5.617202E+11	6.66667	**180**
240	3.693871E+14	0.00000	0.00000	0.15000	2.462580E+15	6.66667	**240**
360	7.099414E+21	0.00000	0.00000	0.15000	4.732943E+22	6.66667	**360**

TABLE C.23

20.00%	Compound Interest Factors						20.00%
	Single Payment		Uniform Payment Series				
n	Compound Amount Factor	Present Worth Factor	Sinking Fund Factor	Capital Recovery Factor	Compound Amount Factor	Present Worth Factor	n
	Find F given P F/P	Find P given F P/F	Find A given F A/F	Find A given P A/P	Find F given A F/A	Find P given A P/A	
1	1.20000	0.83333	1.00000	1.20000	1.00000	0.83333	**1**
2	1.44000	0.69444	0.45455	0.65455	2.20000	1.52778	**2**
3	1.72800	0.57870	0.27473	0.47473	3.64000	2.10648	**3**
4	2.07360	0.48225	0.18629	0.38629	5.36800	2.58873	**4**
5	2.48832	0.40188	0.13438	0.33438	7.44160	2.99061	**5**
6	2.98598	0.33490	0.10071	0.30071	9.92992	3.32551	**6**
7	3.58318	0.27908	0.07742	0.27742	12.91590	3.60459	**7**
8	4.29982	0.23257	0.06061	0.26061	16.49908	3.83716	**8**
9	5.15978	0.19381	0.04808	0.24808	20.79890	4.03097	**9**
10	6.19174	0.16151	0.03852	0.23852	25.95868	4.19247	**10**
11	7.43008	0.13459	0.03110	0.23110	32.15042	4.32706	**11**
12	8.91610	0.11216	0.02526	0.22526	39.58050	4.43922	**12**
13	10.69932	0.09346	0.02062	0.22062	48.49660	4.53268	**13**
14	12.83918	0.07789	0.01689	0.21689	59.19592	4.61057	**14**
15	15.40702	0.06491	0.01388	0.21388	72.03511	4.67547	**15**
16	18.48843	0.05409	0.01144	0.21144	87.44213	4.72956	**16**
17	22.18611	0.04507	0.00944	0.20944	105.93056	4.77463	**17**
18	26.62333	0.03756	0.00781	0.20781	128.11667	4.81219	**18**
19	31.94800	0.03130	0.00646	0.20646	154.74000	4.84350	**19**
20	38.33760	0.02608	0.00536	0.20536	186.68800	4.86958	**20**
21	46.00512	0.02174	0.00444	0.20444	225.02560	4.89132	**21**
22	55.20614	0.01811	0.00369	0.20369	271.03072	4.90943	**22**
23	66.24737	0.01509	0.00307	0.20307	326.23686	4.92453	**23**
24	79.49685	0.01258	0.00255	0.20255	392.48424	4.93710	**24**
25	95.39622	0.01048	0.00212	0.20212	471.98108	4.94759	**25**
30	237.37631	0.00421	0.00085	0.20085	1181.88157	4.97894	**30**
31	284.85158	0.00351	0.00070	0.20070	1419.25788	4.98245	**31**
32	341.82189	0.00293	0.00059	0.20059	1704.10946	4.98537	**32**
33	410.18627	0.00244	0.00049	0.20049	2045.93135	4.98781	**33**
34	492.22352	0.00203	0.00041	0.20041	2456.11762	4.98984	**34**
35	590.66823	0.00169	0.00034	0.20034	2948.34115	4.99154	**35**
36	708.80187	0.00141	0.00028	0.20028	3539.00937	4.99295	**36**
42	2116.47106	0.00047	0.00009	0.20009	10577.35529	4.99764	**42**
48	6319.74872	0.00016	0.00003	0.20003	31593.74358	4.99921	**48**
60	56347.51435	0.00002	0.00000	0.20000	281732.57177	4.99991	**60**
72	5.024001E+05	0.00000	0.00000	0.20000	2.511995E+06	4.99999	**72**
120	3.175042E+09	0.00000	0.00000	0.20000	1.587521E+10	5.00000	**120**
180	1.789057E+14	0.00000	0.00000	0.20000	8.945287E+14	5.00000	**180**
240	1.008089E+19	0.00000	0.00000	0.20000	5.040447E+19	5.00000	**240**
360	3.200727E+28	0.00000	0.00000	0.20000	1.600363E+29	5.00000	**360**

TABLE C.24

25.00%	Compound Interest Factors						25.00%
	Single Payment		Uniform Payment Series				
n	Compound Amount Factor	Present Worth Factor	Sinking Fund Factor	Capital Recovery Factor	Compound Amount Factor	Present Worth Factor	n
	Find F given P F/P	Find P given F P/F	Find A given F A/F	Find A given P A/P	Find F given A F/A	Find P given A P/A	
1	1.25000	0.80000	1.00000	1.25000	1.00000	0.80000	**1**
2	1.56250	0.64000	0.44444	0.69444	2.25000	1.44000	**2**
3	1.95313	0.51200	0.26230	0.51230	3.81250	1.95200	**3**
4	2.44141	0.40960	0.17344	0.42344	5.76563	2.36160	**4**
5	3.05176	0.32768	0.12185	0.37185	8.20703	2.68928	**5**
6	3.81470	0.26214	0.08882	0.33882	11.25879	2.95142	**6**
7	4.76837	0.20972	0.06634	0.31634	15.07349	3.16114	**7**
8	5.96046	0.16777	0.05040	0.30040	19.84186	3.32891	**8**
9	7.45058	0.13422	0.03876	0.28876	25.80232	3.46313	**9**
10	9.31323	0.10737	0.03007	0.28007	33.25290	3.57050	**10**
11	11.64153	0.08590	0.02349	0.27349	42.56613	3.65640	**11**
12	14.55192	0.06872	0.01845	0.26845	54.20766	3.72512	**12**
13	18.18989	0.05498	0.01454	0.26454	68.75958	3.78010	**13**
14	22.73737	0.04398	0.01150	0.26150	86.94947	3.82408	**14**
15	28.42171	0.03518	0.00912	0.25912	109.68684	3.85926	**15**
16	35.52714	0.02815	0.00724	0.25724	138.10855	3.88741	**16**
17	44.40892	0.02252	0.00576	0.25576	173.63568	3.90993	**17**
18	55.51115	0.01801	0.00459	0.25459	218.04460	3.92794	**18**
19	69.38894	0.01441	0.00366	0.25366	273.55576	3.94235	**19**
20	86.73617	0.01153	0.00292	0.25292	342.94470	3.95388	**20**
21	108.42022	0.00922	0.00233	0.25233	429.68087	3.96311	**21**
22	135.52527	0.00738	0.00186	0.25186	538.10109	3.97049	**22**
23	169.40659	0.00590	0.00148	0.25148	673.62636	3.97639	**23**
24	211.75824	0.00472	0.00119	0.25119	843.03295	3.98111	**24**
25	264.69780	0.00378	0.00095	0.25095	1054.79118	3.98489	**25**
30	807.79357	0.00124	0.00031	0.25031	3227.17427	3.99505	**30**
31	1009.74196	0.00099	0.00025	0.25025	4034.96783	3.99604	**31**
32	1262.17745	0.00079	0.00020	0.25020	5044.70979	3.99683	**32**
33	1577.72181	0.00063	0.00016	0.25016	6306.88724	3.99746	**33**
34	1972.15226	0.00051	0.00013	0.25013	7884.60905	3.99797	**34**
35	2465.19033	0.00041	0.00010	0.25010	9856.76132	3.99838	**35**
36	3081.48791	0.00032	0.00008	0.25008	12321.95164	3.99870	**36**
42	11754.94351	0.00009	0.00002	0.25002	47015.77403	3.99966	**42**
48	44841.55086	0.00002	0.00001	0.25001	179362.20343	3.99991	**48**
60	652530.44680	0.00000	0.00000	0.25000	2610117.78720	3.99999	**60**
72	9.495568E+06	0.00000	0.00000	0.25000	3.798227E+07	4.00000	**72**
120	4.257960E+11	0.00000	0.00000	0.25000	1.703184E+12	4.00000	**120**
180	2.778448E+17	0.00000	0.00000	0.25000	1.111379E+18	4.00000	**180**
240	1.813022E+23	0.00000	0.00000	0.25000	7.252089E+23	4.00000	**240**
360	7.719776E+34	0.00000	0.00000	0.25000	3.087910E+35	4.00000	**360**

TABLE C.25

30.00%	Compound Interest Factors						30.00%
	Single Payment		Uniform Payment Series				
n	Compound Amount Factor	Present Worth Factor	Sinking Fund Factor	Capital Recovery Factor	Compound Amount Factor	Present Worth Factor	n
	Find F given P F/P	Find P given F P/F	Find A given F A/F	Find A given P A/P	Find F given A F/A	Find P given A P/A	
1	1.30000	0.76923	1.00000	1.30000	1.00000	0.76923	**1**
2	1.69000	0.59172	0.43478	0.73478	2.30000	1.36095	**2**
3	2.19700	0.45517	0.25063	0.55063	3.99000	1.81611	**3**
4	2.85610	0.35013	0.16163	0.46163	6.18700	2.16624	**4**
5	3.71293	0.26933	0.11058	0.41058	9.04310	2.43557	**5**
6	4.82681	0.20718	0.07839	0.37839	12.75603	2.64275	**6**
7	6.27485	0.15937	0.05687	0.35687	17.58284	2.80211	**7**
8	8.15731	0.12259	0.04192	0.34192	23.85769	2.92470	**8**
9	10.60450	0.09430	0.03124	0.33124	32.01500	3.01900	**9**
10	13.78585	0.07254	0.02346	0.32346	42.61950	3.09154	**10**
11	17.92160	0.05580	0.01773	0.31773	56.40535	3.14734	**11**
12	23.29809	0.04292	0.01345	0.31345	74.32695	3.19026	**12**
13	30.28751	0.03302	0.01024	0.31024	97.62504	3.22328	**13**
14	39.37376	0.02540	0.00782	0.30782	127.91255	3.24867	**14**
15	51.18589	0.01954	0.00598	0.30598	167.28631	3.26821	**15**
16	66.54166	0.01503	0.00458	0.30458	218.47220	3.28324	**16**
17	86.50416	0.01156	0.00351	0.30351	285.01386	3.29480	**17**
18	112.45541	0.00889	0.00269	0.30269	371.51802	3.30369	**18**
19	146.19203	0.00684	0.00207	0.30207	483.97343	3.31053	**19**
20	190.04964	0.00526	0.00159	0.30159	630.16546	3.31579	**20**
21	247.06453	0.00405	0.00122	0.30122	820.21510	3.31984	**21**
22	321.18389	0.00311	0.00094	0.30094	1067.27963	3.32296	**22**
23	417.53905	0.00239	0.00072	0.30072	1388.46351	3.32535	**23**
24	542.80077	0.00184	0.00055	0.30055	1806.00257	3.32719	**24**
25	705.64100	0.00142	0.00043	0.30043	2348.80334	3.32861	**25**
30	2619.99564	0.00038	0.00011	0.30011	8729.98548	3.33206	**30**
31	3405.99434	0.00029	0.00009	0.30009	11349.98112	3.33235	**31**
32	4427.79264	0.00023	0.00007	0.30007	14755.97546	3.33258	**32**
33	5756.13043	0.00017	0.00005	0.30005	19183.76810	3.33275	**33**
34	7482.96956	0.00013	0.00004	0.30004	24939.89853	3.33289	**34**
35	9727.86043	0.00010	0.00003	0.30003	32422.86808	3.33299	**35**
36	12646.21855	0.00008	0.00002	0.30002	42150.72851	3.33307	**36**
42	61040.88153	0.00002	0.00000	0.30000	203466.27175	3.33328	**42**
48	294632.67632	0.00000	0.00000	0.30000	982105.58773	3.33332	**48**
60	6864377.17274	0.00000	0.00000	0.30000	22881253.90915	3.33333	**60**
72	1.599268E+08	0.00000	0.00000	0.30000	5.330895E+08	3.33333	**72**
120	4.711967E+13	0.00000	0.00000	0.30000	1.570656E+14	3.33333	**120**
180	3.234472E+20	0.00000	0.00000	0.30000	1.078157E+21	3.33333	**180**
240	2.220264E+27	0.00000	0.00000	0.30000	7.400879E+27	3.33333	**240**
360	1.046181E+41	0.00000	0.00000	0.30000	3.487270E+41	3.33333	**360**

REFERENCES

Adrian, J. (1978) Construction Accounting. Reston, VA. Reston Publishing.

Adrian, J. and D. J. Adrian. (2006). *Construction Accounting: Financial, Managerial, Auditing, and Tax*. Champaign IL: Stipes Publishing LLC.

American Institute of Certified Public Accountants (2007). *Construction Contractors: AICPA Audit and Accounting Guide*. Durham, NC: AICPA.

Anonymous (2001). *Builder's Guide to Accounting, Revised Edition*. Carlsbad, CA: Craftsman Book Company.

Au, T. and T. P. Au (1992). *Engineering Economics for Capital Investment Analysis*, 2nd edition. Englewood Cliffs, NJ: Prentice Hall.

Ball, M. (2006). *Markets & Institutions in Real Estate & Construction*. Malden, MA: Blackwell Pub.

Barnes, M. (1977). *Measurement in Contract Control: A Guide to the Financial Control of Contracts Using the Civil Engineering Standard Method of Measurement*. London: Institution of Civil Engineers.

Barrie, D. W. and B. S. Paulson (1992). *Professional Construction Management 3rd edition*, New York, NY: McGraw-Hill.

Callan, J. L. (2003). *Construction Accounting Deskbook*, 5th edition. New York, NY: Aspen Law & Business.

Carmichael, D. R. a. R., P (editors) (2003). *Accountants' Handbook*. Hoboken, NJ: John Wiley and Sons.

Clough, R. H. and G. A. Sears (1994). *Construction Contracting*, 6th edition. New York, NY: John Wiley and Sons.

Collier, Keith (1974). *Fundamentals of Construction Estimating and Cost Accounting*. Englewood Cliffs, NJ: Prentice-Hall.

Collier, N. S., C. A. Collier, and D. A. Halperin. (2002). *Construction Funding, The Process of Real Estate Development*, 3rd edition. New York, NY: John Wiley and Sons.

Coombs, W. E. (1989). *Construction Accounting and Financial Management.* New York: McGraw-Hill.

Eppes, B. G. (1984). *Cost Accounting for the Construction Firm.* New York, NY: John Wiley and Sons.

Ganaway, N. (2006). *Construction Business Management: What Every Construction Contractor, Builder & Subcontractor Needs to Know*. Kingston, MA: R.S. Means Co, Inc.

Halpin, D. W. (1985). *Financial and Cost Concepts for Construction Management.* New York, NY: John Wiley and Sons.

Halpin, D. W. (2006) *Construction Management, 3rd edition*, New York, NY: John Wiley and Sons.

Internal Revenue Service (IRS) website: www.irs.gov.

Jackson, I. J. (1986). *Financial Management for Contractors.* New York: McGraw-Hill.

Kenig, M., Coordinator (2004). *Project Delivery Systems for Construction*, Washington, DC: Associated General Contractors of America.

Kenley, R. (1999) "Cash farming in building and construction: a stochastic analysis," *Construction Management and Economics*, 17: 3. Abingdon, Oxfordshire, United Kingdom: Taylor and Francis Group.

Lang, Hans J., and Michael Decoursey (1983). *Profitability Accounting and Bidding Strategy for Engineering and Construction Management.* New York, NY: Van Nostrand Reinhold.

Lucas, P. D. (1984). *Modern Construction Accounting Methods and Controls*. Englewood Cliffs, NJ: Prentice-Hall.

Merna, T. (2002). *Financing Infrastructure Projects*. London, UK: Thomas Telford.

Milliner, M. S. (1988). *Contractor's Business Handbook*. Kingston, MA: R.S. Means Co., Inc.

Miramontes, L. P. and Hugh, L. R. (2004). *Construction Accounting Deskbook.* Chicago, IL: CCH Incorporated.

Navarette, Pablo F. (1995). *Planning Estimating and Control of Chemical Construction Projects*. New York: Marcel Dekker, Inc.

Neil, James N. (1982). *Construction Cost Estimating for Construction Control*, Englewood Cliffs, NJ: Prentice-Hall.

Palmer, W. J., (1995). *Construction Accounting and Financial Management.* New York, NY: McGraw-Hill.

Park, William R. (1979). *Construction Bidding for Profit*, New York, NY: John Wiley and Sons.

Peterson, S. J. (2008). *Construction Accounting and Financial Management*, 2nd edition. Upper Saddle River, NJ: Prentice Hall.

Price, A. D. F. (1999). International Project Accounting (International Construction Management). London, UK: International Labour Office.

RMA Annual Statement Studies, Financial Benchmark Studies, Philadelphia, PA: The Risk Management Association, published annually.

R. S. Means Company, *Building Construction Cost Data*. Kingston, MA: R. S. Means Co., Inc. Published annually.

Rice, H. L. and J. L. Callan (2001). *Construction Accounting Deskbook: Financial, Tax, Accounting, Management, and Legal Answers* (with CD-ROM). Orlando, FL: Harcourt Professional Publishing.

Russell, J. S. (2000). *Surety Bonds for Construction Contracts*. Reston, VA: ASCE Press.

Schexnayder, C. J. (2004). *Construction Management Fundamentals*. Boston, MA: McGraw-Hill Higher Education.

Shinn, E. S. (2008). *Accounting and Financial Management: for Residential Construction*, fifth edition. Washington, DC: NAHB Builderbooks.

Singh, A. (1991) Knowledge Bases for C/SCSC. *Journal of Cost Engineering*, AACE, 33 (6), 39–48.

Troy, Lee. (2007). *Almanac of Business and Industrial Financial Ratios*, Chicago, IL: CCH, Inc. (2008 edition).

U.S. Census Bureau. "Number of Firms, Number of Establishments, Employment, and Annual Payroll by Employment Size of the Enterprise for the United States, All Industries 2004."

Wallace, E. P. (2007). *Construction Guide: Accounting and Auditing*. Chicago, IL: CCH Incorporated.

INDEX